Preface

Air dehumidification is a main task of building's air condition system. Except from air temperature, humidity ratio of indoor air also needs to be kept within a certain range to satisfy needs of human thermal comfort and normal operation of equipment. Desiccant wheel, which has the advantages of handling large air mass flow rate and realizing deep dehumidification, is widely used to process air in the air conditioning systems of commercial and industrial buildings.

Different from condensation dehumidification, which requires low-temperature cooling source to condense vapor from air, air dehumidification with desiccant wheel is powered by high temperature heat source. Therefore, electrical heaters or high temperature steam are usually used. In recent years, renewable energies are playing a more important part in the structure of social energy consumption, especially since the announcement of "carbon peaking and carbon neutrality goals". It has become a hot research topic to replace traditional energies with renewable energies, such as solar energy, waste heat, and heat pump, etc., to drive air dehumidification process of desiccant wheels. However, quality of the above renewable energies is low, which means, to use these energies effectively and efficiently, driving temperature, which is called regeneration temperature, of desiccant wheels needs to be reduced. This is different from traditional energies, efficiency of which is more related to heat quantity rather than heat quality. Therefore, it is essential to investigate desiccant materials, structures of wheel as well as air handling processes that are favorable for the adoption of renewable energies.

In this book, latest dehumidification technologies using desiccant wheels were introduced. Physical properties of solid desiccant materials, structures of desiccant wheels, and air handling processes were suggested to enhance thermal performances. In Chapter 1, dehumidification mechanisms of desiccant wheel and physical properties of regular solid desiccant materials were illustrated. In Chapter 2, three numerical models and two prediction models of desiccant wheels, as well as test validations of these models, were presented. Chapters 3 and 4 were about desiccant wheels, discussing influences of wheel structure, rotation speed and adsorption isotherms on the reduction of regeneration temperature. Chapters 5 to 9 were about air handling processes. In Chapter 5, a reversible desiccant wheel dehumidification and cooling system was designed, and irreversible processes were examined. In Chapter 6, air handling processes with single desiccant wheel were studied, aiming to find the inherent performance influencing factors for the existing ventilation systems through exergy destruction analyses. Then, an advanced system with single-stage desiccant wheel driven by heat pumps was proposed and discussed in Chapter 7. In Chapter 8, dehumidification and cooling systems with multi-stage desiccant wheels were designed, and the influences of the number of stages on regeneration temperature were examined, when water and refrigerant are used as cooling or heating media. In Chapter 9, design criterions of highly efficient desiccant wheel dehumidification systems using traditional energies and renewable energies were discussed.

This book was written based on the author's work on desiccant wheel dehumidification systems from 2010 to 2022. This was also the topic of the author's PhD dissertation in Tsinghua University. I'd like to expvess my appreciate iow for the guidance from Prof. Yi Jiang and Prof. Xiaohua Liu, who were my supervisors in Tsinghua University, as well as Prof. Yunho Hwang, who was the supervisor during my visit in the University of Maryland (College Park). I'd like to thank the University of Science and Technology Beijing for funding this book's publication. Most of the work in this book were accomplished in the University of Science and Technology Beijing. Lastly, I'd like to express my appreciation to the following students, namely, Mengdan Liu, Siqi Wang, Xianzhong Chen, Qiaoxin Zhang, Ming Li, Yuyao Ge and Lu Wang, who contributed a lot in writing and editing this book.

Desiccant wheel dehumidification technologies are hot topics and have been investigated by many scholars. A lot of innovative and advanced opinions, system configurations and devices, which they proposed, have given the author quite a lot of inspirations and directions. If there are something wrong or of difficent opposite opinions in this book, I do appreciate and welcome comments from all sides.

Rang Tu

Aug. 2022

[illegible]

[illegible]

[illegible]

[illegible]

[illegible]

[illegible]

Naijie Hu

Contents

Chapter 1 Dehumidification Mechanisms of Desiccant Wheels

In this chapter, working principles of desiccant wheels used for air dehumidification are briefly illustrated. Properties of desiccant materials are summarized. And regular desiccant wheel dehumidification systems as well as the associated heating and cooling sources are introduced.

1.1 Heat and moisture transfer mechanism in desiccant wheel

Desiccant wheel is a rotor covered with solid desiccant material. It slowly rotates between the processed air and the regenerated air. During the dehumidification process, the processed air operates close to an isenthalpic process; thus, the outlet temperature of the processed air will be very high and assisted cooling must be implemented to cool down the dried processed air before it is introduced into occupied spaces. Schematic and operating principle of a typical rotary wheel system are shown in Fig. 1-1. The desiccant wheel rotates between the regeneration air (RA) and the processed air (PA) to continuously dehumidify the processed air. The processed air and the regeneration air flow in reverse along the wheel's thickness direction. Regeneration air needs to be heated from state A_{rin} to state A_{r1} to regenerate desiccant material. Fig. 1-1 (b) shows that during the dehumidification and regeneration processes, both the processed air and the regeneration air change states near the isenthalpic lines. During dehumidification (regeneration), the air is heated (cooled) and dehumidified (humidified). The processed air after dehumidification needs to be cooled down by extra cooling sources before being supplied into the occupant room[1].

Air handling process of the desiccant wheel is adiabatic, which is different from the inner cooling desiccant bed. For the inner cooling desiccant bed, cold and hot fluids are supplied into the dehumidification bed and the regeneration bed, respectively. Heat and mass are transferred in the same direction. The dehumidification air is cooled and dehumidified, while the regeneration air is heated and humidified.

The fluids in desiccant beds are normally water or refrigerant. DAIKIN Company of Japan has introduced a fresh air treatment device, called DESICA. DESICA combines

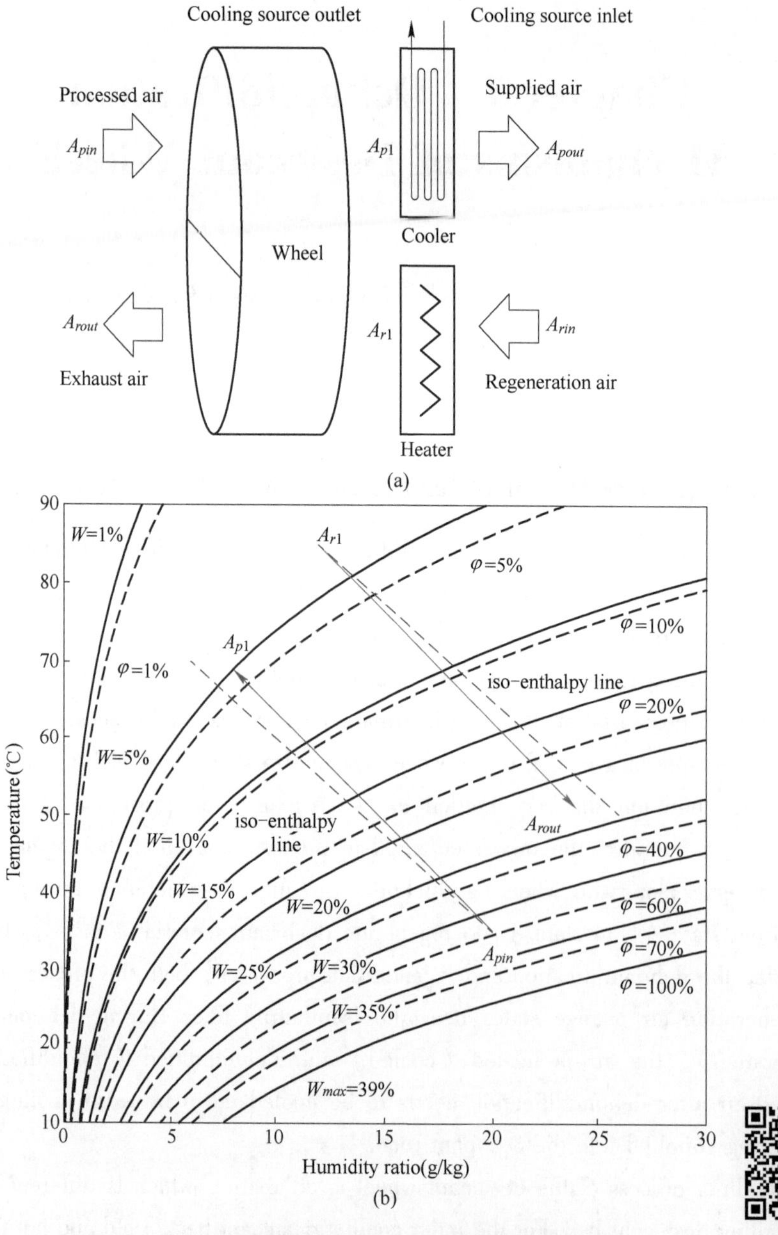

Fig. 1-1 A typical desiccant wheel dehumidification system

(a) Schematic of the system; (b) Air handling process

the refrigeration system with the desiccant system. The adsorbent is attached to fins of the evaporator and the condenser. Schematic of the unit is shown in Fig. 1-2 (a) . In the dehumidification process, the fresh air contacts with the solid desiccant attached to the surface of the evaporator, and the fresh air is dehumidified. The latent heat released during dehumidification is taken by the evaporator, and the fresh air is simultaneously cooled down by the evaporator. During regeneration, the indoor return air contacts with the solid adsorbent attached to the surface of the condenser. The condenser provides heat to increase temperature of the adsorbent, so that water content of the adsorbent rises. Both temperature and humidity ratio of the adsorbent is higher than the regeneration air. The regeneration air is heated and humidified. A typical air handling process of air is shown in Fig. 1-2 (c) . As compared with the desiccant wheel, temperature of the regeneration air does not need to be too high to meet the requirements of regeneration.

However, specific surface area of the desiccant bed is relatively small as compared with the desiccant wheel, and the desiccant layer on the fins' surface is thin. Therefore, bulk density of solid desiccant (mass of solid adsorbent divided by the volume of the desiccant bed) is low. This means that it's fast for the adsorbent material to be fully charged with water or to be regenerated. It usually takes 3 ~ 5minutes for the two heat exchangers to switch between dehumidification mode and regeneration mode. This is realized through changing circulation direction of the refrigerant and flowing paths of the fresh air and the return air, as shown in Fig. 1-2 (b) . This leads to large cold-heat offset of the heating and cooling fluids due to frequently heating and cooling the desiccant bed.

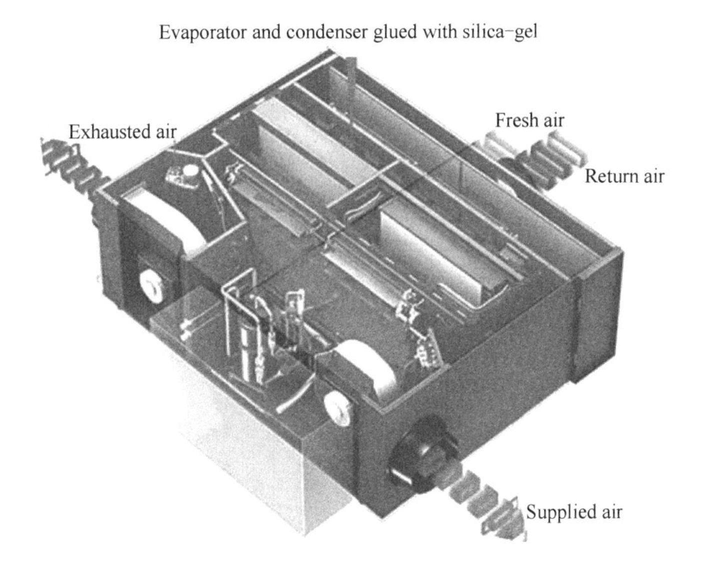

(a)

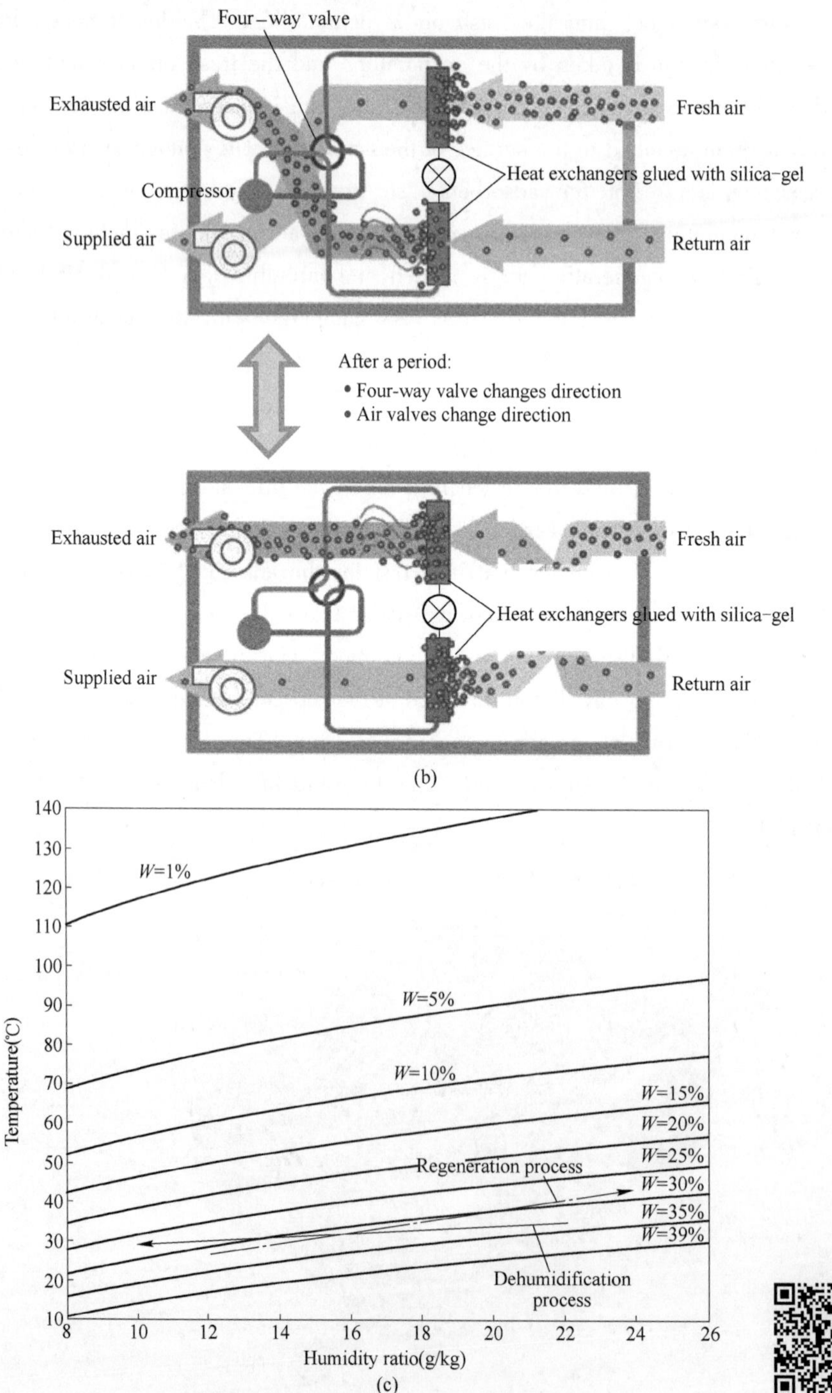

Fig. 1-2 Working principles of DESICA

(a) Schematic of DESICA; (b) Working principle; (c) Air handling process

1.2 Properties of solid desiccants

1.2.1 Introduction of regular solid desiccants

Regular solid desiccants for water vapor adsorption, also called hygroscopic agent, includes silica gel, alumina, molecular sieve, etc. Silica gel ($SiO_2 \cdot xH_2O$) is a non-toxic, odorless, non-corrosive translucent crystal. It is insoluble in water. Porosity is up to 70%[2]. Average density is about 650kg/m^3, and specific surface area is in the range of 100~1000m^2/g. Water uptake capacity is up to 30%.

As shown in Fig. 1-3, there are powder silica gel, particle silica gel, allochroic silica gel and white silica gel. Powder silica gel is mainly used by desiccant wheels and inner-cooling desiccant beds. The powder is glued on walls or fins surfaces, with a thickness of around 2mm. Particle silica gel is mainly used at packed dehumidification beds. And dried allochroic silica gel is blue, while it changes into pink after adsorbing water. Color of the white silica gel doesn't change.

There are numbers of zigzaggy microporous structures inside the adsorbent, with relatively large specific surface area and void fraction[2,3]. Therefore, it has a strong vapor adsorption and water storage ability. According to IUPAC (International Union of Pure and Applied Chemistry) classification[3,4], macropores with pore size greater than 500Å are called macropores. Those with pore size less than 20Å are called micropores. And those with pore size between 500Å and 20Å are called mesopores. The pore structure of silica gel belongs to mesopore or micropore[5]. Macropores and mesopores are the channels of mainstream vapor flow, and micropores mainly play the role of water storage[2,3]. Parameters that describe the internal pore structure are porosity and tortuosity factor. Porosity is defined as the proportion of a material's pore volume to its total volume. The tortuosity factor or zigzag factor is defined as the ratio of the actual path of the channel to the straight path of both ends. Specific surface of silica gel is in the range of 100m^2/g to 1000m^2/g[6,7].

Activated alumina has a strong affinity for water. Under certain operating conditions, air can be dried to the dew point of −70℃, and its regeneration temperature is much lower than the molecular sieve. Aluminum oxide (Al_2O_3) with high microporous particle structure can adsorb water up to 60% of its own weight[5]. The most commonly used activated alumina has a specific surface area of 100 to 415m^2/g[5,6].

Calcium chloride is white crystal and tastes slightly bitter and salty. It has strong moisture adsorption ability and is cheap. However, it becomes calcium chloride solution

Fig. 1-3 Silica gel for vapor adsorption

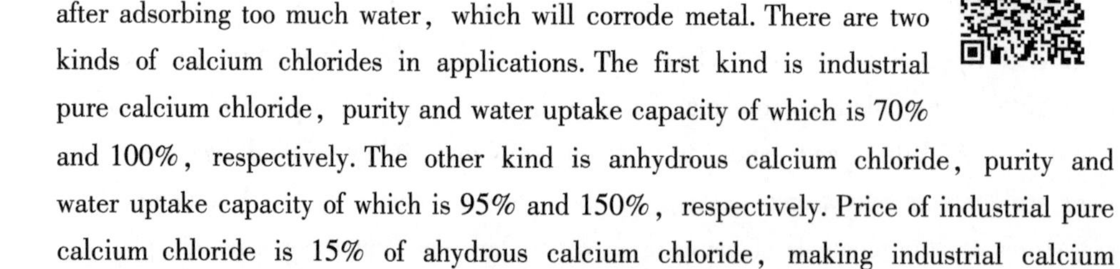

after adsorbing too much water, which will corrode metal. There are two kinds of calcium chlorides in applications. The first kind is industrial pure calcium chloride, purity and water uptake capacity of which is 70% and 100%, respectively. The other kind is anhydrous calcium chloride, purity and water uptake capacity of which is 95% and 150%, respectively. Price of industrial pure calcium chloride is 15% of ahydrous calcium chloride, making industrial calcium chloride more economic.

There are mainly two kinds of Zeolite, namely natural zeolite and synthetic zeolite (molecular sieve). Price of the latter is more expensive. However, its bulk density is higher and heat conduction performance is better. Zeolite is a compound of silicon aluminum acid salt, which has strong hydrophilicity. The hydrophilicity is related to Si/Al ratio. The smaller the ratio is, the better the hydrophilicity will be. Zeolite molecular sieve is a synthetic silicon aluminate crystal. According to crystal framework structures, it can be divided into A type, X type and Y type. Synthetic zeolite (molecular sieve) has relatively uniform micropores, and products with different pore sizes can be produced from it, such as 4A, 5A, 10X and 13X molecular sieve. For the 5A molecular sieve, specific surface area is between 500 and 700m^2/g, porosity is about 0.32, and the maximum water uptake is from 21.5% to 27%. For industrial applications, 20% clay is usually added to the crystalline molecular sieve powders, diameter of which is from 1μm to 10μm.

1.2.2 Adsorption isotherms of solid desiccant materials

Fig. 1-4 shows principles of heat and mass transfer between air and the adsorbent. There is a thin humid air film on the surface of the adsorbent, with which the mainstream air exchange heat and mass. When water vapor diffuses into the desiccant material, it diffuses into the inside through the pore structures. The water vapor and the -OH bond on the surface of the adsorption material combine to form the adsorbed water. There are molecular diffusion, Kudsen diffusion and surface diffusion[2,7].

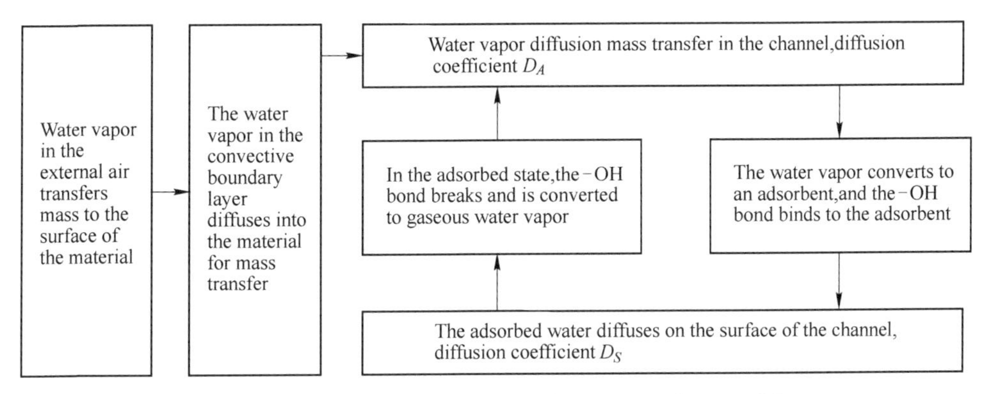

Fig. 1-4 Adsorption process of solid hygroscopic materials

Equations of the above diffusion coefficients are shown in Eqs. (1-1) ~ (1-2)[3,8-11]:

$$D_A = \frac{\varepsilon}{\xi}\left(\frac{1}{D_{AO}} + \frac{1}{D_{AK}}\right)^{-1} \tag{1-1}$$

where $D_{AO} = 1.758 \times 10^{-4} \dfrac{T_d^{1.658}}{p_a}$, $D_{AK} = 97\, a_d \left(\dfrac{T_d}{M}\right)^{0.5}$

$$D_S = \frac{1}{\xi} D_0 \exp(-0.974 \times 10^{-3} r_s / T_d) \tag{1-2}$$

where D_{AO} is the Molecular diffusion coefficient, m^2/s; D_{AK} is the Knudsen diffusion coefficient, m^2/s; D_A is the gas diffusion coefficient, m^2/s; D_S is the surface diffusion coefficient, m^2/s; ε is porosity; ξ is tortuous factor; r_s is the heat of adsorption; D_0 is the surface diffusion constant, equaling to $1.6\times10^{-6} m^2/s$[2,11]; M is the molecular weight of water, kg/kmol; p_a is atmospheric pressure, equaling to 101325Pa; a_d is the average radius of micropores of adsorbent, equaling to 11×10^{-10}m[2,11]; T_d is temperature of the hygroscopic material, K.

Molecular diffusion and Knudsen diffusion are used to describe the diffusion of water vaporin the pore channels. And the surface diffusion is used to describe the diffusion of adsorbed water on the pore surface. When water vapor converses into adsorbed water, a lot of adsorption heat will be released, value of which is slightly higher than the latent heat of vaporization. If the heat is not discharged in time, temperature of the air and the adsorbent would be increased, which is bad for the dehumidification process. Adsorption isotherm of solid hygroscopic material can be used to explained this phenomenon.

The adsorption isotherm of solid hygroscopic material describes the relationship between the water content of the hygroscopic material (W) and the relative humidity (φ_d) of the thin humid air film on the surface of the hygroscopic material under equilibrium state. Eq. (1-3)[2,7,12] is a common equation that describes the relation between φ_d and W for ideal desiccant material:

$$W = \frac{W_{max}}{1 - C + C/\varphi_d} \tag{1-3}$$

where W is water content of the desiccant material, kg/kg. W is calculated through dividing mass of water adsorbed by the material with mass of dry material, as shown in Eq. (1-4)[2,7]:

$$W = \frac{m_{water}}{m_{ad}} \tag{1-4}$$

where W_{max} is the maximum moisture content that can be reached by the desiccant material. W_{max} is measured when it reaches to the equilibrium state with saturation air.

C is shape factor of the desiccant material. Fig. 1-5[2] shows the forms of adsorption isotherms corresponding to different C. With the increase of relative humidity, its equilibrium moisture content increases. When the relative humidity is 100%, the corresponding equilibrium moisture content reaches the maximum.

For type Ⅰ desiccant material, C is less than 1, and the adsorption isotherm is convex. It means that the equilibrium moisture content of the desiccant material can be close to the maximum value when air relative humidity is low. For type Ⅱ desiccant materials, C is equal to 1 and the adsorption isotherm is linear. Water adsorption ability doesn't change with air humidity ratio. For type Ⅲ desiccant material, C is concave. Water adsorption capacity of the desiccant material is poor when air relative humidity is low. However, the water adsorption capacity increases sharply at high air relative humidity. Typical adsorbents for types Ⅰ, Ⅱ and Ⅲ hygroscopic material are molecular sieve, silica gel and activated alumina, respectively[2,7].

For air treatment of commercial buildings or residential buildings, silica gel is a good

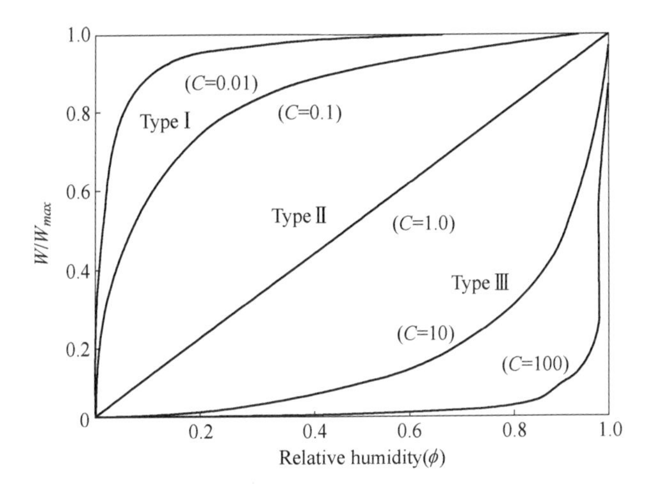

Fig. 1-5 Adsorption isotherms of solid hygroscopic materials

option. Table 1-1 summarizes adsorption isotherms of regular silica gel, also known as RD silica gel. It shows that the adsorption isotherm is generally the relationship between relative humidity and water content. In some formulas, the adsorption isotherm is also related to temperature.

Table 1-1 Adsorption isotherms of silica gel used in the literature

$\varphi_d = 0.0078 - 0.05759W + 24.16554W^2 - 124.478W^3 + 204.226W^4$ [12,14]
$\varphi_d = -0.02833 - 8.18612W - 41.7964W^2 + 82.9974W^3$ [15]
$W = 0.24\,\varphi_d^{2/3}$ [16]
$W = 0.77\varphi_d - 0.38\,\varphi_d^2$ [17,18];
$\varphi_d = (2.112W)^{q_{st}/h_v}(29.91\,P_{vs})^{q_{st}/h_v - 1}$ [19]: q_{st}-adsorption heat, J/kg; h_v-vaporization heat, J/kg
$\varphi_d = \dfrac{0.616238W + 16.7916W^2 - 74.34228W^3 + 116.6834W^4}{1-(t-40)/300}$ [20]
$\varphi_d = S_1tW^2 + S_2tW + S_3W^4 + S_4W^3 + S_5W^2 + S_6W$ [21]
$W = 0.0329 - 0.4113\times10^{-5}t^2 + 0.0105\times10^{-3}\varphi_d^2 + 0.6586\times10^{-6}\varphi_d^3 + 0.7894\times10^{-10}t^3\varphi_d^2 + 0.6747\times10^{-1}$ [22], t is the temperature of the desiccant material, ℃

* $S_1 = -0.04031298$, $S_2 = 0.02170245$, $S_3 = 125.470047$, $S_4 = -72.651229$, $S_5 = 15.5223665$, $S_6 = 0.00842660$.

Under equilibrium state, temperature of the humid air film (t_d) equals to the desiccant material, and relative humidity ratio of the air film (φ_d) can be calculated with Eq. (1-4). Humidity of the air film (ω_d) is calculated with Eq. (1-5)[23]:

$$\frac{\varphi_d}{\omega_d} = 10^{-6}\exp(5279/T_d) - 1.61\varphi_d \tag{1-5}$$

With Eq. (1-4) and Eq. (1-5), relations between ω_d and t_d can be drawn in the psychrometric chart. Shape factor (C) and maximum water capacity (W_{max}) are two parameters relating to adsorption properties of desiccant materials. The adsorption isotherms in Table 1-1 can be simplified as Eq. (1-4) with different C and W_{max}, as shown in Fig. 1-6.

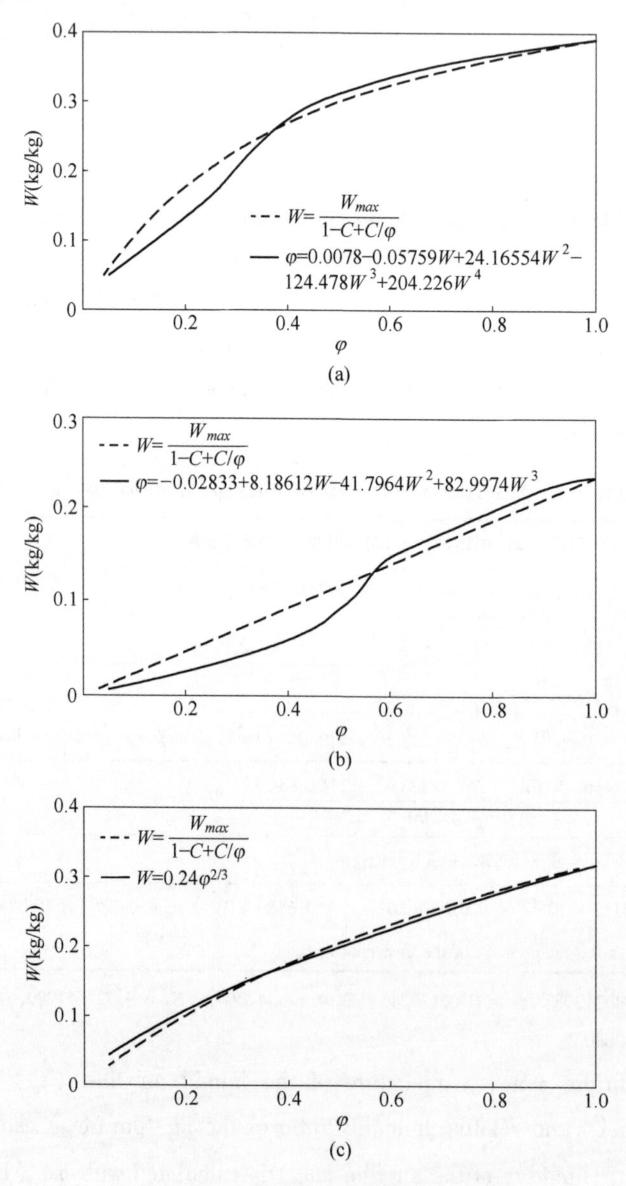

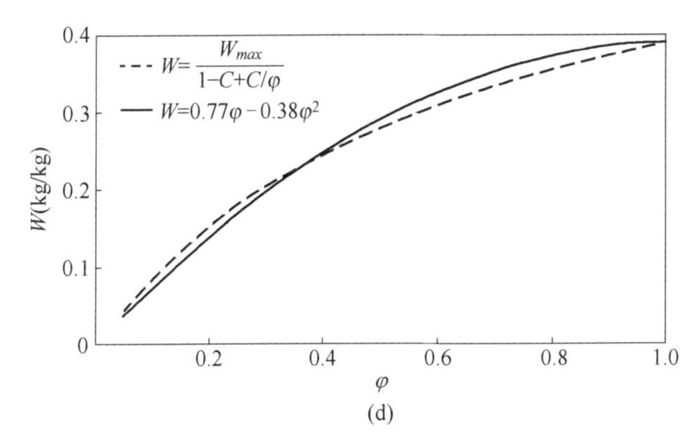

Fig. 1-6 Comparisons of the adsorption isotherm lines drawn from different equations with the universal adsorption isotherm equation

(a) W_{max} = 0.39kg/kg, C = 0.3; (b) W_{max} = 0.316kg/kg, C = 1; (c) W_{max} = 0.24kg/kg, C = 0.55; (d) W_{max} = 0.39kg/kg, C = 0.4

Combining the adsorption isotherm equations and the Clapeyron equation[23] shown in Eq. (1-5), the equilibrium humidity ratio of the desiccant (ω_d), which is influenced by t_d and W/W_{max}, can be obtained, shown as Eq. (1-6):

$$\omega_d = \frac{W/W_{max}}{10^{-6}\exp\left(\frac{5294}{t_d + 273.15}\right)\left(\frac{1}{C} + \frac{C-1}{C}\frac{W}{W_{max}}\right) - 1.61\frac{W}{W_{max}}} \tag{1-6}$$

The adsorption process is accompanied with the release of adsorption heat. Table 1-2 lists the adsorption heat equations given in literature. Adsorption heat (q_{st}, J/kg) is generally a function of W and temperature (t_d), and its value is generally greater than the latent heat of vaporization (h_v, J/kg).

Table 1-2 Adsorption heat of silica-gel in the literature

$q_{st} = h_v[1 + 0.2843\exp(-10.28W)]$ [14,17,24]
$q_{st} = (-0.02t^3 + t^2 - 2386.2t + h_v) + (-375.867 - 550\log_{10}W + 420W) \times 10^3$ [14,17,18]
$q_{st} = \begin{cases} -12400W + 3500, & W \leqslant 5\% \\ -1400W + 2900, & W > 5\% \end{cases}$ [14,17]
$q_{st} = 2650000$ [2,18,22]

Nomenclature

A area

a_d the average radius of micropores of adsorbent, 11×10^{-10}m

C	shape factor of the desiccant material
D_A	the gas diffusion coefficient, m^2/s
D_{AO}	the diffusion coefficient of Molecular diffusion, m^2/s
D_{AK}	the diffusion coefficient of Knudsen diffusion, m^2/s
DE	a direct evaporative cooler
DW	the desiccant wheel
D_0	the surface diffusion constant, 1.6×10^{-6} m^2/s
D_S	the diffusion coefficient of surface diffusion, m^2/s
h_v	the latent heat of vaporization, J/kg
M	the molecular weight of water, kg/kmol
PA	processed air
p_a	atmospheric pressure, 101325Pa
q_{st}	adsorption heat, J/kg
RA	regeneration air
r_s	the heat of adsorption
T_d	temperature of the desiccant material, K
t	temperature, ℃
t_{reg}	regeneration temperature, ℃
W	the water content
W_{max}	maximum water content

Greek symbols

ε	porosity, dimensionless
ξ	tortuous factor
φ_d	the relative humidity
ω_d	the equilibrium humidity ratio of the desiccant

Subscripts

A	air
ad	adsorbent
d	desiccant
max	maximum
pin	processed airinlet
$pout$	processed airoutlet
rin	regeneration airinlet
$rout$	regeneration airoutlet
v	vaporization

References

[1] R Tu, X H Liu, Y Jiang. Analysis of heat and mass exchange process and performance comparison of different solid dehumidification methods [J]. Chemical Industry Journal, 2013, 64 (6): 1939-1947.

[2] L Z Zhang. Dehumidification Technology [M]. Beijing: Chemical Industry Press, 2005.

[3] X Li, Z Li. Effects of pore sizes of porous silica gels on desorption activation energy of water vapour [J]. Applied Thermal Engineering, 2006, 27 (5): 869-876.

[4] J Rouquerol, D Avnir, C W Fairbridge, D H Everett, J M Haynes, N Pernicone, J D F Ramsay, K S W Sing, K K Unger. Recommendations for the characterization of porous solids (Technical Report) [J]. Pure and Applied Chemistry, 2013, 66 (8): 1739-1758.

[5] C X Jia. Study on enhanced dehumidification Mechanism and Application of Silica gel composite desiccant [D]. Shanghai Jiaotong University, 2006.

[6] Y H Niu, S B Xiu, L Q Zhang, X M Hu, Y K Li. Research progress of solid dehumidifying materials for air conditioning [J]. Applied Chemical Industry, 2018, 47 (11): 2464-2468.

[7] Y P Zhang, L Z Zhang, X H Liu, J H Mo. Mass transfer in the built environment [M]. Beijing: China Architecture and Architecture Press, 2006.

[8] K J Sladek, E R Gilliland, R F Buddour. Diffusion on Surfaces. Ⅱ. Correlation of Diffusivities of Physically and Chemically Adsorbed Species [J]. Ind. Eng. Chem. Fundam, 1974, 13 (2): 100-105.

[9] Z H Ye. Chemical adsorption separation process [M]. Beijing: China Petrochemical Press, 1992.

[10] P Majumdar. Heat and mass transfer in composite desiccant pore structures for dehumidification [J]. Sol Energy, 1998, 62 (1): 1-10.

[11] R Tu, X H Liu, Y Jiang. Performance analysis of a new kind of heat pump-driven outdoor air processor using solid desiccant [J]. Renewable Energy, 2013, 57: 101-110.

[12] J D Chung, D Y Lee. Effect of desiccant isotherm on the performance of desiccant wheel [J]. Int. J. Refrig, 2009, 32 (4): 720-726.

[13] R Narayanan, W Y Saman, S D White, M Goldsworthy. Comparative study of different desiccant wheel designs [J]. Appl. Therm. Eng, 2011, 31 (10): 1613-1620.

[14] A A Pesaran. Air dehumidification in packed silica gel bed [D]. CA: University of California Los Angeles, 1980.

[15] A K Ramzy, R Kadoli. Improved utilization of desiccant material in packed bed dehumidifier using composite particles [J]. Renew. Energy, 2011, 36 (2): 732-742.

[16] A Kodama, T Hirayama, M Goto, T Hirose, R E Critoph. The use of psychometric charts for the optimisation of a thermal swing desiccant wheel [J]. Appl. Therm. Eng, 2001, 21 (16): 1657-1674.

[17] A Pesaran. Moisture Transport in Silica Gel Particle Beds [D]. CA: University of California,

Los Angeles, 1983.

[18] C R Ruivo, J J Costa, A R Figueiredo. Analysis of simplifying assumptions for the numerical modeling of the heat and mass transfer in a porous desiccant medium [J]. Numer. Heat Tr. A-Appl, 2006, 49 (9): 851-872.

[19] E Van Den Bulk, J W, Mitchell, S A Klein. Design theory for rotary heat and mass exchangers-Ⅱ. Effeciveness-number-of-transfer-units method for rotary hcat and mass exchangers [J]. Int. J. Heat Mass Tran, 1985, 28 (8): 1587-1595.

[20] J Jeong, S Yamaguchi, K Saito, S. Kawai. Performance analysis of four-partition desiccant wheel and hybrid dehumidification air-conditioning system [J]. Int. J. Refrig, 2010, 33 (3): 496-509.

[21] M Dupont, B Celestine, J Merigoux, B Brandan. Desiccant solar air conditioning in tropical climate: Ⅰ-dynamic experimental and numerical studies of silica gel and activated alumina [J]. Sol Energy, 1994, 52 (6): 509-517.

[22] P Majumdar, W M Worek. Combined heat and mass transfer in a porous adsorbent [J]. Energy, 1989, 14 (3): 161-175.

[23] L Z Zhang. Total Heat Recovery: Heat and Moisture Recovery from Ventilation Air [M]. New York: Nova Science Publisher, 2008.

[24] J Y San. Heat and mass transfer in a two-dimensional cross-flow regenerator with a solid conduction effect [J]. Int. J. Heat Mass Tran, 1993, 36 (3): 633-643.

Chapter 2 Numerical Models of the Desiccant Wheel

In this chapter, three mathematic models and two prediction models of desiccant wheels are introduced. The mathematic models are based on coupled heat and mass transfer processes between air and desiccant material. The three mathematic models are two-dimensional two-diffusion model, one-dimensional two-diffusion model, and one-dimensional non-diffusion model. The two prediction models are built based on multiple regression methods and artificial neural networks, respectively.

2.1 Coupled heat and mass transfer equations

2.1.1 Working principles of desiccant wheels

In the desiccant wheel, the heat and mass transfer processes between adsorbent and humid air is shown in Fig. 2-1. Cross-section of a single air channel is sinusoidal, with a height of a and a width of b. P and d_h are the wet cycle and hydraulic diameter of a single channel. Air enters the channel at a speed of u_a. For a single channel, facial area of the air flow part and the solid part is A_a and A_d, respectively. And the proportion of the air flowing part is $f=A_a/(A_a+A_d)$. The solid part includes the substrate and the adsorbent, in which the volume ratio of the hygroscopic material to the total solid part is x. Thickness of the desiccant wheel is L, and the area ratio of the dehumidification zone to the regeneration zone on the windward surface is A_r. The A_r is generally 3 : 1 or 1 : 1[1]. Air flow rate ratio between the dehumidification air and the regeneration air (F_r) normally equals to A_r.

Mathematical model of a desiccant wheel is established for the control unit as shown in Fig. 2-1, based on the following assumptions[2,3]:

(1) Thermal properties of heat and mass transfer between the desiccant wall and moist air are constant throughout the flow channel.

(2) The material of the air passage wall is uniform, and the coating thickness of the silica gel desiccant wheel on the air passage is uniform.

(3) The heat loss from the desiccant wheel to the environment is ignored, and the

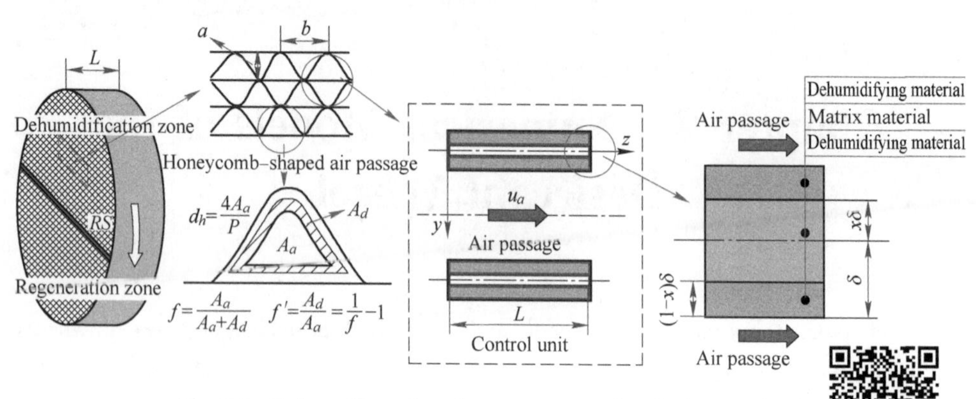

Fig. 2-1 Schematic and main parameters of a desiccant wheel

inlet parameters of the air are the same in the angle direction.

(4) The airflow through the air passage of the desiccant wheel is incompressible, and there is no leakage of air streams.

(5) The heat conduction and mass diffusion of the airflow in the flow direction are ignored. The heat and mass transfer in the airflow is only caused by convection, and the radiation heat transfer in the model is ignored.

(6) The effect of the centrifugal force on the running process of the desiccant wheel is ignored. And the flow and the heat and mass transfer in the angle direction are different, but the same in the radial direction.

2.1.2 Two-Dimensional-Double-Diffusion model

For the two-dimensional-double-diffusion model[2], heat conduction and mass diffusion in y and z directions, as shown in Fig. 2-1, are considered. The governing equations can be obtained from the above physical model, as shown in Eqs. (2-1) ~ (2-4).

Heat transfer differential equation:

$$\frac{1}{u_a}\frac{\partial t_a}{\partial \tau}+\frac{\partial t_a}{\partial z}=\frac{4h}{\rho_a c_{pa} u_a d_h}(t_d-t_a) \tag{2-1}$$

Mass transfer differential equation:

$$\frac{1}{u_a}\frac{\partial \omega_a}{\partial \tau}+\frac{\partial \omega_a}{\partial z}=\frac{4h_m}{\rho_a c_{pa} u_a d_h}(\omega_d-\omega_a) \tag{2-2}$$

Energy conservation equation:

$$\begin{aligned}\rho_d\left(c_{pd}+\frac{\rho_{ad}x}{\rho_d}c_{pw}W\right)\frac{\partial t_d}{\partial \tau}+x\rho_{ad}c_{pw}t_d\frac{\partial W}{\partial \tau}\\=k_d\left(\frac{\partial^2 t_d}{\partial z^2}+\frac{\partial^2 t_d}{\partial y^2}\right)+r_s x\rho_{ad}\frac{\partial W}{\partial \tau}+\frac{4h}{d_h f}(t_a-t_d)\end{aligned} \tag{2-3}$$

Mass conservation equation:

$$\varepsilon\rho_a \frac{\partial \omega_d}{\partial \tau} + \rho_{ad}\frac{\partial W}{\partial \tau} = D_S\rho_{ad}\left(\frac{\partial^2 W}{\partial z^2} + \frac{\partial^2 W}{\partial y^2}\right) + \varepsilon\rho_a D_A\left(\frac{\partial^2 \omega_d}{\partial z^2} + \frac{\partial^2 \omega_d}{\partial y^2}\right) + \frac{4h_m}{xd_h f'}(\omega_a - \omega_d) \quad (2\text{-}4)$$

where, D_A and D_S can be calculated with Eqs. (2-1) ~ (2-2); f' is the ratio of solid windward area to air windward area of a single passage, and f is the proportion of air passage area on the windward surface to the total area. The calculation formula of f and f' is shown in Eq. (2-5). ρ_d and c_{pd} are the weighted density and weighted specific heat capacity of solid hygroscopic materials after considering the substrate and adsorbent materials. The calculation formula is shown in Eq. (2-6) and Eq. (2-7).

$$f = \frac{A_a}{A_a + A_d},\ f' = \frac{A_d}{A_a} = \frac{1}{f} - 1 \quad (2\text{-}5)$$

$$\rho_d = x\rho_{ad} + (1 - x)\rho_m \quad (2\text{-}6)$$

$$c_{pd} = x^* c_{pad} + (1 - x^*)c_{pm} \quad (2\text{-}7)$$

x and x^* are respectively the volume ratio and mass ratio of solid adsorbent in solid material to the total material. The relationship between the two is shown in Eq. (2-8).

$$x^* = x\rho_d/[x\rho_{ad} + (1 - x)\rho_m] \quad (2\text{-}8)$$

According to the comparison between heat transfer and mass transfer, the relation between convective heat transfer coefficient and mass transfer parameter load is expressed with Eq. (2-9).

$$h_m = \frac{h}{c_{pa} \cdot Le} \quad (2\text{-}9)$$

where, c_{pa} is the specific heat capacity of air at constant pressure; *Le* is the Lewis number. For h, it is calculated with Eq. (2-10)[4,5].

$$h = \frac{Nu \cdot \lambda_a}{d_h} \quad (2\text{-}10)$$

where, λ_a is the thermal conductivity of air; d_h is the hydraulic diameter of a sinusoidal channel, which can be calculated with Eq. (2-11)[2-5], where P is the wet perimeter of a single passage, m:

$$d_h = \frac{4A_a}{P} \quad (2\text{-}11)$$

It is pointed out in literature[6,7] that Nu is mainly related to the ratio between the height of a single air passage and the width of a single air passage (a/b), which can be calculated with Eq. (2-12)[8-10]:

$$Nu_T = 1.1791 \times [1 + 2.7701(a/b) - 3.1901(a/b)^2 - 1.9975(a/b)^3 - 0.4966(a/b)^4]$$
$$Nu_H = 1.903 \times [1 + 0.455(a/b) + 1.2111(a/b)^2 - 1.6805(a/b)^3 + 0.7724(a/b)^4 - 1.228(a/b)^5] \quad (2\text{-}12)$$
$$Nu = (Nu_T + Nu_H)/2$$

Boundary conditions and initial conditions are expressed as follows:

$$-k_d \frac{\partial t_d}{\partial y}\Big|_{y=\delta} = h(t_d - t_a)$$
$$\left(-\rho_a D_A \frac{\partial W}{\partial y} - \rho_d D_S \frac{\partial W}{\partial y}\right)\Big|_{y=\delta} = h(W - \omega_a)$$
$$\frac{\partial W}{\partial y}\Big|_{y=0} = \frac{\partial W}{\partial z}\Big|_{z=0} = \frac{\partial W}{\partial z}\Big|_{z=L} = 0$$
$$\frac{\partial t_d}{\partial z}\Big|_{y=0} = \frac{\partial t_d}{\partial z}\Big|_{z=0} = \frac{\partial t_d}{\partial z}\Big|_{z=L} = 0$$
$$t_d\Big|_{z=0\sim L}^{\tau=0} = t_{d,\ ini},\quad W\Big|_{z=0\sim L}^{\tau=0} = W_{ini},\quad t_a\Big|_{z=0\sim L}^{\tau=0} = t_{a,\ ini},\quad \omega_a\Big|_{z=0\sim L}^{\tau=0} = \omega_{a,\ ini} \quad (2\text{-}13)$$

2.1.3 One-Dimensional-Double-Diffusion model

Since the wall thickness is very thin, which is around 0.2mm, it can be considered that temperature and water content of the hygroscopic material are uniform in the wall thickness direction. So that the two-dimensional-double-diffusion model can be simplified into an one-dimensional-double-diffusion model, which only considers heat conduction and vapor diffusion in the wheel thickness direction.

Governing equations of the one-dimensional-double-diffusion model can be referred to Eqs. (2-14) ~ (2-17).

Heat transfer differential equation:

$$\frac{1}{u_a}\frac{\partial t_a}{\partial \tau} + \frac{\partial t_a}{\partial z} = \frac{4h}{\rho_a c_{pa} u_a d_h}(t_d - t_a) \quad (2\text{-}14)$$

Mass transfer differential equation:

$$\frac{1}{u_a}\frac{\partial \omega_a}{\partial \tau} + \frac{\partial \omega_a}{\partial z} = \frac{4h_m}{\rho_a c_{pa} u_a d_h}(\omega_d - \omega_a) \quad (2\text{-}15)$$

Energy conservation equation:

$$\rho_d\left(c_{pd} + \frac{\rho_{ad} x}{\rho_d} c_{pw} W\right)\frac{\partial t_d}{\partial \tau} + x\rho_{ad} c_{pw} t_d \frac{\partial W}{\partial \tau} = k_d \frac{\partial^2 t_d}{\partial z^2} + r_s x \rho_{ad}\frac{\partial W}{\partial \tau} + \frac{4h}{d_h f}(t_a - t_d) \quad (2\text{-}16)$$

Mass conservation equation:

$$\varepsilon\rho_a \frac{\partial\omega_d}{\partial\tau} + \rho_{ad}\frac{\partial W}{\partial\tau} = D_S\rho_{ad}\frac{\partial^2 W}{\partial z^2} + \varepsilon\rho_a D_A \frac{\partial^2\omega_d}{\partial z^2} + \frac{4h_m}{xd_h f'}(\omega_a - \omega_d) \quad (2\text{-}17)$$

The boundary conditions and initial conditions are given in Eq. (2-18), where β is the angle, including the initial temperature and initial moisture content of the desiccant wheel, and the initial temperature and humidity of the air:

$$t_{a,p}\Big|_{z=0}^{\beta=0\sim 2\pi A_r/(1+A_r)} = t_{pin},\quad \omega_{a,p}\Big|_{z=0}^{\beta=0\sim 2\pi A_r/(1+A_r)} = \omega_{pin}$$

$$t_{a,r}\Big|_{z=L}^{\beta=0\sim 2\pi A_r/(1+A_r)} = t_{rin},\quad \omega_{a,r}\Big|_{z=L}^{\beta=0\sim 2\pi A_r/(1+A_r)} = \omega_{rin}$$

$$\frac{\partial W}{\partial z}\Big|_{z=0}^{\beta=0\sim 2\pi} = \frac{\partial W}{\partial z}\Big|_{z=L}^{\beta=0\sim 2\pi} = 0,\quad \frac{\partial t_d}{\partial z}\Big|_{z=0}^{\beta=0\sim 2\pi} = \frac{\partial t_d}{\partial z}\Big|_{z=L}^{\beta=0\sim 2\pi} = 0$$

$$t_d\Big|_{z=0\sim L}^{\tau=0} = t_{d,ini},\quad W\Big|_{z=0\sim L}^{\tau=0} = W_{ini},\quad t_a\Big|_{z=0\sim L}^{\tau=0} = t_{a,ini},\quad \omega_a\Big|_{z=0\sim L}^{\tau=0} = \omega_{a,ini}$$

(2-18)

2.1.4 One-Dimensional-Non-Diffusion model

Ignoring the energy term $\left(x\rho_{ad}c_{pw}t_d\frac{\partial W}{\partial\tau}\right)$, which is caused by mass change of the adsorbed water, the thermal conductivity term $\left(k_d\frac{\partial^2 t_d}{\partial z^2}\right)$ of desiccant material and substrate in Eq. (2-16), the variation of water vapor in the desiccant material $\left(\varepsilon\rho_a\frac{\partial\omega_d}{\partial\tau}\right)$, gas diffusion of water vapor in the desiccant material $\left(\varepsilon\rho_a D_A\frac{\partial^2\omega_d}{\partial Z^2}\right)$ and the surface diffusion of adsorbed water in the desiccant material $\left(D_S\rho_{ad}\frac{\partial^2 W}{\partial Z^2}\right)$ in Eq. (2-17), Eqs. (2-16) ~ (2-17) can be simplified into Eqs. (2-19) ~ (2-20), respectively:

Simplified energy conservation equation:

$$\rho_d\left(c_{pd} + \frac{\rho_{ad}x}{\rho_d}c_{pw}W\right)\frac{\partial t_d}{\partial\tau} = r_s x\rho_{ad}\frac{\partial W}{\partial\tau} + \frac{4h}{d_h f}(t_a - t_d) \quad (2\text{-}19)$$

Simplified mass conservation equation:

$$\rho_{ad}\frac{\partial W}{\partial\tau} = \frac{4h}{x\,d_h f}(\omega_a - \omega_d) \quad (2\text{-}20)$$

Boundary conditions and initial conditions are expressed as Eq. (2-18) .

2.2 Validations of the mathematic models

2.2.1 Test bench

Tests of the desiccant wheel were carried out in a psychrometric room. For the first test, schematic of the test rig is shown in Fig. 2-2.

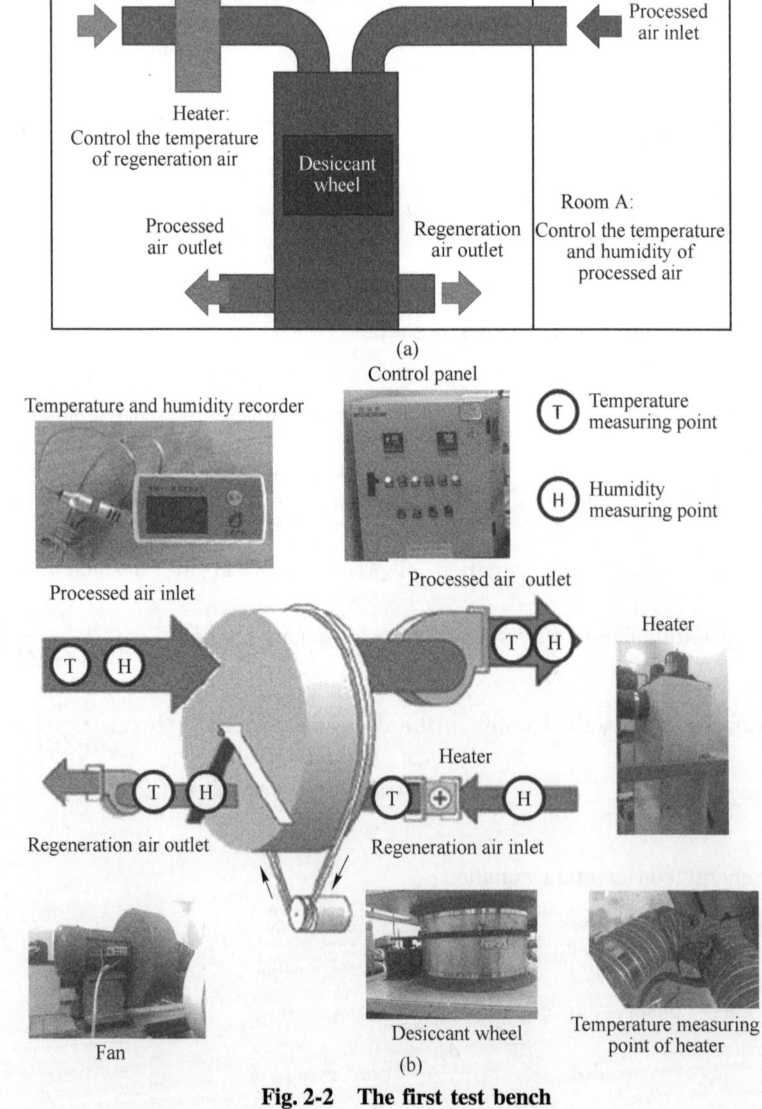

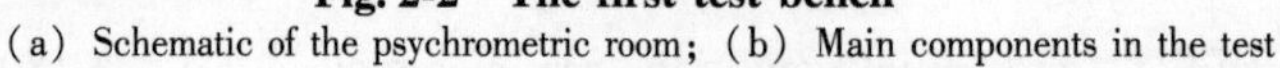

Fig. 2-2 The first test bench

(a) Schematic of the psychrometric room; (b) Main components in the test

The psychrometric room has two rooms. One is used for controlling the inlet temperature and moisture content of the processed air (Room A). The other room is used to control the inlet moisture content of the regenerated air (Room B), inlet temperature of which is regulated by an electric heater placed in the air inlet duct. Diameter of the desiccant wheel is 370mm, and the thickness is 200mm. One quarter of its windward side is used for regeneration and the other three quarters are used for dehumidification. Temperature and relative humidity of the processed air and the regeneration air at the intet and outlet of the wheel were measured. The temperature and humidity recorder (WSZY-1) were used to record the temperature and humidity, and the errors were ± 0.3℃ and ± 2% relative humidity. The air flow rate was obtained through measuring the air duct area and wind speed. Air speed was measured with a hot bulb anemometer (FB-1), error of which is ± 5%. Air outlet state under different control conditions and working conditions were obtained in the tests.

For the second experiment, the tested desiccant wheels was made of polymer. Four desiccant wheels with different length and radius were tested. The average height and width of the sinusoidal air channel were 1.28mm and 2.83mm, respectively. The cross section of the wheel was evenly divided between the process air and the regeneration air. And the two streams of air were designed to have the same mass flow rate, which was 0.1kg/s. Under each working condition the outlet temperature and humidity ratio of the air were recorded for different rotational speeds. The experiment setup and results were detailed introduced in Cao et al.'s work[11].

2.2.2 Validations

The one-dimensional-two-diffusion wodel was validated with the above test results. Simulations were carried out under the parameter settings same to the tests, which are listed in Table 2-1. The experimental results and energy and mass unbalance ratios are shown in Table 2-2. The unbalance rates were maintained within 20%. Simulation results and the experimental results were compared as shown in Fig. 2-3. The experimental results were in good agreement with the simulation results. So that the heat and mass transfer model described in this section can be used to simulate the performance of the desiccant wheel. Taking the 4th and 8th operating points as examples, the experimental results and simulation results of the air inlet and outlet state are shown in Fig. 2-4.

Table 2-1 Parameter settings in the simulation

ξ	λ_d [W/(m · ℃)]	q_{st} (J/kg)	ρ_{ad} (kg/m³)	x	c_{pad}[J/(kg · ℃)]	M_v(kg/kmol)
2.8	0.22	2.65×10^{6}	1129	0.7	920	18
ε	W_{max} (kg/kg)	D_0 (m²/s)	ρ_m (kg/m³)	a_d (m)	c_{pm}[J/(kg · ℃)]	P_0 (Pa)
0.7	0.39	1.6×10^{-6}	625	11×10^{-10}	880	100325

Table 2-2 Experimental results and simulation results

No.	Processed air inlet			Regeneration air inlet			Experimental outlet condition				unbalance ratio		The simulation results			
	$G_{a,p}$	$t_{a,p,in}$	$d_{a,p,in}$	$G_{a,r}$	$t_{a,r,in}$	$d_{a,r,in}$	$t_{a,p,out}$	$d_{a,p,in}$	$t_{a,r,out}$	$d_{a,r,in}$	Energy	Mass	$d_{a,p,in}$	$t_{a,r,out}$	$d_{a,r,in}$	$d_{a,p,in}$
1	610	35.2	21.5	170	62.3	10.3	43.8	18.4	36.5	21.5	16.4%	0.7%	42.7	18.8	34.9	19.8
2	610	35.0	17.9	170	61.8	10.2	41.7	15.6	37.3	18.7	1.9%	3.0%	42.0	15.6	36.4	18.5
3	610	35.1	15.4	170	61.8	10.3	40.9	13.8	38.1	17.3	13.9%	21.9%	41.6	13.4	38.1	17.6
4	610	35.1	23.0	170	61.3	9.8	41.9	19.7	34.7	20.6	9.0%	8.8%	42.7	20.2	33.7	19.7
5	610	35.0	22.9	170	60.9	9.3	41.9	19.4	34.5	20.2	6.6%	13.2%	42.6	20.1	33.4	19.3
6	610	35.2	25.7	150	69.0	9.7	42.8	22.3	34.9	25.1	10.3%	11.4%	43.8	22.6	33.6	22.2
7	610	35.2	25.7	150	78.6	9.1	44.6	22.1	35.6	26.8	12.5%	20.9%	45.8	22.1	34.9	23.9
8	610	35.2	25.7	150	88.8	9.3	46.6	21.8	36.5	28.6	12.8%	21.7%	47.8	21.5	36.6	26.3
9	610	35.2	25.7	150	98.0	10.3	48.5	21.6	37.4	30.7	12.0%	22.4%	49.5	21.1	38.3	29.0
10	610	35.1	25.7	150	97.9	9.1	50.5	20.9	37.1	30.2	2.9%	8.1%	49.6	21.0	37.8	28.1

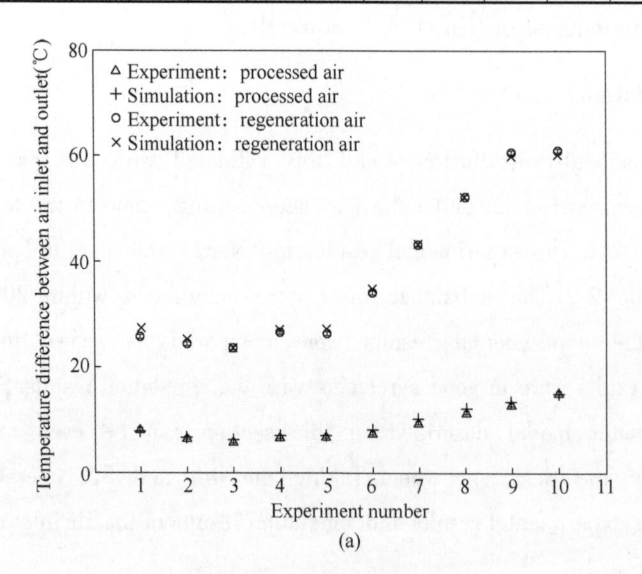

(a)

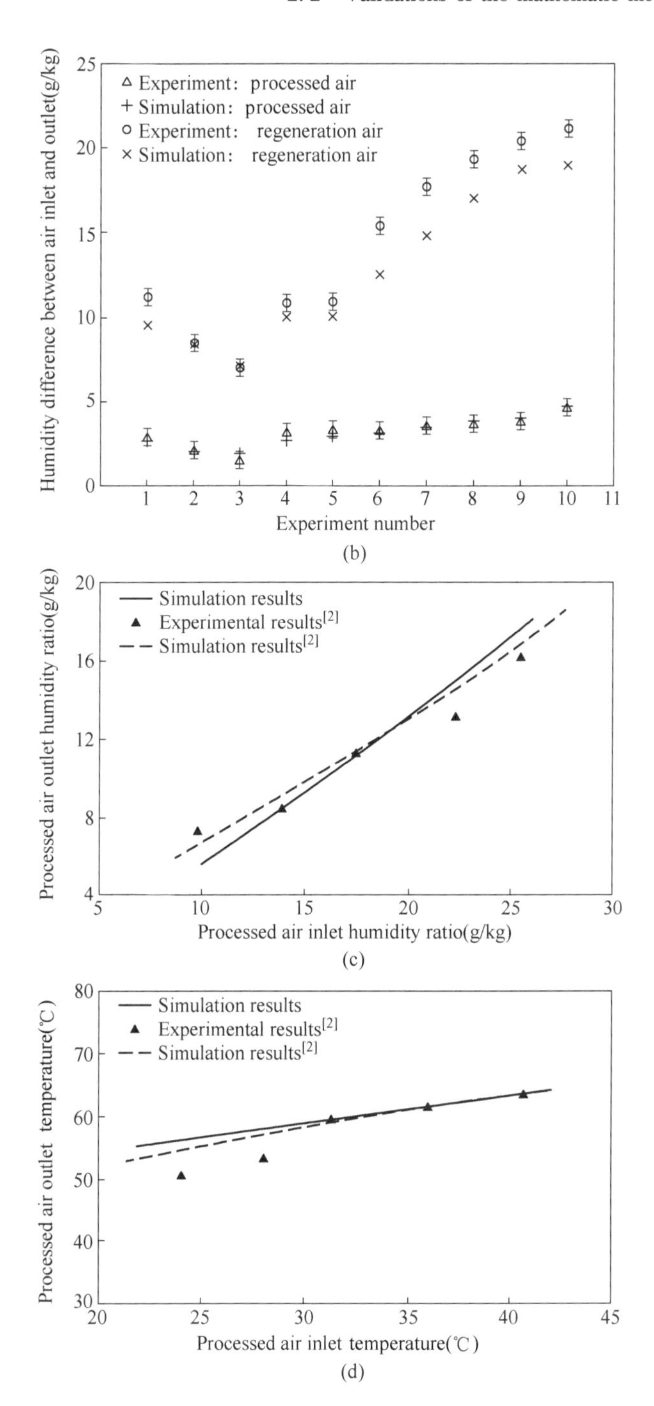

Fig. 2-3 Comparison of experimental and simulated results

(a) Temperature difference between inlet and outlet of the processed air; (b) Humidity difference between inlet and outlet of the processed air; (c) Outlet humidity of the processed air; (d) Outlet temperature of the processed air

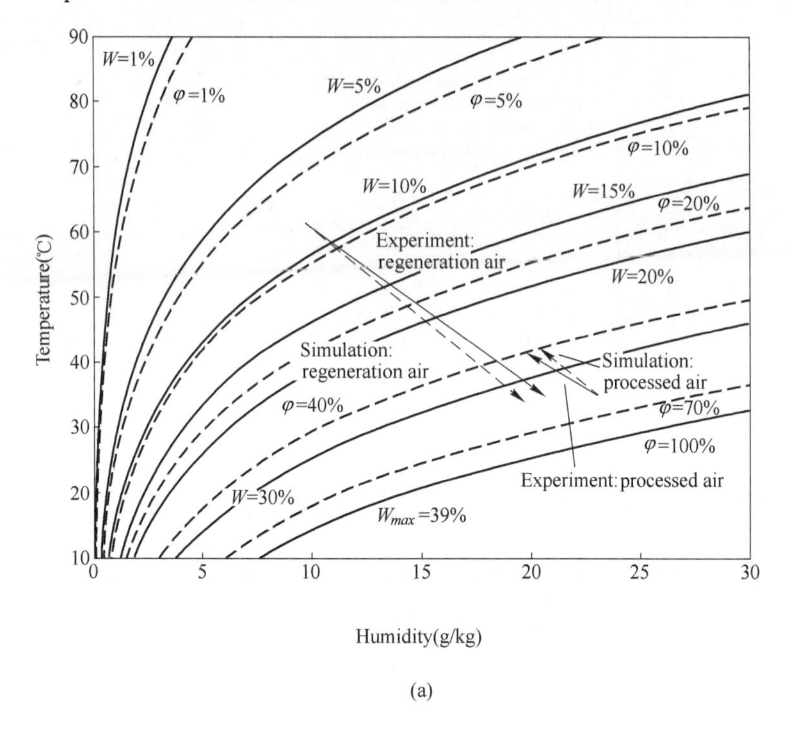

(a)

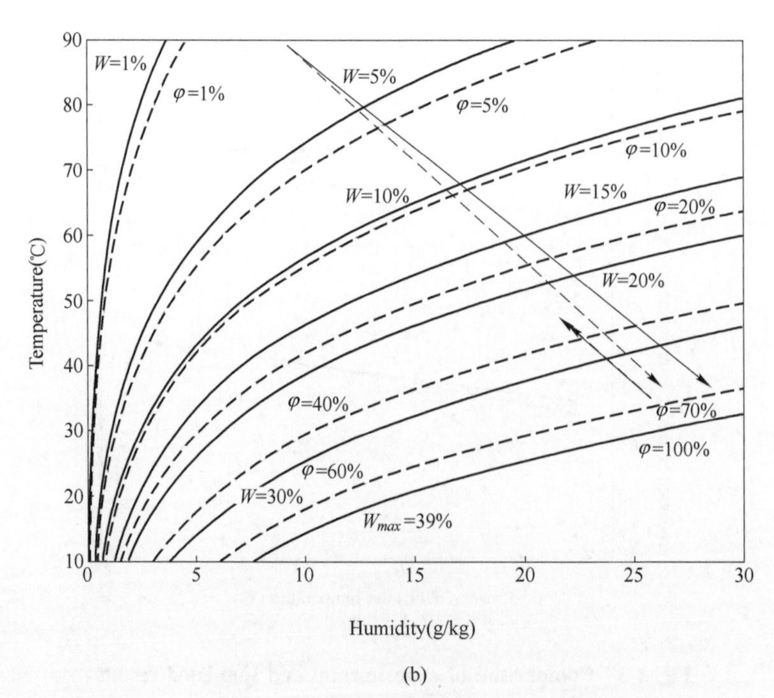

(b)

Fig. 2-4 Comparison of experimental and simulated results

(a) Working condition 4; (b) Working condition 8

For the second experiment, test results of one desiccant wheel under five groups of working conditions, which are shown in Fig. 2-5, were used for model validations. The diameter and length of the desiccant wheel are 150mm and 175mm, respectively. Parameters used in the mathematical model are listed in Table 2-3.

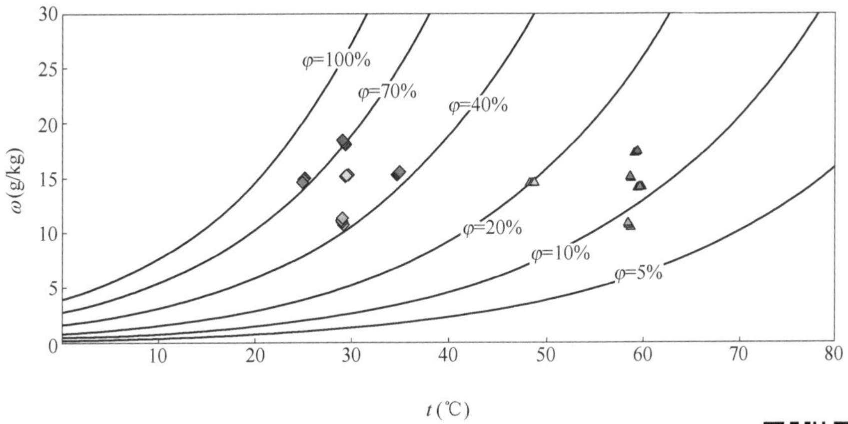

Fig. 2-5 Five groups of test conditions shown on the psychrometric chart

Table 2-3 Parameters used in the model

f	k_d [kW/(m·℃)]	r_s (kJ/kg)	ρ_{ad} (kg/m^3)	x	c_{pad} [kJ/(kg·℃)]	Mol (kg/kmol)
0.1765	0.00022	2.65×10^3	1129	0.7	0.92	18
σ	W_{max} (kg/kg)	C	ρ_d (kg/m^3)	a (m)	c_{pd} [kJ/(kg·℃)]	d_h (m)
0.7	0.39	0.5	978	11×10^{-10}	0.912	0.0012

The simulation results of process air outlet temperature and relative humidity changing with rotational speed under the five working conditions were compared with the corresponding experiment results. The detailed information can be referred to Fig. 2-6. The differences between the simulation results and the test results are shown in Fig. 2-7. It is demonstrated that the deviations of process air outlet temperature and relative humidity were within ± 5% and ± 12%, respectively. Therefore, it was concluded that this mathematical model could predict the experiment results well. The following discussion were based on the simulation results of this model.

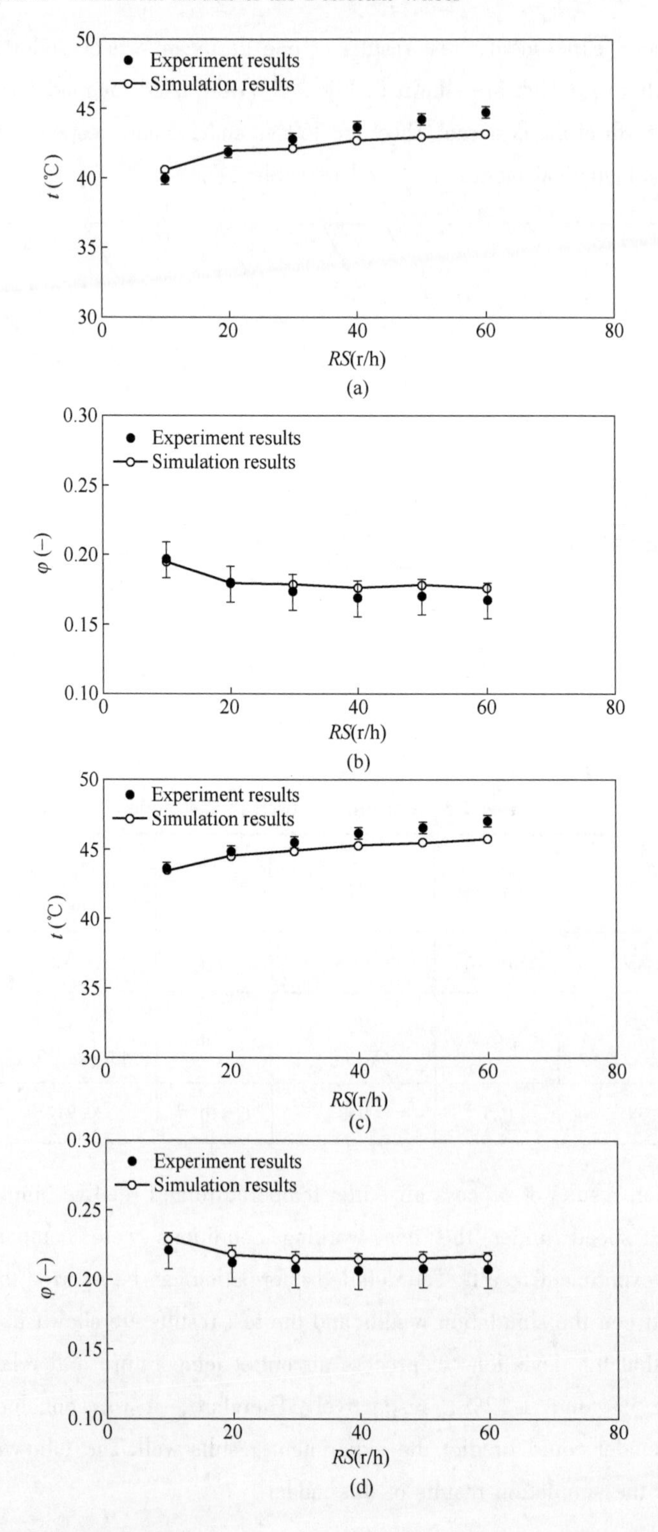

(a)
(b)
(c)
(d)

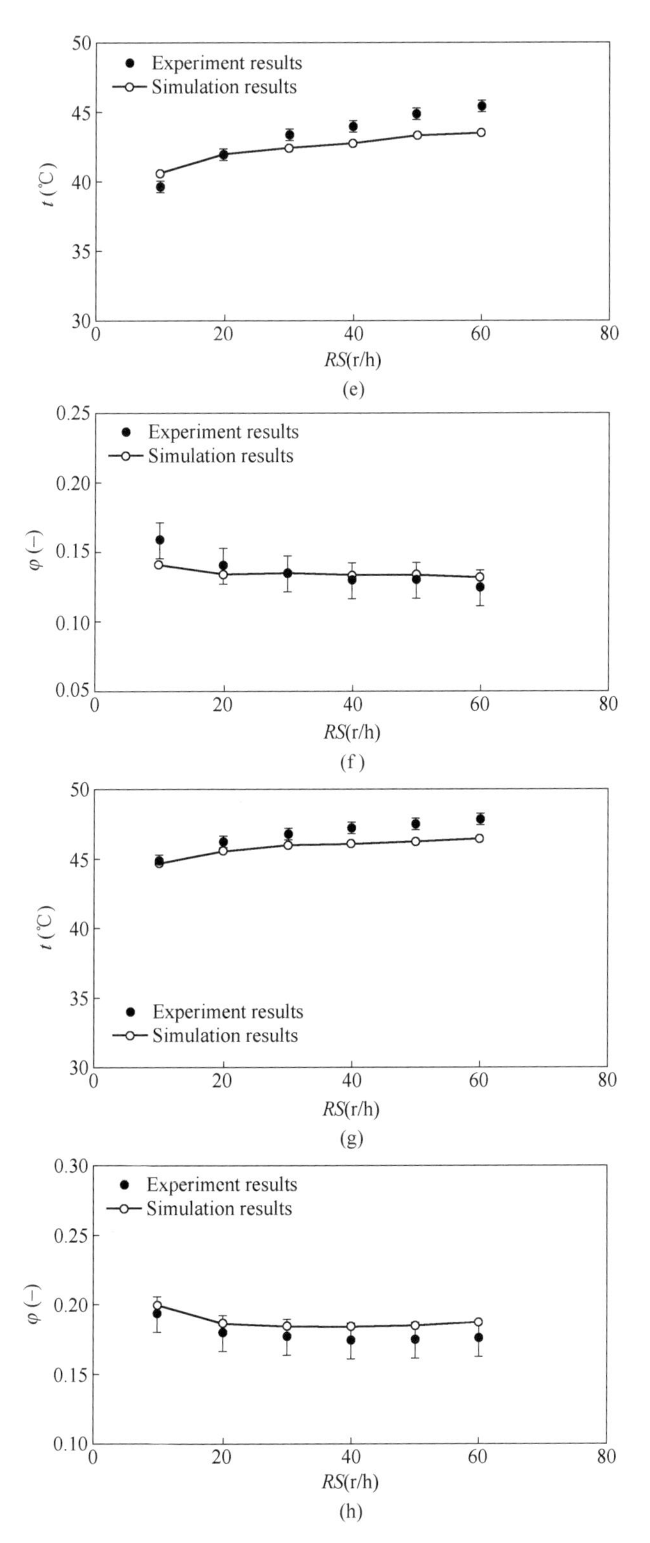
50
45
40
35
30
t (℃)
Experiment results
Simulation results
0
20
40
60
80
RS(r/h)
(e)
0.25
0.20
0.15
0.10
0.05
φ (–)
Experiment results
Simulation results
RS(r/h)
(f)
t (℃)
Experiment results
Simulation results
RS(r/h)
(g)
0.30
0.25
0.20
0.15
0.10
φ (–)
Experiment results
Simulation results
RS(r/h)
(h)

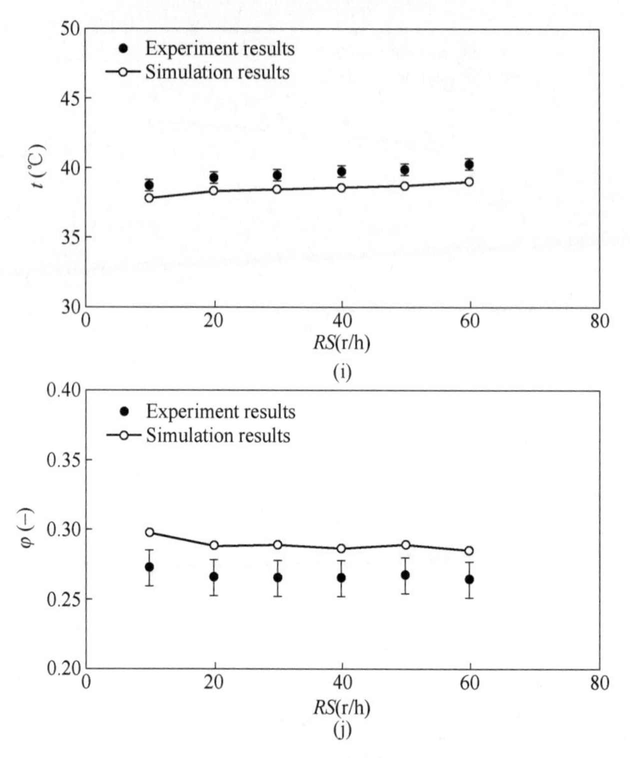

Fig. 2-6 Comparison of process air outlet temperature and relative humidity from experiment and simulation

(a) P1 and R1: temperature; (b) P1 and R1: relative humidity; (c) P2 and R2: temperature; (d) P2 and R2: relative humidity; (e) P3 and R3: temperature; (f) P3 and R3: relative humidity; (g) P4 and R4: temperature; (h) P4 and R4: relative humidity; (i) P5 and R5: temperature; (j) P5 and R5: relative humidity

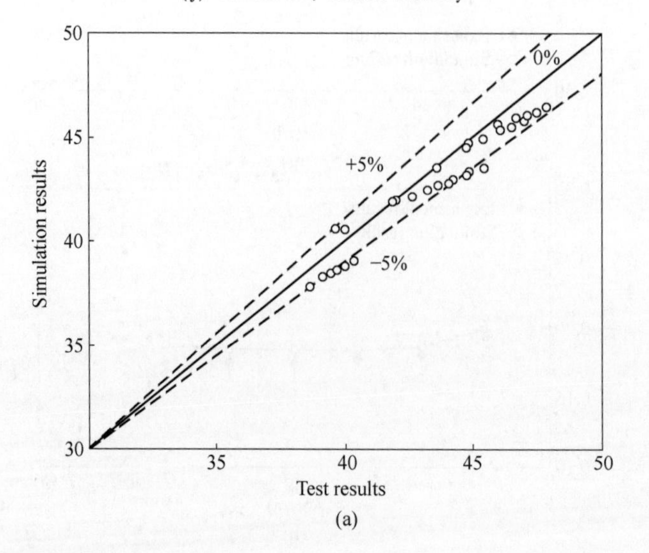

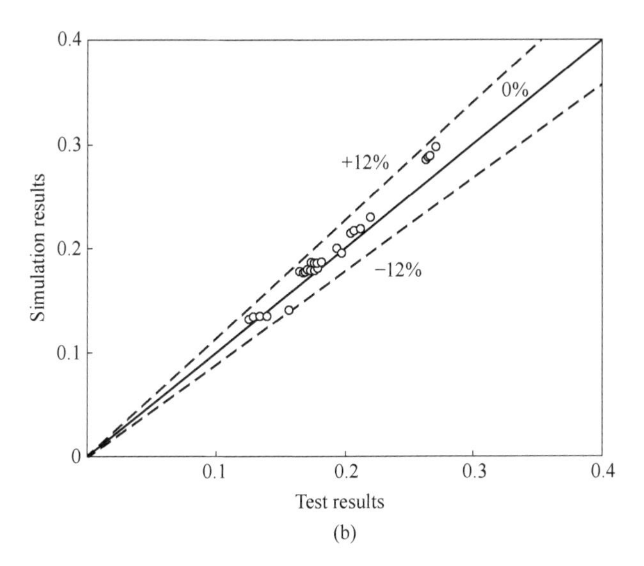

Fig. 2-7 Deviations between results from experiment and simulation

(a) Process air outlet temperature (℃); (b) Process air outlet relative humidity (-)

2.3 Dimensionless mathematic model

In this section, dimensionless numerical model was given based on the one-dimensional-non-diffusion desiccant model described above.

Outlet air temperature and humidity calculated from the one-dimensional double-diffusion model (model 1) and one-dimensional-non-diffusion model (model 2) are compared. The results are presented in Fig. 2-8 (a) and Fig. 2-8 (b), respectively. The deviations of the results calculated using the two models are within ±1%. Therefore, the accuracy of the simplified model is acceptable.

2.3.1 Dimensionless form

In order to make the calculation model more universal, dimensionless parameters of length, temperature, moisture content and angle are degined, as shown in Eq. (2-21):

$$Z^* = \frac{z}{L},\ \theta = \frac{t - t_{pin}}{t_{rin} - t_{pin}},\ \vartheta = \frac{\omega}{\omega_{pin}},\ W^* = \frac{W}{W_{max}},\ \alpha = \tau \cdot RS/3600 \quad (2\text{-}21)$$

where L is desiccant wheel's thickness, m; t_{rin} is the inlet temperature of regeneration air, ℃; t_{pin} is the inlet temperature of the processed air, ℃; W_{max} is the maximum moisture content of the dehumidifying material, kg water/kg dry desiccant; RS is the speed of the desiccant wheel, r/h; and τ is the running time, s.

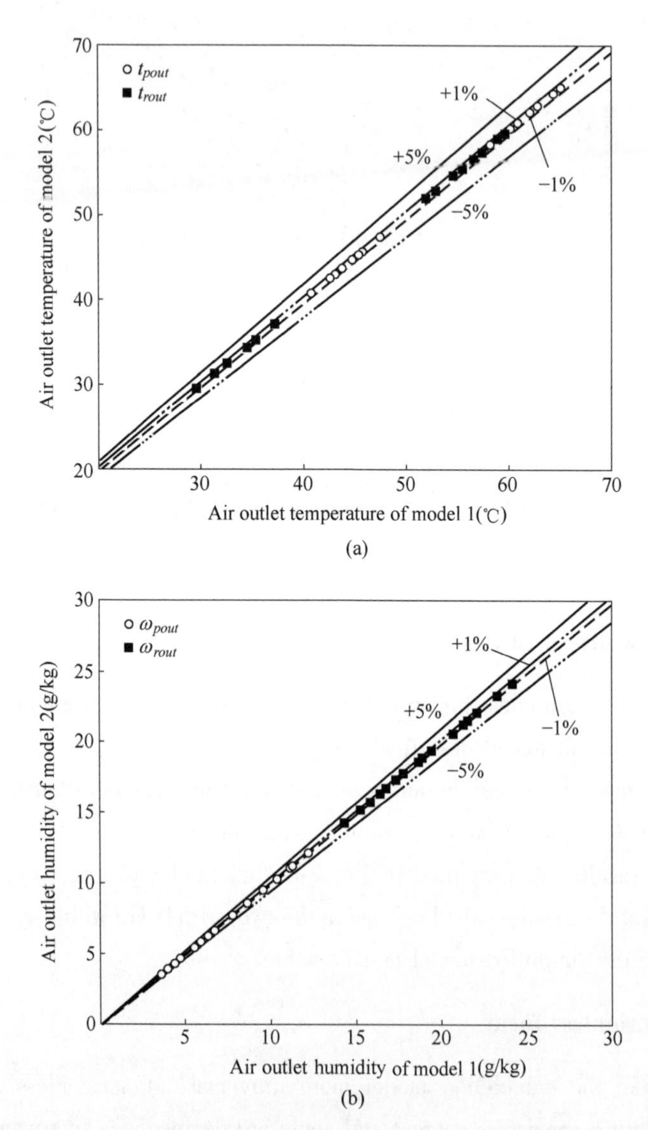

Fig. 2-8 Comparisons of the one-dimensional-non-diffusion model with the one-dimensional-double-diffusion model

(a) Temperature; (b) Humidity

The simplified model is nondimensionalized using the above dimensionless parameters, and the following dimensionless governing equations can be obtained. Eq. (2-22) is the dimensionless heat transfer differential equation:

$$\frac{1}{3600}\frac{L\cdot RS}{u_a}\frac{\partial\theta_a}{\partial\alpha}+\frac{\partial\theta_a}{\partial Z^*}=\frac{hLP}{\rho_a c_{pa}u_a A_a}(\theta_d-\theta_a) \tag{2-22}$$

Eq. (2-23) is the dimensionless mass transfer differential equation.

$$\frac{1}{3600}\frac{L\cdot RS}{u_a}\frac{\partial\vartheta_a}{\partial\alpha}+\frac{\partial\vartheta_a}{\partial Z^*}=\frac{h_m LP}{\rho_a u_a A_a}(\vartheta_d-\vartheta_a) \tag{2-23}$$

Eq. (2-24) is the dimensionless energy conservation equation.

$$\begin{aligned}&\left(1+\frac{A_d\rho_{ad}c_{pw}x\,W_{max}}{A_d\rho_d c_{pd}}W^*\right)\frac{\partial\theta_d}{\partial\alpha}\\&=\frac{A_d\rho_{ad}x\,W_{max}r_s}{A_d\rho_d c_{pd}(t_{rin}-t_{pin})}\frac{\partial W^*}{\partial\alpha}+\frac{hP}{A_d\rho_d c_{pd}RS/3600}(\theta_a-\theta_d)\end{aligned} \tag{2-24}$$

Eq. (2-25) is the dimensionless mass conservation equation.

$$\frac{\partial W^*}{\partial\alpha}=\frac{Ph_m\,\omega_{pin}}{xA_d\rho_{ad}\,W_{max}(\vartheta_a-\vartheta_d)} \tag{2-25}$$

2.3.2 Main dimensionless criteria parameters

The dimensionless parameters in Eqs. (2-22) ~ (2-25) are defined as shown in Eqs. (2-26) ~ (2-32):

The physical interpretations of the seven dimensionless criterion numbers are explained as follows. As expressed by Eq. (2-26), $\Psi=\Psi^*\cdot 3600$, where Ψ^* is the time required for air to pass through the wheel (L/u_a) divided by the time required for the wheel to rotate once (3600/RS).

$$\Psi = 3600\,\Psi^*$$

$$\Psi^*=\frac{L/u_a}{3600/RS}=\frac{1}{3600}\frac{L\cdot RS}{u_a} \tag{2-26}$$

As expressed by Eq. (2-27) and Eq. (2-28), NTU and NTU_m are the numbers of transfer units of the heat and mass transfer processes, respectively:

$$NTU=\frac{hF}{G_a\rho_a c_{pa}}=\frac{h(LP)}{(u_a A_a)\rho_a c_{pa}} \tag{2-27}$$

$$NTU_m=\frac{h_m F}{G_a\rho_a}=\frac{h_m(LP)}{(u_a A_a)\rho_a}=\frac{h_m}{h}\cdot NTU\cdot c_{pa} \tag{2-28}$$

where F is the heat transfer area of the control unit (m^2). As expressed by Eq. (2-29), the sensible heat ratio (SHR) is the maximum sensible heat capacity of water ($Q_{w,sh,max}$) divided by the maximum sensible heat of the solid part ($Q_{d,sh,max}$).

$$SHR = \frac{Q_{w,\ sh,\ max}}{Q_{d,\ sh,\ max}} = \frac{(\rho_{ad}x)\ W_{max}c_{pw}(t_{rin} - t_{pin})\ (3600/RS)}{\rho_d c_{pd}(t_{rin} - t_{pin})/(3600/RS)} = \frac{\rho_{ad}c_{pw}xW_{max}}{\rho_d c_{pd}} \tag{2-29}$$

As expressed by Eq. (2-30), the heat transfer ratio (HTR) is the maximum heat transfer rate ($Q_{ht,max}$) divided by the maximum sensible heat of the solid part ($Q_{d,sh,max}$).

$$HTR = \frac{Q_{ht,\ max}}{Q_{d,\ sh,\ max}} = \frac{hP(t_{rin} - t_{pin})}{A_d\rho_d c_{pd}(t_{rin} - t_{pin})\ /(3600/RS)} = \frac{4h}{f'd_h\rho_d c_{pd}RS/3600} \tag{2-30}$$

As expressed by Eq. (2-31), the latent heat ratio (LHR) is the maximum adsorption heat ($Q_{w,adh,max}$) divided by the maximum sensible heat of the solid part ($Q_{d,sh,max}$).

$$LHR = \frac{Q_{w,\ adh,\ max}}{Q_{d,\ sh,\ max}} = \frac{(\rho_{ad}x)\ W_{max}r_s/(3600/RS)}{\rho_d c_{pd}(t_{rin} - t_{pin})/(3600/RS)} = \frac{\rho_{ad}xW_{max}r_s}{\rho_d c_{pd}(t_{rin} - t_{pin})} \tag{2-31}$$

As expressed by Eq. (2-32), the mass transfer ratio (MTR) is the maximum mass transfer rate ($M_{mt,max}$) divided by the maximum water in the solid part ($M_{w,max}$).

$$MTR = \frac{M_{mt,\ max}}{M_{w,\ max}} = \frac{h_m P(\omega_{pin} - 0)}{(xA_d\rho_{ad})\ W_{max}/(3600/RS)} = \frac{4h_m\omega_{pin}f}{x(1-f)d_h\rho_{ad}W_{max}RS/3600} \tag{2-32}$$

The dimensionless equations in Eqs. (2-22) ~ (2-25) can be further simplified, as shown in Eqs. (2-33) ~ (2-36):

$$\frac{1}{3600}\Psi\frac{\partial\theta_a}{\partial\alpha} + \frac{\partial\theta_a}{\partial Z^*} = NTU(\theta_d - \theta_a) \tag{2-33}$$

$$\frac{1}{3600}\Psi\frac{\partial\vartheta_a}{\partial\alpha} + \frac{\partial\vartheta_a}{\partial Z^*} = NTU_m(\vartheta_d - \vartheta_a) \tag{2-34}$$

$$(1 + SHR \cdot W^*)\frac{\partial\theta_d}{\partial\alpha} = LHR\frac{\partial W^*}{\partial\alpha} + HTR(\theta_a - \theta_d) \tag{2-35}$$

$$\frac{\partial W^*}{\partial\alpha} = MTR(\vartheta_a - \vartheta_d) \tag{2-36}$$

The corresponding dimensionless boundary conditions are shown in Eq. (2-37).

$$\theta_a \Big|_{z^*=0}^{\alpha=0\sim A_r/(1+A_r)} = \theta_{pin},\quad \vartheta_a \Big|_{z^*=0}^{\alpha=0\sim A_r/(1+A_r)} = \vartheta_{pin}$$

$$\theta_a \Big|_{z^*=1}^{\alpha=A_r/(1+A_r)\sim 1} = \theta_{rin},\quad \vartheta_a \Big|_{z^*=1}^{\alpha=A_r/(1+A_r)\sim 1} = \vartheta_{rin}$$

$$\frac{\partial W^*}{\partial z^*}\Big|_{z^*=0}^{\alpha=0\sim1}=0,\ \frac{\partial W^*}{\partial z^*}\Big|_{z^*=1}^{\alpha=0\sim1}=0,\ \frac{\partial \theta_d}{\partial z^*}\Big|_{z^*=0}^{\alpha=0\sim1}=0,\ \frac{\partial \theta_d}{\partial z^*}\Big|_{z^*=1}^{\alpha=0\sim1}=0 \tag{2-37}$$

The dimensionless initial conditions are shown in Eq. (2-38) .

$$\theta_d\Big|_{z^*=0\sim1}^{\alpha=0}=\theta_{d,\ ini},\ W^*\Big|_{z^*=0\sim1}^{\alpha=0}=W^*_{ini}$$

$$\theta_a\Big|_{z^*=0\sim1}^{\alpha=0}=\theta_{a,\ ini},\ \vartheta_a\Big|_{z^*=0\sim1}^{\alpha=0}=\vartheta_{a,\ ini} \tag{2-38}$$

It is observed that there are fixed constants such as ρ_a, c_{pa} and c_{pw} in the definition of dimensionless parameters, and there are clear relationships among some parameters, such as A_a, A_d and f; h and h_m. In this part, the dimensionless parameters involved are divided into the basic dimensionless number and the extended dimensionless number, and the extended dimensionless number can be expressed by the basic dimensionless number and constant.

$$NTU_m=\frac{1}{c_{pa}\cdot Le}\cdot NTU \tag{2-39}$$

$$HTR=\frac{fc_{pa}\rho_a}{(1-f)\rho_{ad}c_{pw}xW_{max}}\cdot\frac{NTU\cdot SHR}{\Psi} \tag{2-40}$$

$$LHR=\frac{1}{t_{rin}-t_{pin}}\cdot\frac{r_s}{c_{pw}}\cdot SHR \tag{2-41}$$

$$MTR=\omega_{pin}\cdot\frac{\rho_a f}{x(1-f)\rho_{ad}W_{max}}\cdot\frac{NTU}{\Psi} \tag{2-42}$$

Therefore, under the condition of fixed desiccant material, structure ratio (x), inlet air temperature difference ($t_{rin}-t_{pin}$) and air humidity (ω_{pin}), 3 basic dimensionless numbers (Ψ, NTU and SHR) can be used to represent the remaining 4 extended dimensionless numbers, and the 7 key dimensionless numbers of the desiccant wheel can be simplified to 3 basic dimensionless numbers.

2.4 Prediction models of desiccant wheels

In this section, the silica gel, which is widely used for air dehumidification, was used as the desiccant material for example.

2.4.1 Value ranges of the 3 basic dimensionless numbers

The values of the parameters for the desiccant wheels, air, water, desiccant material, and matrix material are listed in Table 2-4.

Table 2-4 Values of parameters for desiccant wheels, air, water, desiccant material, and matrix material

Category		Value
Desiccant wheels	L(m)	0.1~0.4[2,12-14]
	RS(r/h)	8~18[2,15]
	u_a(m/s)	1~4[2,14]
	δ(mm)	0.3[16], 0.45[17], 0.1[18], 0.15[17-20]
	A_r	1[17]
	f	0.816[17]
	x	0.75[18]
Desiccant material	ε	0.75[18], 0.7[17-20]
	ρ_{ad} (kg/m^3)	1129[2,7,18]
	λ_{ad} [W/(m · K)]	0.20[2,18], 0.175[17], 0.22[21]
	r_s(J/kg)	2650000[2,21,22]
	W_{max} (kg/kg)	0.39[2,14,21]
	c_{pad} [J/(kg · K)]	921[17,20,21]
Matrix material	c_{pm} [J/(kg · K)]	880[2,21]
	ρ_m (kg/m^3)	793[15], 625[14]
Water	c_{pw} [J/(kg · K)]	4200[2,15]
Air	c_{pa} [J/(kg · K)]	1003[2,15]
	ρ_a (kg/m^3)	1.15[2,15]
	λ_a [W/(m · K)]	0.23[2,15,23]

For different a/b values provided in the literature[7,15,16], Nu, d_h, and h were calculated (Table 2-5). f is fixed at 0.816 (Table 2-5). d_h ranges between 0.97 and 2.14mm, and h ranges between 21.9 and 48.9W/(m^2 · K).

Table 2-5 Air passage parameters

a/b (mm/mm)	d_h (mm)	Nu	h [W/(m^2 · K)]
1.2/2.4	0.97	2.07	48.9
1.4/2.6	1.11	2.04	42.3
1.34/4.35	1.21	2.03	38.6
1.9/3.4	1.49	2.02	31.2

Continued Table 2-5

a/b (mm/mm)	d_h (mm)	Nu	h [W/(m^2 · K)]
2.5/4.2	1.92	1.97	23.7
2.7/5	2.14	2.04	21.9
1.8/3.2	1.41	2.02	32.9
2.0/5.0	1.72	2.08	27.9
1.75/3.5	1.47	2.07	32.3
1.74/4.35	1.53	2.08	31.4

Based on the values listed in Table 2-4 and Table 2-5, the value ranges of the three basic criterion numbers were calculated, as listed in Table 2-6.

Table 2-6 Value ranges of the three basic criterion numbers

Category	Value range	References [14, 23]	Reference [16]
Ψ	0.2~3.2	2.4	0.92~1.2
NTU	9.46~21.088	14.228	10.94~15.2
SHR	0.992~1.654	1.444	1.08~1.6

2.4.2 Original data calculated from the dimensionless model

Based on the value ranges listed in Table 2-6, five values were selected for each of the three basic criterion numbers (Table 2-7). In total, there were 125 scenarios, considering different combinations of the three basic criterion numbers.

Table 2-7 Values of three basic criterion numbers for simulation study

Category	Value				
SHR	0.992	1.1575	1.323	1.444	1.654
NTU	9.46	12.367	14.228	18.181	21.088
Ψ	0.2	0.95	1.7	2.45	3.2

Precooling A_{pro} is an effective method for reducing t_{rin} [20,23]. In addition, the return air is often adopted as A_{reg} [23,24]. In this study, the inlet air states were selected based on the two aspects. t_{pin} was fixed at 25℃, which could be achieved with free cooling. ω_{rin} was fixed at 12g/kg, which satisfied the thermal comfort requirements of residential and commercial buildings[25]. The values of the other air inlet states are listed in Table 2-8.

Table 2-8 Air inlet parameters for simulation study

Category	Value
t_{rin} (℃)	50~90[2,12-14,23,24]
ω_{rin} (g/kg)	12[2,12,13,23,24]
t_{pin} (℃)	25[2,12,13,23,24]
ω_{pin} (g/kg)	14~20[2,12,13,23,24]

ω_{pin} ranged from 14g/kg to 20g/kg, and four values (14, 16, 18 and 20g/kg) were selected for the simulation study. t_{rin} ranged from 50℃ to 90℃, and five values (90, 80, 70, 60 and 50℃) were selected for the simulation study. Therefore, 20 working conditions were evaluated, considering different combinations of t_{rin} and ω_{pin}. Finally, 2500 cases were designed, considering the 20 working conditions and 125 scenarios. The output data of 2500 groups, namely θ_{pout}, ϑ_{pout}, θ_{rout} and ϑ_{rout}, were calculated.

2.4.3 Prediction model using multiple regression method

The multiple regression method was used to derive formulas for the fixed working conditions. Two working conditions were selected for the analysis. For the first working condition (Case 1), t_{rin} was 90℃, ω_{rin} was 12g/kg, t_{pin} was 25℃, and ω_{pin} was 20g/kg. For the second working condition (Case 2), t_{rin} was 50℃, ω_{rin} was 12g/kg, t_{pin} was 25℃, and ω_{pin} was 14g/kg. For each working condition, there were 125 groups of output data related to the three basic criterion numbers. The relations of the four dimensionless output parameters with the three basic criterion numbers were regressed.

The goodness-of-fit was evaluated using the coefficient of determination (R^2), R^2_{adj} and the root-mean-square error ($RMSE$), as expressed by Eqs. (2-43) ~ (2-45), respectively:

$$R^2 = 1 - \frac{\sum (Y_{act} - Y_{pre})^2}{\sum (Y_{act} - Y_{ave})^2} \tag{2-43}$$

$$R^2_{adj} = 1 - \frac{(1 - R^2)(n - 1)}{n - p - 1} \tag{2-44}$$

$$RMSE = \sqrt{\frac{1}{n}\sum (Y_{act} - Y_{pre})^2} \tag{2-45}$$

where Y_{act}, Y_{pre} and Y_{ave} are the target, predicted, and average target values, respectively, n is the number of samples, and p is the number of features. The R^2 and

R^2_{adj} were smaller than 1. The higher the R^2 and R^2_{adj} values, the better the goodness-of-fit. The *RMSE* is the error between Y_{act} and Y_{pre}. The lower the *RMSE* value, the better the goodness-of-fit.

θ_{pout} in Case 1 was used as an example, and the formulas were regressed in four steps. The steps for the multiple regression calculations are depicted in the flow diagram shown in Fig. 2-9.

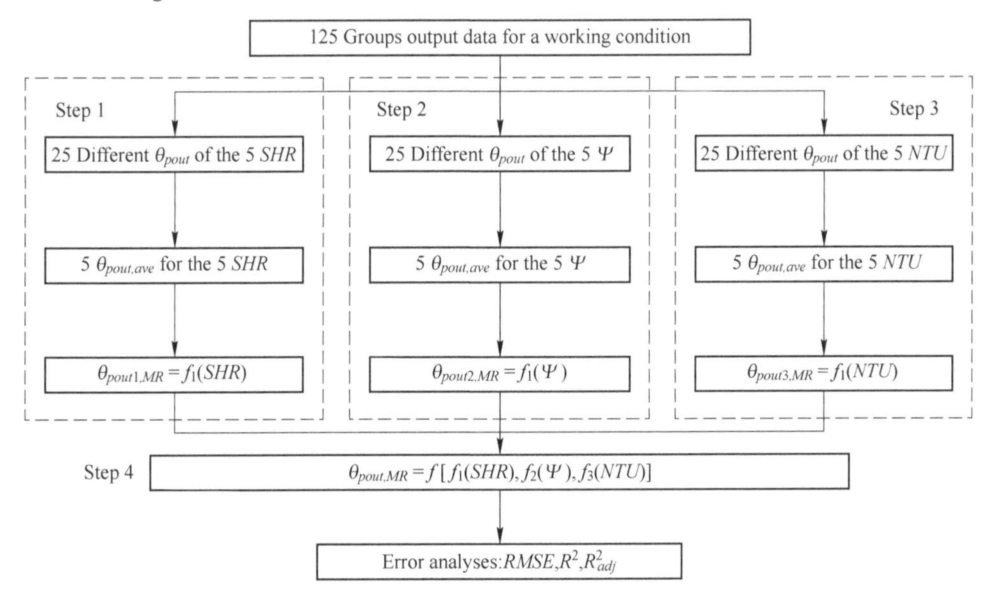

Fig. 2-9 Flow diagram of multiple regression method using $\theta_{pout,MR}$ as example

Step 1: Derive the regression equation expressing the relationship between θ_{pout} and *SHR* as $\theta_{pout1,\ MR} = f_1(\mathrm{SHR})$

As shown in Fig. 2-10 (a), all 25 θ_{pout} (shown as black squares) for the same *SHR* were averaged. Thus, for the five *SHR* values, five $\theta_{pout,\ ave}$ were obtained (shown as stars in Fig. 2-10 (a)). The regression equation expressing the relationship between $\theta_{pout1,\ MR}$ and *SHR* is as follows.

$$\theta_{pout1,\ MR} = f_1(SHR) = -164.1079 \cdot \exp\left(\frac{-SHR}{0.15462}\right) + 0.8078 \qquad (2\text{-}46)$$

The dot-dash curve depicted in Fig. 2-10 (a) represents the prediction result ($\theta_{pout1,MR}$) calculated using Eq. (2-46). If an error of ± 10% is added to $\theta_{pout1,MR}$, full line and dashed can be drawn. All the black squares were within the line, indicating that the errors between $\theta_{pout1,\ MR}$ and θ_{pout} for the same *SHR* were within ±10%.

Step 2: Derive the regression equation expressing the relationship between θ_{pout} and Ψ

as $\theta_{pout2,\ MR}=f_2(\Psi)$

Step 3: Derive the regression equation expressing the relationship between θ_{pout} and NTU as $\theta_{pout3,\ MR}=f_3(NTU)$

Similar to Step 1, $\theta_{pout2,\ MR}=f_2(\Psi)$ and $\theta_{pout3,\ MR}=f_3(NTU)$ can be derived in Steps 2 and 3, respectively. The results obtained following the two steps are presented in Fig. 2-10 (b) and (c), respectively, and the regression equations are expressed by Eqs. (2-47) and (2-48), respectively.

$$\theta_{pout2,\ MR}=f_2(\Psi)=-0.10385\cdot\exp\left(\frac{-\Psi}{2.88429}\right)+0.78736 \tag{2-47}$$

$$\theta_{pout3,\ MR}=f_3(NTU)=0.13842\exp\frac{-NTU}{8.89275}+0.69758 \tag{2-48}$$

Step 4: Derive the final regression equation expressing the correlation between θ_{pout} and f_1, f_2 and f_3 as $\theta_{pout,\ MR}=f\,[f_1(SHR),\ f_2(\Psi),\ f_3(NTU)]$

Based on Eqs. (2-46) ~ (2-48), the final regression formula can be derived, as expressed by Eq. (2-49).

$$\begin{aligned}\theta_{pout,\ MR}&=\theta_{pout1,\ MR}+\theta_{pout2,\ MR}+\theta_{pout3,\ MR}-1.4516\\&=-164.1079\cdot\exp\left(\frac{-SHR}{0.15462}\right)-0.10385\cdot\exp\left(\frac{-\Psi}{2.88429}\right)+\\&\quad 0.13842\cdot\exp\left(\frac{-NTU}{8.89275}\right)+0.84114\end{aligned} \tag{2-49}$$

The $RMSE$, R^2 and R^2_{adj} of the formulas were 1.054×10^{-2}, 0.9946, and 0.9893, respectively (Fig. 2-10 (d)). The results indicated a good fitting effect.

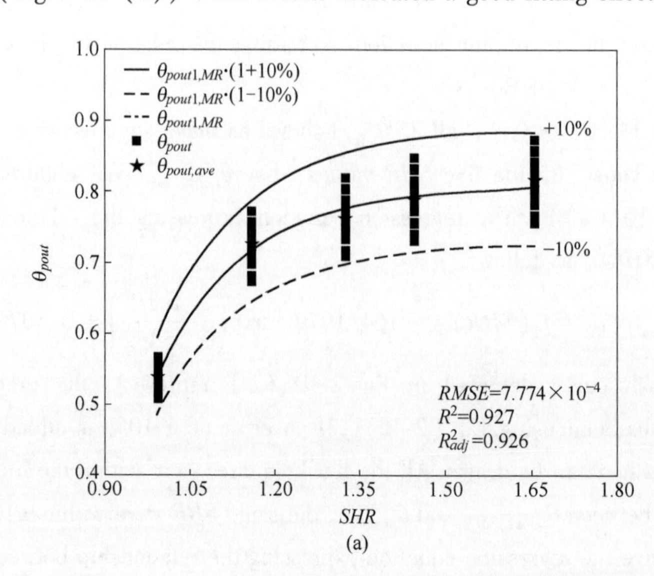

(a)

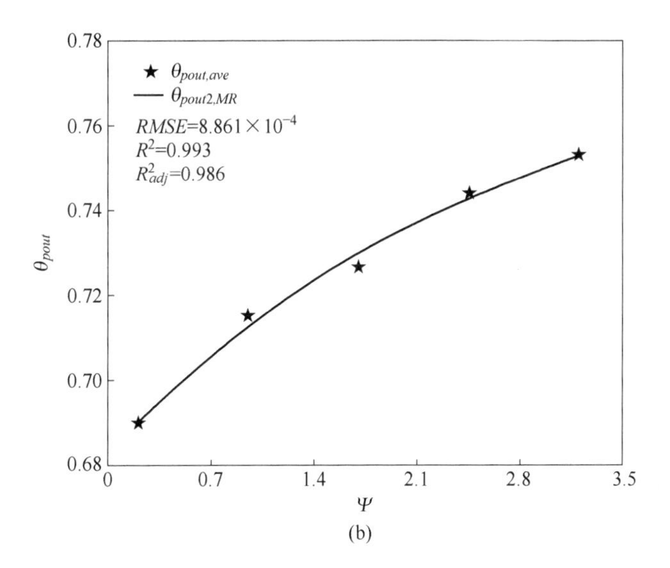

0.78
0.76
0.74
0.72
0.70
0.68
$\theta_{pout,ave}$
$\theta_{pout2,MR}$
$RMSE$=8.861×10^{-4}
R^2=0.993
R^2_{adj}=0.986
θ_{pout}
0
0.7
1.4
2.1
2.8
3.5
Ψ

(b)

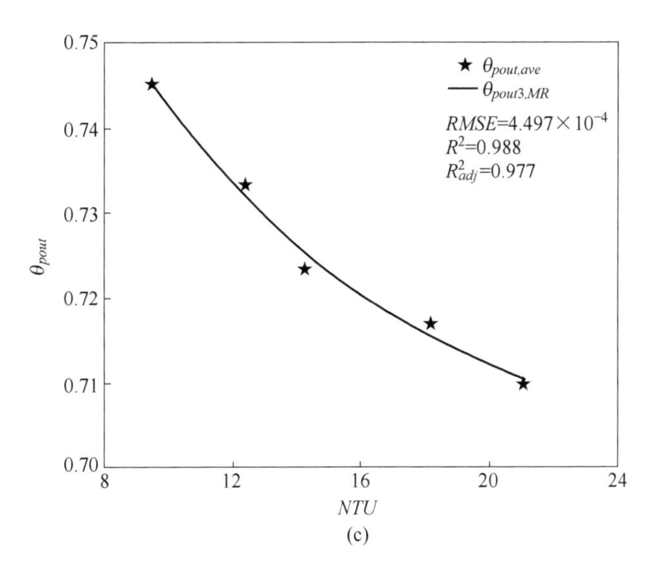

0.75
0.74
0.73
0.72
0.71
0.70
$\theta_{pout,ave}$
$\theta_{pout3,MR}$
$RMSE$=4.497×10^{-4}
R^2=0.988
R^2_{adj}=0.977
θ_{pout}
8
12
16
20
24
NTU

(c)

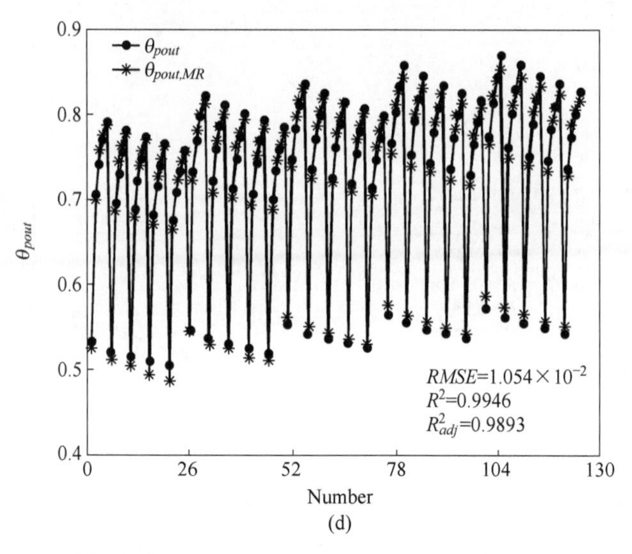

(d)

Fig. 2-10 Four steps to regress formula of θ_{pout}

(a) $\theta_{pout1,\ MR} = f_1(SHR)$; (b) $\theta_{pout2,\ MR} = f_2(\Psi)$;
(c) $\theta_{pout3,\ MR} = f_3(NTU)$; (d) $\theta_{pout,\ MR} = f\,[f_1(SHR),\ f_2(\Psi),\ f_3(NTU)]$

By following the same steps described above, the multiple regression formulas of ϑ_{pout}, θ_{rout} and ϑ_{rout} can be obtained, as expressed by Eqs. (2-50) ~ (2-52), respectively.

$$\vartheta_{pout,\ MR} = 0.73689 \cdot SHR^2 - 1.8449 \cdot SHR + 0.21426 \cdot \exp\left(\frac{-NTU}{7.95844}\right) + 0.08867 \cdot \exp\left(\frac{-\Psi}{2.59879}\right) + 1.44354 \tag{2-50}$$

$$\theta_{rout,\ MR} = 587.14356 \cdot \exp\left(\frac{-SHR}{0.12819}\right) - 0.1245 \cdot \exp\left(\frac{-NTU}{8.91438}\right) + 0.10238 \cdot \exp\left(\frac{-\Psi}{2.91691}\right) + 0.16886 \tag{2-51}$$

$$\vartheta_{rout,\ MR} = -0.75028 \cdot SHR^2 + 1.85284 \cdot SHR - 0.22124 \cdot \exp\left(\frac{-NTU}{8.00056}\right) - 0.09168 \cdot \exp\left(\frac{-\Psi}{2.78305}\right) + 0.16811 \tag{2-52}$$

For Case 2, the regression formulas are expressed by Eqs. (2-53) ~ (2-56).

$$\theta_{pout,\ MR} = -0.11223 \cdot \exp\left(\frac{-\Psi}{2.84148}\right) + 0.13064 \cdot \exp\left(\frac{-NTU}{8.10751}\right) -$$

$$1.17431 \cdot \exp\left(\frac{-SHR}{0.81112}\right) + 1.03096 \tag{2-53}$$

$$\vartheta_{pout,\ MR} = 0.06066 \cdot \exp\left(\frac{-\Psi}{2.86565}\right) + 0.13668 \cdot \exp\left(\frac{-NTU}{8.15043}\right) -$$

$$1.89011 \cdot \exp\left(\frac{-SHR}{5.75038}\right) + 2.14354 \tag{2-54}$$

$$\theta_{rout,\ MR} = 0.09183 \cdot \exp\left(\frac{-\Psi}{2.43083}\right) - 0.11206 \cdot \exp\left(\frac{NTU}{8.5483}\right) +$$

$$1.22534 \cdot \exp\left(\frac{SHR}{0.55592}\right) + 0.12028 \tag{2-55}$$

$$\vartheta_{rout,\ MR} = -0.04884 \cdot \exp\left(\frac{-\Psi}{2.4702}\right) - 0.14026 \cdot \exp\left(\frac{-NTU}{8.27211}\right) +$$

$$2.32973 \cdot \exp\left(\frac{-SHR}{6.11397}\right) - 0.68421 \tag{2-56}$$

The goodness-of-fit values for the two cases are listed in Table 2-9.

Table 2-9 Goodness-of-fit values for Cases 1 and cases 2

Category		*RMSE*	R^2	R^2_{adj}
Case 1	$\theta_{pout,\ MR}$	1.054×10^{-2}	0.9946	0.9893
	$\vartheta_{pout,\ MR}$	9.251×10^{-2}	0.8786	0.772
	$\theta_{rout,\ MR}$	1.289×10^{-2}	0.9947	0.9894
	$\vartheta_{rout,\ MR}$	9.854×10^{-2}	0.8871	0.7869
Case 2	$\theta_{pout,\ MR}$	6.747×10^{-3}	0.9956	0.9913
	$\vartheta_{pout,\ MR}$	8.996×10^{-3}	0.9897	0.9795
	$\theta_{rout,\ MR}$	5.077×10^{-3}	0.9958	0.9912
	$\vartheta_{rout,\ MR}$	9.93×10^{-3}	0.9906	0.9814

The errors between the simplified model and regression formulas for both cases are shown in Fig. 2-11, respectively.

For Case 1, the errors of θ_{pout} and ϑ_{rout} were within ±5%, and the errors of ϑ_{pout} and θ_{rout} were within ±15%. For Case 2, the errors of θ_{pout}, ϑ_{pout}, θ_{rout} and ϑ_{rout} were within ±5%. The regression formulas were highly accurate for predicting the four dimensionless outlet states of air.

Similarly, the results for Case 3 (t_{rin} = 70℃, ω_{rin} = 12g/kg, t_{pin} = 25℃, and ω_{pin} = 16g/kg) are expressed by Eqs. (2-57)~(2-60).

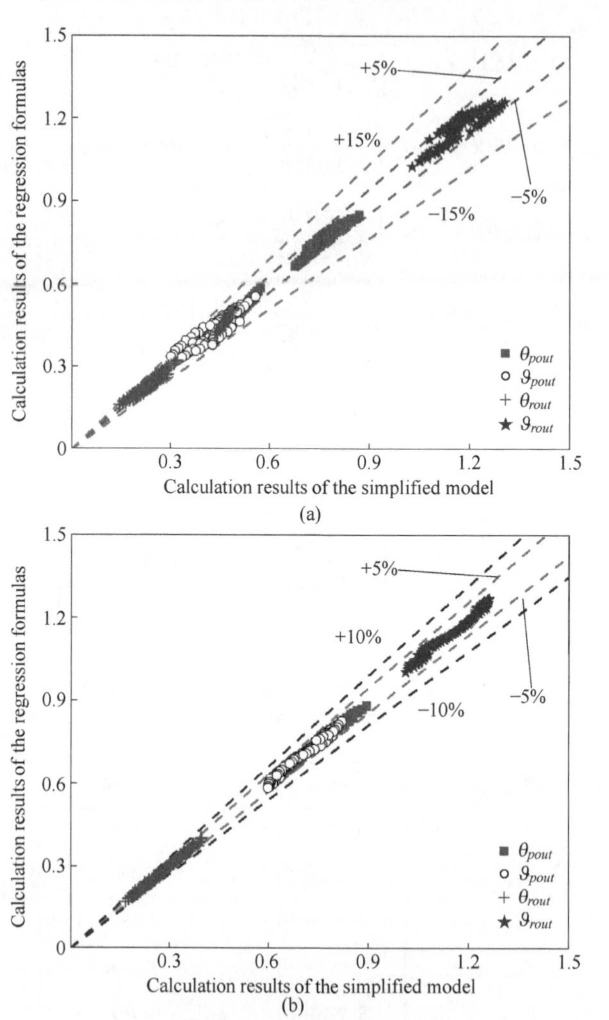

Fig. 2-11 Fitting errors

(a) Case 1; (b) Case 2

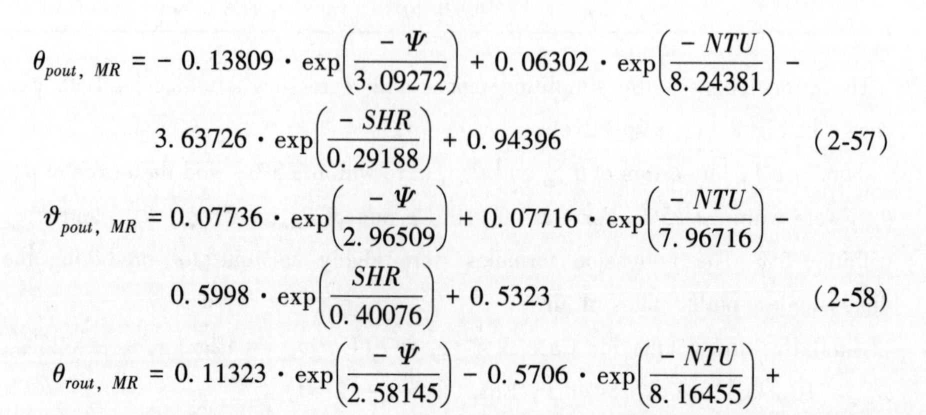

$$\theta_{pout,\ MR} = -0.13809 \cdot \exp\left(\frac{-\Psi}{3.09272}\right) + 0.06302 \cdot \exp\left(\frac{-NTU}{8.24381}\right) - 3.63726 \cdot \exp\left(\frac{-SHR}{0.29188}\right) + 0.94396 \tag{2-57}$$

$$\vartheta_{pout,\ MR} = 0.07736 \cdot \exp\left(\frac{-\Psi}{2.96509}\right) + 0.07716 \cdot \exp\left(\frac{-NTU}{7.96716}\right) - 0.5998 \cdot \exp\left(\frac{SHR}{0.40076}\right) + 0.5323 \tag{2-58}$$

$$\theta_{rout,\ MR} = 0.11323 \cdot \exp\left(\frac{-\Psi}{2.58145}\right) - 0.5706 \cdot \exp\left(\frac{-NTU}{8.16455}\right) +$$

$$80.11523 \cdot \exp\left(\frac{-SHR}{0.14108}\right) + 0.12793 \quad (2\text{-}59)$$

$$\vartheta_{rout,\ MR} = -0.0651 \cdot \exp\left(\frac{-\Psi}{2.56672}\right) - 0.07756 \cdot \exp\left(\frac{-NTU}{8.09942}\right) -$$

$$0.00823 \cdot \exp\left(\frac{SHR}{0.62939}\right) + 1.14252 \quad (2\text{-}60)$$

As $\theta_{pout,\ MR}$ and $\vartheta_{pout,\ MR}$ were known, the heat and mass balance principle was applied to calculate the outlet states of the regeneration air ($\theta_{rout,\ prin}$ and $\vartheta_{rout,\ prin}$) using Eqs. (2-61) ~ (2-62), respectively.

$$\theta_{rout,\ prin} = 1 - \theta_{pout,\ MR} \quad (2\text{-}61)$$

$$\vartheta_{rout,\ prin} = \vartheta_{rin} + 1 - \vartheta_{pout,\ MR} \quad (2\text{-}62)$$

$\theta_{rout,\ prin}$ and $\vartheta_{rout,\ prin}$ were compared with $\theta_{rout,\ MR}$ and $\vartheta_{rout,\ MR}$, respectively. The results are presented in Figs. 2-12 (a) and (b). The maximum errors for Cases 1 and 2 were within ±10% and ±15%, respectively. When the dimensionless output values were converted into values with dimensions, the maximum errors of the two cases are within ±5%, as shown in Figs. 2-12 (c) and (d).

In summary, the multiple regression formulas reflect the three basic criterion numbers and accurately calculate the outlet states of A_{pro} and A_{reg}. However, the above formulas were regressed for the fixed air inlet states. If air inlet states are adopted as input parameters, the sample data size becomes large, making it difficult to regress the formulas. However, this problem can be solved using an artificial neural network method, which is suitable for processing big data. In the next part, different air inlet conditions are considered input variables, and a BPNN is used to predict the air outlet states.

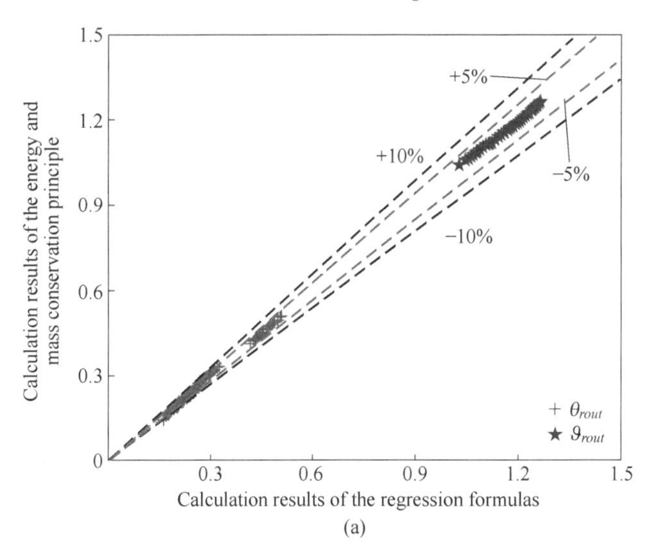

(a)

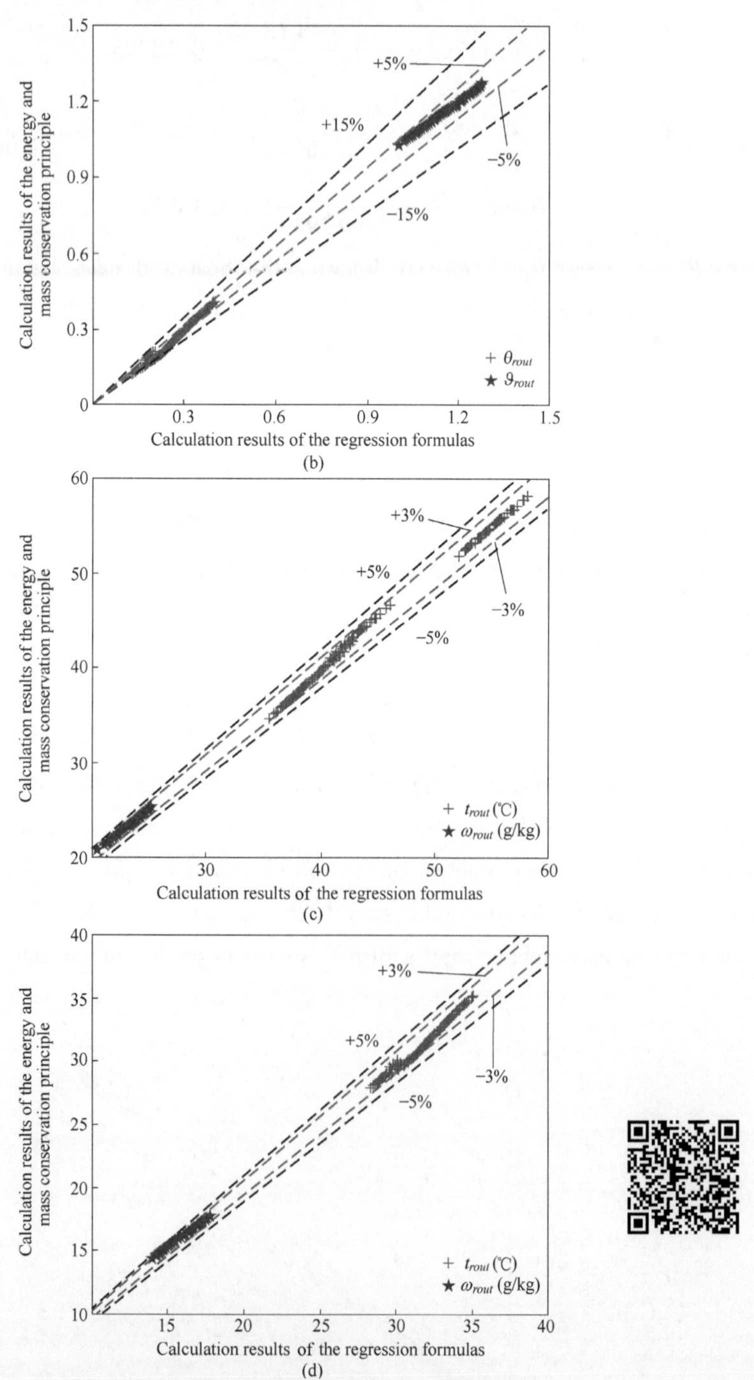

Fig. 2-12 Comparisons of outlet states of A_{reg} calculated using formulas and heat and mass balance principles

(a) Case 1: Dimensionless outlet humidity ratio and temperature of A_{reg};

(b) Case 2: Dimensionless outlet humidity ratio and temperature of A_{reg};

(c) Case 1: Outlet humidity ratio and temperature of A_{reg}; (d) Case 2: Outlet humidity ratio and temperature of A_{reg}

2.4.4 Prediction model using artificial neural network method

The regression formulas are presented in Section 2.3.1. In this subsection, t_{rin} and ω_{pin} are selected as variables, and the BPNN is used to predict the dimensionless air outlet states. Analyses based on the 2500 groups of the original data are discussed.

The BPNN is an error neural network based on BP with three layers: the input, hidden, and output layers. It is one of the most widely used neural network models for performance predictions[24]. The *RMSE*, R^2, R^2_{adj}, and *MSE* were used as the evaluation parameters. MSE is the mean square error (range $[0, +\infty)$), and it was calculated using Eq. (2-63).

$$MSE = \frac{1}{n}\sum (Y_{act} - Y_{pre})^2 \tag{2-63}$$

When the predicted value is equal to the actual value, $MSE = 0$. A larger *MSE* indicates a more significant error, and a smaller *MSE* indicates a better prediction effect.

The number of hidden layer neurons was determined using Eq. (2-64) proposed by Huang et al. [26]:

$$K = \sqrt{m_1 + m_2} + m_3 \tag{2-64}$$

Where K, m_1 and m_2 are the number of neurons in the hidden layer, the number of input parameters, and the number of output parameters, respectively. m_3 is an empirical constant ranging between 1 and 10. In this subsection, five inputs (θ_{rin}, ϑ_{pin}, Ψ, NTU, SHR) and four outputs (θ_{pout}, ϑ_{pout}, θ_{rout}, and ϑ_{rout}) were considered. Therefore, $m_1 = 5$ and $m_2 = 4$.

The *MSE* and training epochs of the different hidden neurons are listed in Table 2-10. The results obtained using the different learning methods are listed in Table 2-11.

Table 2-10 *MSE* and training epochs of different hidden layer neurons

Number of hidden neurons	*MSE*	Epoch of best validation performance
4	1.402×10^{-3}	84
5	6.7701×10^{-4}	77
6	3.8986×10^{-4}	34
7	4.3788×10^{-4}	51
8	3.7416×10^{-4}	50
9	3.17164×10^{-4}	97

Table 2-11 *MSE* and training epochs of different learning methods

Learning method	*MSE*	Epoch of best validation performance
Levenberg-Marquardt	3.7017×10^{-4}	34
Bayesian regularization	4.96×10^{-4}	166
Scaled conjugate gradient	8.6936×10^{-4}	138

Fig. 2-13 shows the corresponding neural network structure when $m_3 = 3$.

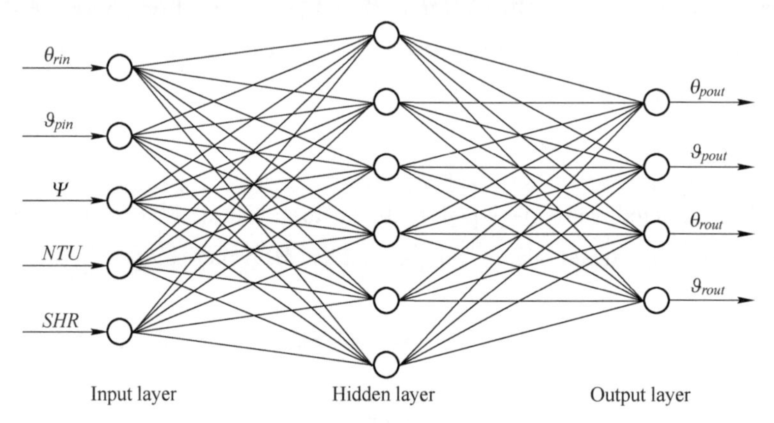

Fig. 2-13 BPNN structure

The parameter settings of the BPNN used in this study are listed in Table 2-12.

Table 2-12 Parameter settings of BPNN

Category	Value
Total number of samples	2500
Number of samples for training	1600
Number of samples for validation	400
Number of samples for testing and predicting	500
Data division	Random
Number of hidden neurons	6
Training algorithm	Levenberg-Marquardt[27]
Learning rate	0.01
Minimum error	1×10^{-6}
Number of iterations	3000
Normalized function	Mapminmax

When the number of neurons in the hidden layer was six, the BPNN in the training process achieved the best validation performance at the 34th epoch, with an *MSE* of 3.7017×10^{-4} (Fig. 2-14).

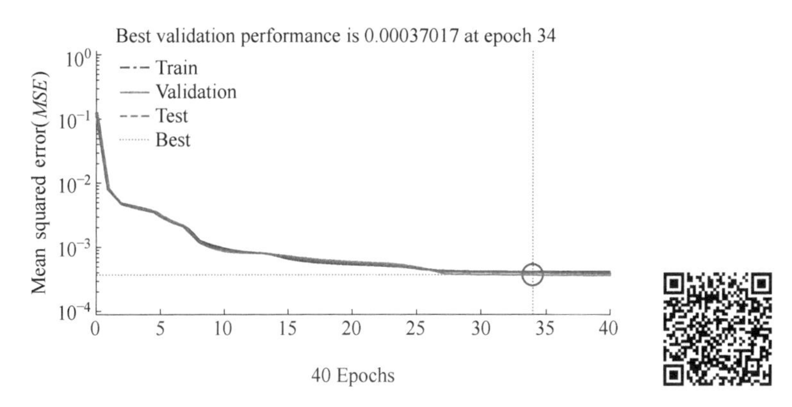

Fig. 2-14 Variation in *MSE* with number of iterations

Fig. 2-15 shows the R^2 of the model during the training, validation, testing, and all stages. The corresponding R^2 values were 0.98835, 0.98708, 0.98856 and 0.98817, respectively. The BPNN model exhibited high accuracy and reliability for predicting the air outlet states of desiccant wheels.

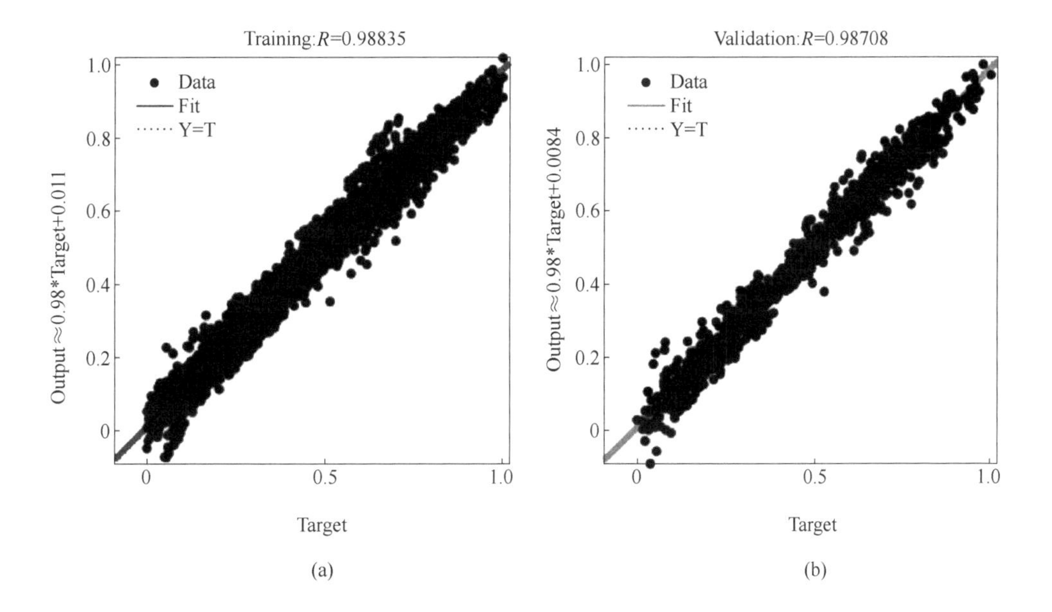

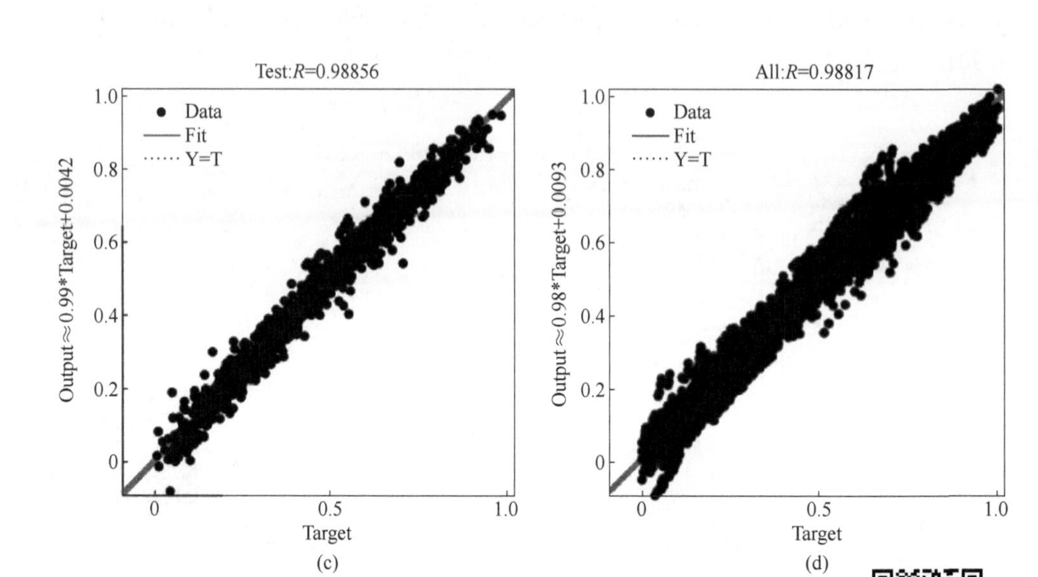

Fig. 2-15 Accuracy of BPNN method

(a) Training; (b) Validation; (c) Testing; (d) All

The prediction errors are shown in Fig. 2-16. The prediction errors of θ_{pout}, ϑ_{pout}, θ_{rout} and ϑ_{rout} were mainly within the ranges of ±5%, ±10%, ±10%, and ±5%, respectively. Overall, 95% of the errors were within ±10%, and the maximum error was less than 15%.

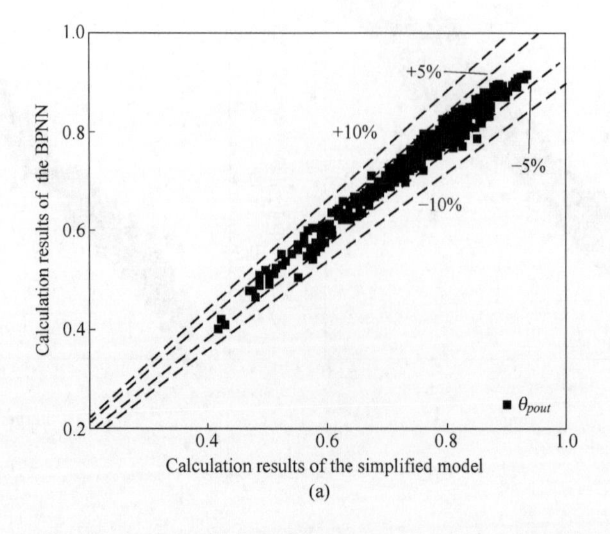

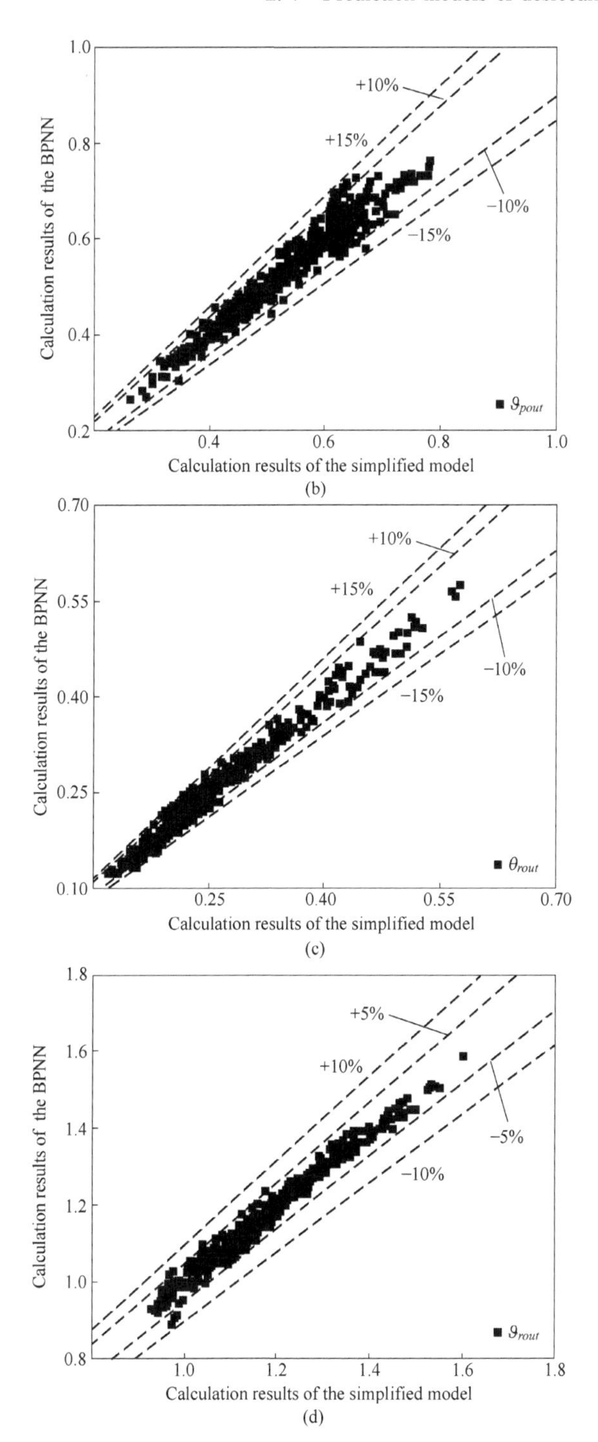

Fig. 2-16 Prediction errors of BPNN

(a) θ_{pout}; (b) ϑ_{pout}; (c) θ_{rout}; (d) ϑ_{rout}

The goodness-of-fit values are listed in Table 2-13. The *MSE* ranged from 1.574×10^{-4} to 6.801×10^{-4}, and the *RMSE* ranged from 1.254×10^{-2} to 2.608×10^{-2}. The minimum R^2 and R_{adj}^2 values were 0.9693 and 0.9691, respectively. These results indicate that the prediction model has high precision.

Table 2-13 Goodness-of-fit parameters of BPNN prediction

Parameter	*MSE*	*RMSE*	R^2	R_{adj}^2
$\theta_{pout,\ BP}$	2.297×10^{-4}	1.516×10^{-2}	0.9875	0.9874
$\vartheta_{pout,\ BP}$	6.801×10^{-4}	2.608×10^{-2}	0.9693	0.9691
$\theta_{rout,\ BP}$	1.574×10^{-4}	1.254×10^{-2}	0.989	0.9889
$\vartheta_{rout,\ BP}$	4.34×10^{-4}	2.083×10^{-2}	0.9881	0.988

Cases 1 and 2 described in Section 4 were used as examples to compare the results of the BPNN model with those calculated using the regression formulas. The comparisons are depicted in Fig. 2-17. Overall, the difference between the two prediction methods was within ±10%.

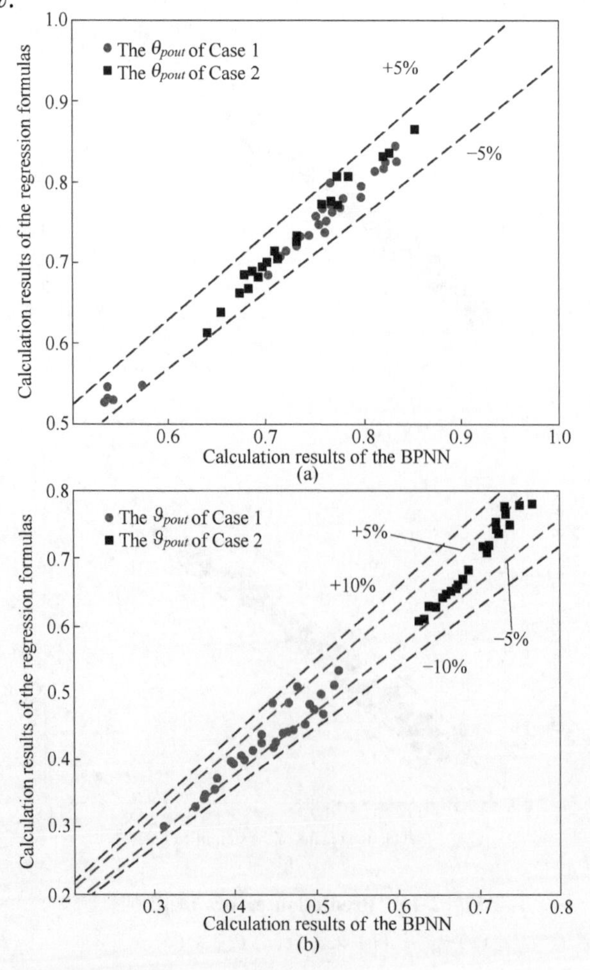

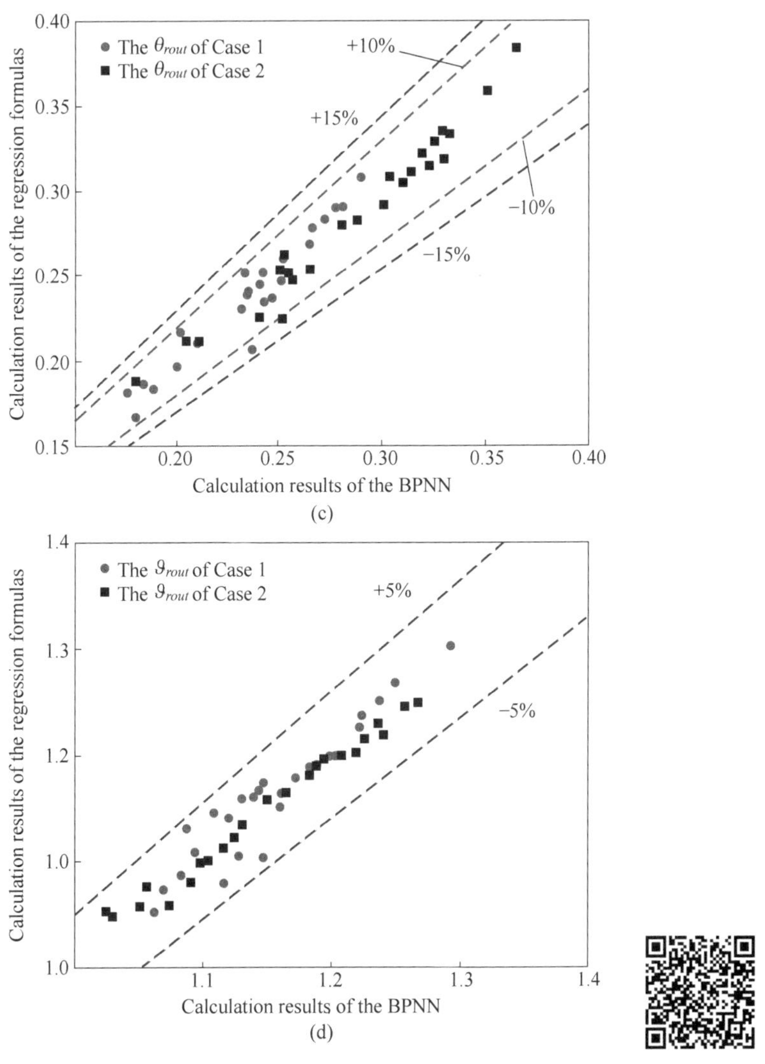

Fig. 2-17 Comparisons between BPNN and prediction formulas

(a) θ_{pout}; (b) ϑ_{pout}; (c) θ_{rout}; (d) ϑ_{rout}

Nomenclature

A	area
a	pore radius, m
C	shape factor of desiccant material
c_p	specific heat, kJ/(kg · K)

d_h	hydraulic diameter, m
D_A	ordinary diffusion coefficient, m^2/s
D_{AO}	ordinary diffusion coefficient, m^2/s
D_{AK}	Knudsen diffusion coefficient, m^2/s
D_0	surface diffusion constant, m^2/s
D_S	surface diffusion coefficient, m^2/s
h_m	mass transfer coefficient, $kg/(m^2 \cdot s)$
h_v	heat of vaporization, kJ/kg
k	thermal conductivity, $kW/(m \cdot K)$
L	wheel thickness, m
Le	Lewis number
m	mass flow rate, kg/s
M	mass of desiccant wheel, kg
Nu	Nusselt number
P_a	standard atmospheric pressure, Pa
P_{vs}	saturated vapor pressure, Pa
P	wet cycle, m
R	regeneration air
r_s	adsorption or desorption heat, kJ/kg
RS	rotational speed
t	Celsius temperature, ℃
x	volume ratio of adsorption material
x^*	mass ratio of adsorption material
L	thickness, m
f	area ratio, dimensionless
τ	time, s
u	velocity, m/s
R^2	the coefficient of determination
Y_{act}	the target values
Y_{pre}	the predicted values
Y_{ave}	the average target values
n	the number of samples
p	the number of features
K	the number of neurons in the hidden layer
m_1	the number of input parameters
m_2	the number of output parameters
m_3	an empirical constant ranging between 1 and 10

Greek symbols

ω	humidity ratio, g/kg
φ	relative humidity ratio
τ	time
ρ	density, kg/m^3
σ	porosity
λ	thermal conductivity
β	the angle
Ψ^*	required time

Subscripts

a	air
ave	average
ad	adsorption material
c	cold
d	desiccant material
p	process
reg	regeneration
in	inlet
out	outlet
max	maximum
min	minimum
w	water
DW	desiccant wheel

References

[1] Q Y Lu. Practical heating and air conditioning design manual [M]. Beijing: China Architecture and Architecture Press, 2015.

[2] L Z Zhang. Dehumidification Technology [M]. Beijing: Chemical Industry Press, 2005.

[3] R Tu, X H Liu, Y Jiang. Performance analysis of a new kind of heat pump-driven outdoor air processor using solid desiccant [J]. Renew. Energy, 2013, 57: 101-110.

[4] D A Stefano, M J Cesare, M Luca. Simulation, performance analysis and optimization of desiccant wheels [J]. Energy and Buildings, 2010, 42 (9): 1386-1393.

[5] X M Zhang, Z P Ren, F M Mei. Heat Transfer [M]. Beijing: China Architecture and Architecture Press, 2007.

[6] D Charoensupaya, W M Worek. Parametric study of an open-cycle adiabatic, solid, desiccant cooling system [J]. Energy, 1988, 13 (9): 739-747.

[7] Y F Ding, J Ding, Y T Fang, X X Yang. Study on heat and mass transfer performance of

modified Silica gel desiccation wheel [J]. Journal of Guangzhou University (Natural Science), 2006, 5 (3): 80-85.

[8] J D Chung, D Y Lee, S M Yoon. Optimization of desiccant wheel speed and area ratio of regeneration to dehumidification as a function of regeneration temperature [J]. Solar Energy, 2009, 83 (5): 625-635.

[9] C R Ruivo, J J Costa, A R Figueiredo. Analysis of simplifying assumptions for the numerical modeling of the heat and mass transfer in a porous desiccant medium [J]. Numer. Heat Tr. A-Appl, 2006, 49 (9): 851-872.

[10] S Kakac, R K Shah, W Aung. Handbook of single phase convective heat transfer [M]. New York: John Wiley & Sons, 1987.

[11] T Cao, H Lee, Y H Hwang, R Radermacher, H-H Chun. Experimental investigations on thin polymer desiccant wheel performance [J]. Int. J. Refrig, 2014 (44): 1-11.

[12] X Y Sun, Y J Dai, T S Ge, Y Zhao, R Z Wang. Experimental and comparison study on heat and moisture transfer characteristics of desiccant coated heat exchanger with variable structure sizes [J]. Appl. Therm. Eng, 2018, 137: 32-46.

[13] G G Ilis, H Demir. Influence of bed thickness and particle size on performance of microwave regenerated adsorption heat pump [J]. Int. J. Heat Mass Transf, 2018, 123: 16-24.

[14] R Tu, R Z Wang, T S Ge. Moisture uptake dynamics on desiccant-coated, water-sorbing heat exchanger [J]. Int. J. Therm. Science, 2018, 126: 13-22.

[15] D Charoensupaya, W M Worek. Parametric study of an open-cycle adiabatic, solid, desiccant cooling system [J]. Energy, 1988, 13: 739-747.

[16] L Z Zhang, H X Fu, Q R Yang, J C Xu. Performance comparisons of honeycomb-type adsorbent beds (wheels) for air dehumidification with various desiccant wall materials [J]. Energy, 2014, 65: 430-440.

[17] T S Ge, F Ziegler, R Z Wang. A mathematical model for predicting the performance of a compound desiccant wheel (A model of compound desiccant wheel) [J]. Appl. Therm. Eng, 2010, 30 (8-9): 1005-1015.

[18] L Z Zhang, J L Niu. Performance comparisons of desiccant wheels for air dehumidification and enthalpy recovery [J]. Appl. Therm. Eng, 2002, 22 (12): 1347-1367.

[19] J D Chung, D Y Lee. Effect of desiccant isotherm on the performance of desiccant wheel [J]. Int. J. Refrig, 2009, 32 (4): 720-726.

[20] J D Chung, D Y Lee, S M Yoon. Optimization of desiccant wheel speed and area ratio of regeneration to dehumidification as a function of regeneration temperature [J]. Sol. Energy, 2009, 83 (5): 625-635.

[21] P Majumdar. Heat and Mass transfer in composite desiccant pore structures for dehumidification [J]. Sol. Energy, 1998, 62 (1): 1-10.

[22] P Majumdar. W M Worek. Combined heat and mass transfer in a porous adsorbent [J]. Energy, 1989, 14 (3): 161-175.

[23] R Tu Y H Hwang, T Cao, M D Hou, et al. Investigation of adsorption isotherms and rotational speeds for low temperature regeneration of desiccant wheel systems [J]. Int. J. Refrig, 2017, 86: 495-509.

[24] V K Mishraa, R P Singh, R K Das. Performance prediction of solid desiccant rotary system using artificial neural network [J]. IOP Conf, 2018, 404 (1) .

[25] Ministry of Housing and Urban-Rural Development of the People' s Republic of China. GB 50736—2012 Code for design of heating, ventilation, and air conditioning for civil buildings [S]. Beijing: China Architecture & Building Press, 2012.

[26] H Y Huang, J M Zhu, S Z Li. Research on application of BP neural network based on genetic algorithm optimization in stock index prediction [J]. Journal of Yunnan university, 2017, 39 (3): 350-355.

[27] Z Q Cheng, G Ke-Nan, Z B Kan. Prediction of dehumidification system energy consumption based on Elman neural network [J]. Comp. Eng. Design, 2014, 35 (2): 677-680.

Chapter 3 Lowering Regeneration Temperature of Desiccant Wheels

This chapter analyzes factors that influence regeneration temperature of desiccant wheels, such as facial area ratio of dehumidification side to regeneration side (A_r), air flow rate ratio of the processed air to the regeneration air (F_r), and air handling processes. Exergy analyzes were carried out based on temperature and humidity ratio fields of air and desiccant wheel. Optimized values of A_r and F_r, were provided and effective air handling processed were proposed.

3.1 Exergy analysis of the desiccant wheel

Operating principles of a typical rotary wheel dehumidification system were illustrated in Chapter 1. Regeneration air is heated to a high temperature to make sure that the processed air can be dehumidified to the required humidity ratio. Heat and mass transfer are driven independently by temperature and humidity ratio differences, which are also the cause of exergy destructions. It is essential to carry out exergy analyses on desiccant wheels to find out relations between exergy destructions and regeneration temperature. So that effective ways can be proposed to reduce exergy destructions and the regeneration temperature.

3.1.1 Exergy balance for the rotary wheel

Exergy of humid air per kilogram of dry air (ex) under the atmospheric pressure can be described with Eqs. (3-1) ~ (3-3), representing the theoretical maximum work that can be obtained when the air reaches the dead state (T_0, ω_0). The two terms on the right side of Eq. (3-1) are thermal exergy (ex_{th}) and chemical exergy (ex_{ch}), respectively.

$$ex(T_a,\ \omega_a) = ex_{th}(T_a) + ex_{ch}(\omega_a) \tag{3-1}$$

$$ex_{th}(T_a) = c_{pa}T_0\left(\frac{T_a}{T_0} - 1 - \ln\frac{T_a}{T_0}\right) \tag{3-2}$$

$$ex_{ch}(\omega_a) = R_a T_0 \left[(1 + 1.608\omega_a) \ln \frac{1 + 1.608\omega_0}{1 + 1.608\omega_a} + 1.608\omega_a \ln \frac{\omega_a}{\omega_0} \right] \quad (3\text{-}3)$$

The exergy diagram of humid air is shown in Fig. 3-1, and the dead point (T_0, ω_0) is selected as the saturated state under ambient air temperature. The influence of T_a on thermal exergy (ex_{th}) and that of ω_a on chemical exergy (ex_{ch}) are shown in Figs. 3-2 (a) and (b), respectively. Farther T_a or ω_a is from the dead point, the higher the ex_{th} *or* ex_{ch} will be.

For the air handling process in the rotary wheel, the processed air is dried and heated, while the regeneration air is cooled and humidified along the air flow direction. Thus, exergy of the processed air increases after dehumidification, and exergy of the regeneration air decreases after regeneration, as shown in Fig. 3-2. The exergy balance equation in the desiccant wheel is written as Eq. (3-4). The exergy provided by the regeneration air is equal to the exergy obtained by the processed air plus exergy destruction (ΔEx).

$$\dot{m}_r(ex_{r1} - ex_{rout}) = \dot{m}_p(ex_{p1} - ex_{pin}) + \Delta Ex \quad (3\text{-}4)$$

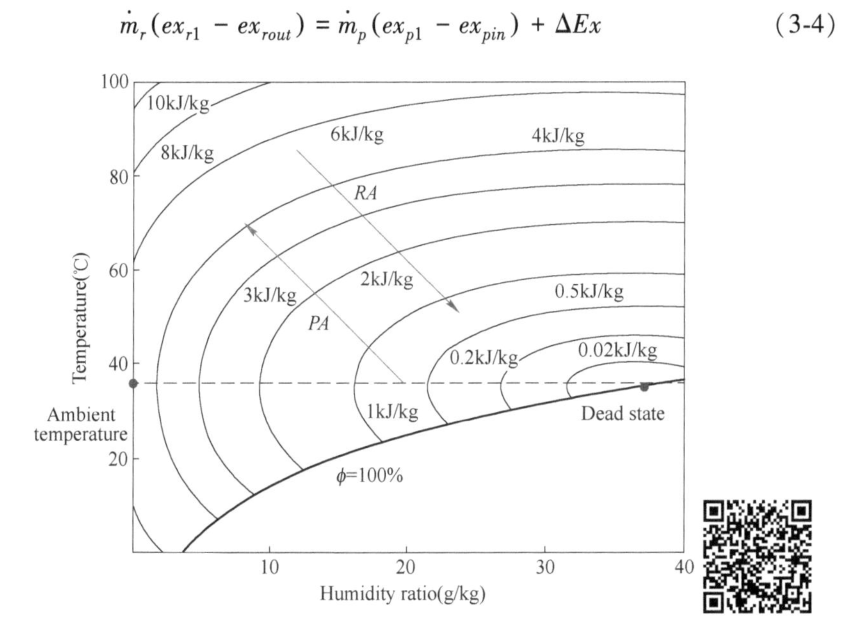

Fig. 3-1 Exergy diagram of humid air

The numbers in the subscripts of the air states are shown in Fig. 3-1; $\dot{m}_r$ and $\dot{m}_p$ represent the mass flow rates of the regeneration air and the processed air, respectively. The exergy destruction (ΔEx) includes the heat and mass transfer exergy destruction inside the desiccant wheel (ΔEx_{tr}) and the air mixing exergy destruction

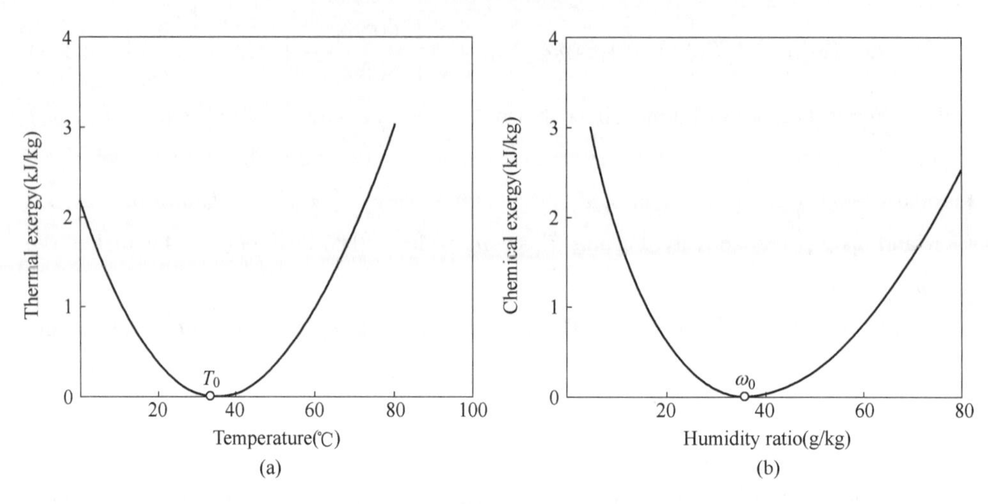

Fig. 3-2　Exergy

(a) Thermal exergy; (b) Chemical exergy

(ΔEx_{mix}) at the outlet of the wheel for the regeneration air and processed air. Eq. (3-4) can be rewritten as Eq. (3-5), where the exergy obtained or provided by the processed air or the regeneration air is divided into thermal exergy changes ($ex_{p1,\ th} - ex_{pin,\ th}$ or $ex_{r1,\ th} - ex_{rout,\ th}$) and chemical exergy changes ($ex_{p1,\ th} - ex_{pin,\ ch}$ or $ex_{r1,\ ch} - ex_{rout,\ ch}$).

$$\dot{m}_r[(ex_{r1,\ th} - ex_{rout,\ th}) + (ex_{r1,\ ch} - ex_{rout,\ ch})] \\ = \dot{m}_p[(ex_{p1,\ th} - ex_{pin,\ th}) + (ex_{p1,\ ch} - ex_{pin,\ ch})] + \Delta Ex \tag{3-5}$$

The energy and mass conservation equations of the two streams of air can be written as Eq. (3-6) and Eq. (3-7), respectively.

$$\dot{m}_r c_{pa}(t_{ar1} - t_{arout}) = \dot{m}_p c_{pa}(t_{ap1} - t_{apin}) \tag{3-6}$$

$$\dot{m}_r(\omega_{ar1} - \omega_{arout}) = \dot{m}_p(\omega_{ap1} - \omega_{apin}) \tag{3-7}$$

According to Eqs. (3-5) ~ (3-7), when the air inlet states and the air flow rates of the processed air and regeneration air are fixed, under the same humidity ratio removal rate of the processed air, the chemical exergy changes of the processed air ($ex_{p1,\ ch} - ex_{pin,\ ch}$) and the regeneration air ($ex_{r1,\ ch} - ex_{rout,\ ch}$) are fixed. Under these circumstances, when $ex_{p1,\ th} - ex_{pin,\ th}$ and the exergy destruction (ΔEx) increase, higher thermal exergy provided by the regeneration air ($ex_{r1,\ th} - ex_{rout,\ th}$) will be needed. Higher $ex_{r1,\ th} - ex_{rout,\ th}$ means higher inlet thermal exergy of the regeneration air ($ex_{r1,\ th}$). The regeneration temperature is usually higher than T_0, and a higher $ex_{r1,\ th}$ means a higher regeneration temperature, as seen in Fig. 3-2 (a). Thus, to reduce the regeneration temperature, both the exergy destruction (ΔEx) and the thermal exergy obtained by the processed air ($ex_{r1,th} - ex_{rout,th}$) during dehumidification should be reduced.

3.1.2 Exergy efficiency of dehumidification

Thermal exergy change of the processed air in the desiccant wheel is due to a temperature increase, and the chemical exergy change of the processed air is due to a decrease in humidity ratio. The latter is consistent with the main aim of an air-conditioning system; thus, the exergy efficiency for the desiccant wheel can be defined as Eq. (3-8).

$$\eta_{ex} = \frac{\dot{m}_p(ex_{p1,ch} - ex_{pin,ch})}{\dot{m}_r(ex_{r1} - ex_{rout})}$$

$$= \frac{\dot{m}_p(ex_{p1,ch} - ex_{pin,ch})}{\dot{m}_p[(ex_{p1,th} - ex_{pin,th}) + (ex_{p1,ch} - ex_{pin,ch})] + \Delta Ex} \tag{3-8}$$

According to Eq. (3-8), when the working conditions are fixed, a higher exergy efficiency is obtained when ΔEx and $ex_{p1,th} - ex_{pin,th}$ are lower, which also leads to a lower regeneration temperature.

The working conditions and the design parameters of the desiccant wheel are listed in Table 3-1. Silica gel is a commonly used adsorption material for dehumidification, which was adopted as the adsorption material in this study, and the iso-water content line can be drawn in a psychometric chart. The equilibrium humidity ratio of the desiccant (ω_d) can be calculated if the desiccant temperature and water content are known. The wheel is 200mm thick, with a diameter of 1m. Ambient air is used as the processed air; the inlet states are 35℃ and 20g/kg, which represent typical cities with hot and humid climate in China, such as Beijing, Guangzhou and Shanghai. The air flow rate is 2500m^3/h. Indoor exhaust air is used as the regeneration air, with an inlet humidity ratio of 12g/kg. The supplied air humidity ratio is fixed at 11g/kg, which is lower than that of the indoor air, and the humidity ratio difference is used to remove the indoor moisture load. The following discussions are all based on these parameters (listed in Table 3-1).

Table 3-1 Working conditions and desiccant wheel design parameters

Working conditions	Desiccant wheel design parameters
Processed air inlet: 35.0℃, 20.0g/kg	Radius = 0.5m, Thickness = 0.2m
Regeneration air inlet: 26.0℃, 12.0g/kg	$G_p = 2500m^3/h$
Supplied air: 11.0g/kg	Adsorption material: silica gel

Firstly, effects of A_r on the performance of the wheel were examined. Scenarios of $A_r = 1$ and $A_r = 3$ were taken as examples, and the air velocity was the same for the two streams of air. Therefore, $F_r = A_r$. The results are shown in Fig. 3-3.

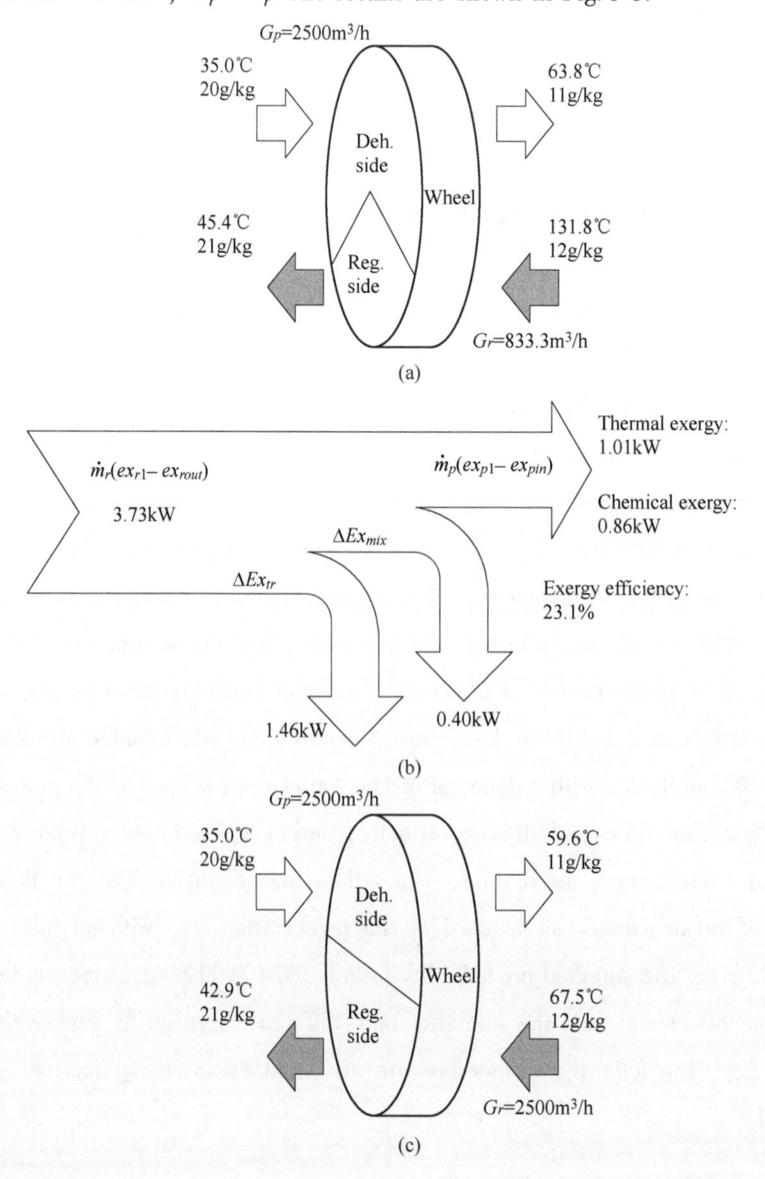

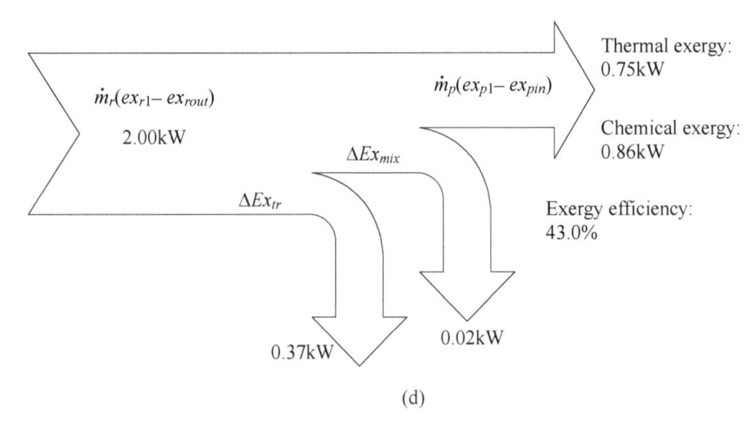

(d)

Fig. 3-3 Exergy flow chart for different A_r values ($A_r = F_r$)

(a) Sketch of wheel when $A_r = 3$; (b) Exergy flow chart when $A_r = 3$;

(c) Sketch of wheel when $A_r = 1$; (d) Exergy flow chart when $A_r = 1$

When $A_r = 3$, the regeneration area is only a quarter of the whole wheel area, and the regeneration air flow rate is one third of that of the processed air. The required regeneration temperature is 131.8℃. ΔEx is 1.86kW, of which ΔEx_{tr} is 1.46kW and ΔEx_{mix} is 0.40kW. Exergy obtained by the processed air is 1.87kW, of which the chemical exergy change is 0.86kW and the thermal exergy change is 1.01kW. Hence, exergy provided by the regeneration air is 3.73kW, and η_{ex} is 23.1%.

When $A_r = 1$, the regeneration area is half of the whole wheel area, and the regeneration air flow rate is equal to that of the processed air. The required regeneration temperature is only 67.5℃. ΔEx is 0.39kW, of which ΔEx_{tr} is 0.37kW and ΔEx_{mix} is 0.02kW. The exergy obtained by the processed air is 1.61kW, of which the chemical exergy change is 0.86kW and the thermal exergy change is 0.75kW. Therefore, exergy provided by the regeneration air is only 2.00kW, and η_{ex} is 43.0%.

Comparing the two cases, significant exergy differences can be seen mainly interms of exergy destruction in the desiccant wheel and thermal exergy obtained by the processed air. When A_r changes from 3 to 1, exergy destruction ΔEx decreases from 1.86kW to 0.39kW, and the thermal exergy obtained by the processed air decreases from 1.01kW to 0.75kW, resulting in a lower regeneration temperature and higher exergy efficiency. The influencing factors of ΔEx and the thermal exergy obtained by the processed air will be analyzed in Section 3.2 and Section 3.3, respectively.

3.2 Decreasing exergy destruction

Exergy destruction includes heat and mass transfer exergy destruction (ΔEx_{tr}) and air mixing exergy destruction (ΔEx_{mix}) . According to Fig. 3-3, ΔEx_{tr} is the dominant part of exergy destruction. Therefore, this section mainly focuses on the factors that influence heat and mass transfer exergy destruction.

3.2.1 Factors influencing heat and mass transfer exergy destruction

The driving forces of heat and mass transfer during the dehumidification and regeneration processes in the rotary wheel are the temperature and humidity ratio differences (Δt *and* $\Delta\omega$, respectively) between the air (t_a and ω_a) and the desiccant (t_d and ω_d). The temperature and humidity ratio fields of the air and the desiccant can be simulated using the one-dimensional-double-diffusion numerical model introduced in Chapter 2.

The wheel is simplified into a two-dimensional chart and divided into several grids, as shown in Fig. 3-4. The thickness is divided into n grids, and the angle is divided into k grids. Heat and mass transfer exergy destruction is the sum of the exergy destruction in each grid, as shown in Eq. (3-9).

$$\Delta Ex_{tr} = \sum_{i,j} \Delta Ex_{tr,h}^{i,j} + \sum_{i,j} \Delta Ex_{tr,m}^{i,j} \tag{3-9}$$

In each grid, when heat is transferred from T_a to T_d with a sensible heat transfer capacity of $\delta Q^{i,j}$ (as illustrated in Fig. 3-4), heat transfer exergy destruction can be written as Eq. (3-10), where dF is the heat and mass transfer area in a grid.

$$\Delta Ex_{tr,h}^{i,j} = \left(\frac{\partial ex_{th}}{\partial T} \Big|_{T_a^{i,j}} - \frac{\partial ex_{th}}{\partial T} \Big|_{T_d^{i,j}} \right) \frac{\delta Q^{i,j}}{c_{pm}}, \quad \delta Q^{i,j} = (T_a^{i,j} - T_d^{i,j}) h \mathrm{d}F \tag{3-10}$$

Combining this equation with Eq. (3-2), $\Delta Ex_{tr,h}^{i,j}$ can be written as Eq. (3-11); F is the total heat and mass transfer area in the wheel, and d$F=F/(hk)$.

$$\Delta Ex_{tr,h}^{i,j} = T_0 \frac{(T_a^{i,j} - T_d^{i,j})^2}{T_a^{i,j} \cdot T_d^{i,j}} h \mathrm{d}F = T_0 \frac{hF}{nk} \cdot \frac{(\Delta T^{i,j})^2}{T_a^{i,j} \cdot T_d^{i,j}} \tag{3-11}$$

Similarly, when massis transferred from ω_a to ω_d with a mass transfer capacity of $\delta M^{i,j}$, mass transfer exergy destruction in each grid can be written as Eq. (3-12).

$$\Delta Ex_{tr,m}^{i,j} = \left(\frac{\partial ex_{ch}}{\partial \omega} \Big|_{\omega_a^{i,j}} - \frac{\partial ex_{ch}}{\partial \omega} \Big|_{\omega_d^{i,j}} \right) \delta M^{i,j}, \quad \delta M^{i,j} = (\omega_a^{i,j} - \omega_d^{i,j}) h_m \mathrm{d}F \tag{3-12}$$

Combining this equation with Eq. (3-3), mass transfer exergy destruction in each grid of the wheel can be written as follows:

$$\Delta Ex_{tr,h}^{i,j} = 1.608 R_a T_0 \left(\ln \frac{\omega_a^{i,j}}{1 + 1.608\omega_a^{i,j}} - \ln \frac{\omega_d^{i,j}}{1 + 1.608\omega_d^{i,j}} \right) (\omega_a^{i,j} - \omega_d^{i,j}) h_m \mathrm{d}F \tag{3-13}$$

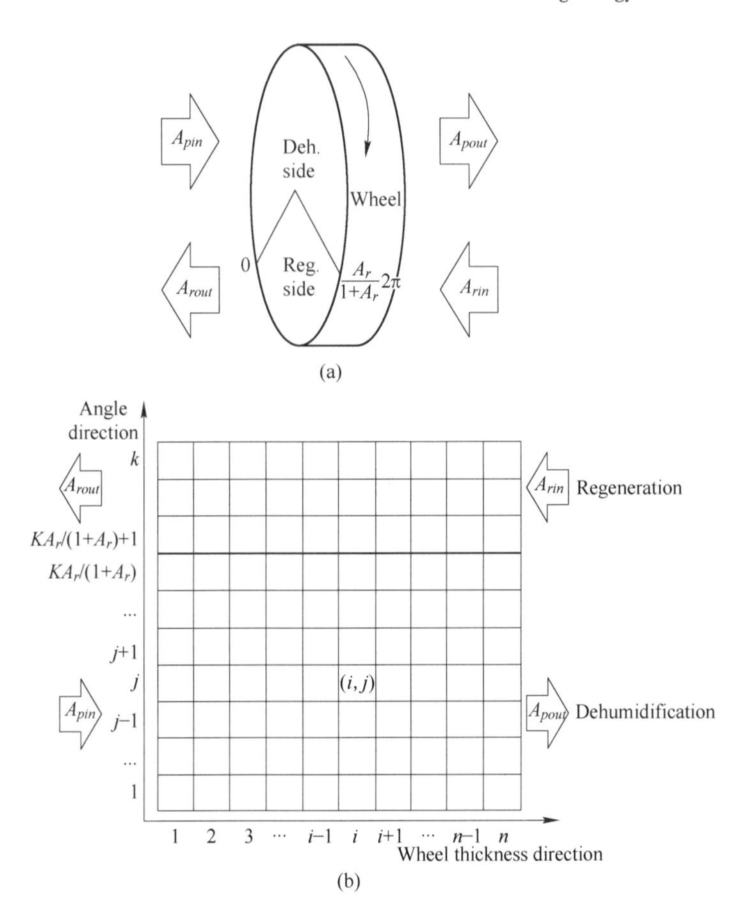

Fig. 3-4 Grid division of desiccant wheel

(a) Sketch of desiccant wheel; (b) Grid division of desiccant wheel

According to Zhang, Eq. (3-13) can be rewritten as Eq. (3-14), where $T_{a,\,dew}^{i,\,j}$ and $T_{d,\,dew}^{i,\,j}$ represent the dew point temperatures of air and the desiccant, respectively, as shown in Fig. 3-5.

$$\Delta Ex_{tr,\,h}^{i,\,j} = T_0 r\left(\frac{1}{T_{a,\,dew}^{i,\,j}} - \frac{1}{T_{d,\,dew}^{i,\,j}}\right)(\omega_a^{i,\,j} - \omega_d^{i,\,j})h_m \mathrm{d}F = T_0 r\,\frac{h_m F}{nk}\cdot\frac{\Delta T_{dew}^{i,\,j}\Delta\omega^{i,\,j}}{T_{a,\,dew}^{i,\,j}\cdot T_{d,\,dew}^{i,\,j}} \tag{3-14}$$

Eq. (3-14) can be approximated as Eq. (3-15), where S_l is the slope of the saturated line ($S_l = \Delta T_{\mathrm{dew}}/\Delta\omega$). S_l can be regarded as a constant in such a narrow air handling range, as shown in Fig. 3-5.

$$\Delta Ex_{tr,\,h}^{i,\,j} \approx T_0 r\,\frac{h_m F}{nk}S_l\cdot\frac{(\Delta\omega^{i,\,j})^2}{T_{a,\,dew}^{i,\,j}\cdot T_{d,\,dew}^{i,\,j}} \tag{3-15}$$

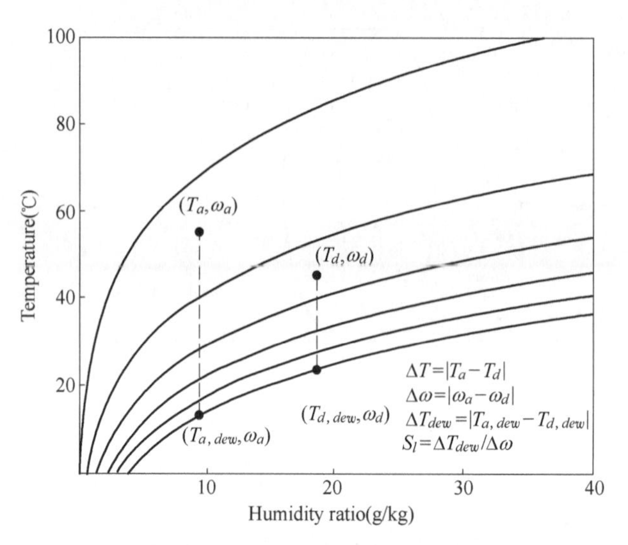

Fig. 3-5 Symbols in the equations

The exergy destruction in the regeneration side (Reg. side), the dehumidification side (Deh. side), and the whole wheel is the sum of $\Delta Ex_{tr,h}^{i,j}$ and $\Delta Ex_{tr,m}^{i,j}$ of all the grids in each part. For the Deh. side, j ranges from 1 to $kA_r/(1+A_r)$, and i ranges from 1 to n. For the Reg. side, j ranges from $kA_r/(1+A_r)+1$ to k, and i ranges from 1 to n. For the whole wheel, j ranges from 1 to k, and i ranges from 1 to n.

The temperatures in Eqs. (3-11) ~ (3-15) are measured in Kelvin (K), calculated by adding 273. 15K to the Celsius temperature. Taking the whole wheel as an example, the heat and mass transfer exergy destruction can be approximated as Eq. (3-16) and Eq. (3-17), respectively, where $\overline{T}_a$ and $\overline{T}_d$ represent the average temperature of the air and the desiccant, respectively, and $\overline{T}_{a,dew}$ and $\overline{T}_{d,dew}$ represent the average dew point temperature of the air and the desiccant, respectively.

$$\Delta Ex_{tr,h} \approx \frac{T_0}{\overline{T}_a \overline{T}_d} \frac{Q^2}{\frac{hF}{nk}} \frac{\sum_{j=1}^{k}\sum_{i=1}^{n}(\Delta T^{i,j})^2}{\left(\sum_{j=1}^{k}\sum_{i=1}^{n}\Delta T^{i,j}\right)^2} = \frac{T_0}{\overline{T}_a \overline{T}_d} \frac{Q^2}{hF}\zeta_t \tag{3-16}$$

$$\Delta Ex_{tr,m} \approx \frac{T_0 S_l h_v}{\overline{T}_{a,dew} \overline{T}_{d,dew}} \frac{m^2}{\frac{h_m F}{nk}} \frac{\sum_{j=1}^{k}\sum_{i=1}^{n}(\Delta \omega^{i,j})^2}{\left(\sum_{j=1}^{k}\sum_{i=1}^{n}\Delta \omega^{i,j}\right)^2} = \frac{T_0 S_l h_v}{\overline{T}_{a,dew} \overline{T}_{d,dew}} \frac{M^2}{h_m F}\zeta_\omega \tag{3-17}$$

In Eq. (3-16) and Eq. (3-17), ζ_t and ζ_ω are unmatched coefficients; their definitions are written as Eq. (3-18) and Eq. (3-19), respectively. The unmatched coefficients can be used to describe uniformity of temperature differences (the driving force for heat transfer) and humidity ratio differences (the driving force for mass transfer) for the whole wheel.

$$\zeta_t = nk\sum_{i=1}^{n}\sum_{j=1}^{k}\Delta t^2 \Big/ \Big(\sum_{i=1}^{n}\sum_{j=1}^{k}\Delta t\Big)^2 \tag{3-18}$$

$$\zeta_\omega = nk\sum_{i=1}^{n}\sum_{j=1}^{k}\Delta \omega^2 \Big/ \Big(\sum_{i=1}^{n}\sum_{j=1}^{k}\Delta \omega\Big)^2 \tag{3-19}$$

If the driving force for heat transfer (or mass transfer) is uniform in the desiccant wheel, i. e., $\Delta T^{i,\,j}$ = const (or $\Delta\omega^{i,\,j}$ = const), the unmatched coefficient ζ_t (or ζ_ω) will be 1; otherwise, ζ_t (or ζ_ω) will be greater than 1. The larger the value of ζ_t (or ζ_ω), the less uniform the heat (or mass) transfer driving force within the desiccant wheel.

According to Eq. (3-16) and Eq. (3-17), it can be seen that when the heat and mass transfer capacities are fixed, the heat and mass transfer exergy destruction (ΔEx_{tr}) is influenced not only by the heat and mass transfer capacities (hF and $h_m F$), but also by the unmatched coefficients (ζ_t and ζ_ω). Therefore, when the heat and mass transfer capacities are determined (approximately proportional to the inputs of the desiccant wheel), it is important to increase uniformity to reduce the exergy destruction (ΔEx_{tr}).

Similarly, the exergy destruction for the Deh. side and the Reg. side can be written as Eq. (3-20) and Eq. (3-21), respectively, with j ranging from1 to $kA_r/(1+A_r)$ in the former and from $kA_r/(1+A_r)$ +1 to k in the latter.

$$\Delta Ex_{tr,\,h,\,deh} \approx \frac{T_0}{\overline{T}_a\overline{T}_d}\frac{Q_{deh}^2}{hF_{deh}}\zeta_{t,\,deh},\quad \Delta Ex_{tr,\,m,\,deh} \approx \frac{T_0 S_l h_v}{\overline{T}_{a,\,dew}\overline{T}_{d,\,dew}}\frac{m_{deh}^2}{h_m F_{deh}}\zeta_{\omega,\,deh} \tag{3-20}$$

$$\Delta Ex_{tr,\,h,\,reg} \approx \frac{T_0}{\overline{T}_a\overline{T}_d}\frac{Q_{reg}^2}{hF_{reg}}\zeta_{t,\,reg},\quad \Delta Ex_{tr,\,m,\,reg} \approx \frac{T_0 S_l h_v}{\overline{T}_{a,\,dew}\overline{T}_{d,\,dew}}\frac{m_{reg}^2}{h_m F_{reg}}\zeta_{\omega,\,reg} \tag{3-21}$$

$|Q_{deh}|$ is equal to $|Q_{reg}|$, $|Q_{deh}| + |Q_{reg}| = |Q|$, and $F_{deh} + F_{reg} = F$. Because $\Delta Ex_{tr,\,h} = \Delta Ex_{tr,\,h,\,deh} + \Delta Ex_{tr,\,h,\,reg}$, the relationship among the unmatched coefficients of the dehumidification side ($\zeta_{t,\,deh}$), the regeneration side ($\zeta_{t,\,reg}$), and the whole wheel (ζ_t) can be written as Eq. (3-22); $\zeta_{\omega,\ deh}$, $\zeta_{\omega,\ reg}$, and ζ_ω have asimilar relationship.

$$\frac{4}{A_r + 1}\zeta_t \approx \frac{1}{A_r}\zeta_{t,\ deh} + \zeta_{t,\ reg} \tag{3-22}$$

According to Fig. 3-3, the area ratio (A_r) and the air flow rate ratio (F_r) between the processed air and regeneration air both have significant influences on exergy destruction. The influences of A_r and F_r on both uniformity and exergy destruction will be discussed in the following section.

3.2.2 Influence of A_r and F_r on uniformity and exergy destruction

The design parameters of the desiccant wheel used in this study and its working conditions can be seen in Table 3-1. The flow rate of the processed air is fixed at 2500m^3/h. Indoor exhaust air is used for regeneration, so its air flow rate can vary. Three cases are considered in order to investigate the influence of A_r and F_r.

(1) Case 1: $A_r = 1$, variable F_r ($G_r = 800 \sim 2500$ m^3/h, $G_p = 2500$ m^3/h)

Fig. 3-6 shows the effect of regeneration air flow rate G_r on the unmatched coefficients of Δt and $\Delta\omega$, as well as the required regeneration temperature, exergy efficiency, and exergy destruction of each part.

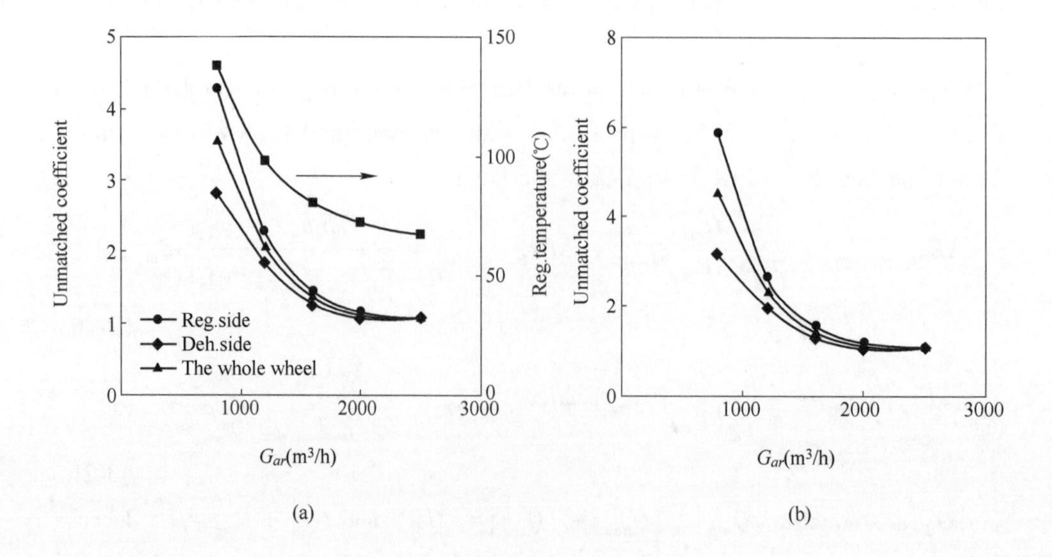

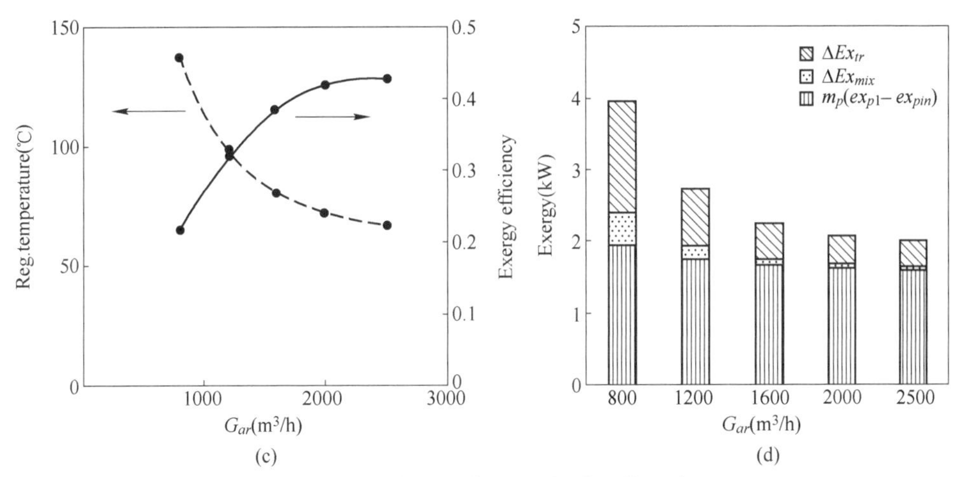

Fig. 3-6 Simulation results for Case 1

(a) Unmatched coefficient of Δt and regeneration temperature; (b) Unmatched coefficient of Δd;

(c) Regeneration temperature and exergy efficiency; (d) Exergy destructions

In this case, $A_r = 1$, and according to Eq. (3-22), the value of ζ_t is in between the values of $\zeta_{t,\ deh}$ and $\zeta_{t,\ reg}$. As G_r approaches G_p, the uniformity improves, the required regeneration temperature decreases from 137.8℃ to 67.5℃, and the exergy efficiency increases from 21.7% to 43.0%. As shown in Fig. 3-7 (d), the exergy obtained by the processed air is almost unchanged; the performance improvement is due to the decrease of the exergy destruction, which benefits from the reduction of the unmatched coefficient as F_r approaches 1.

Fig. 3-7 shows the average t_a, t_d, Δt, ω_a, ω_d and $\Delta\omega$ in the angle direction at the dehumidification side (Deh. side) and regeneration side (Reg. side) as a function of the wheel thickness when regeneration air flow rate $G_r = 800\text{m}^3/\text{h}$ (air flow rate ratio of processed air to regeneration air $F_r = 3.13$) and $G_r = 2500\text{m}^3/\text{h}$ ($F_r = 1$).

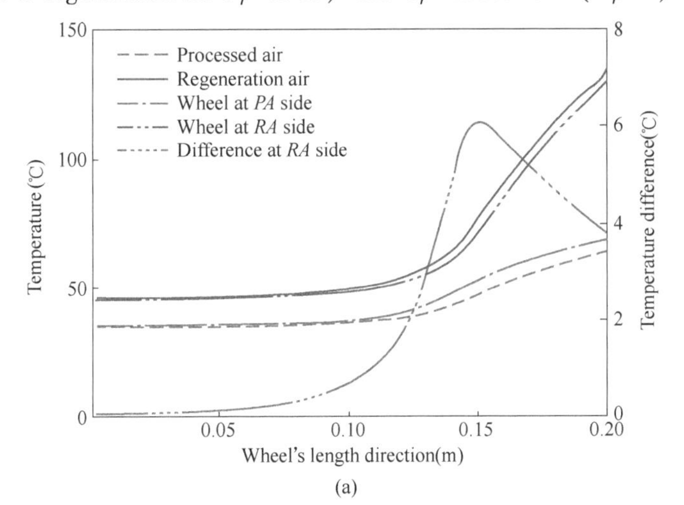

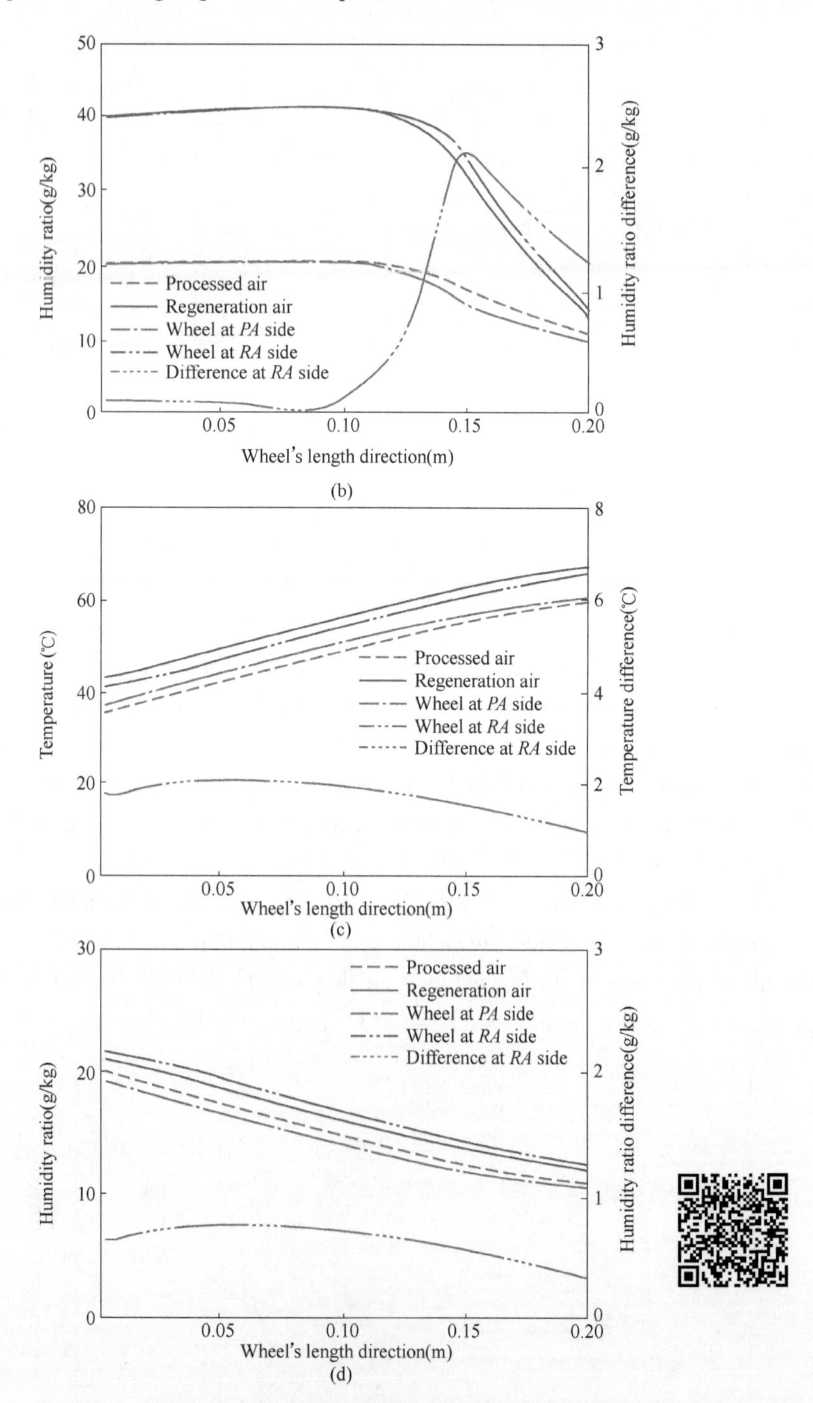

Fig. 3-7 Simulation results for Case 1

(a) Temperature (difference) distribution, $G_{ar}=800\text{m}^3/\text{h}$; (b) Humidity ratio (difference) distribution, $G_{ar}=800\text{m}^3/\text{h}$; (c) Temperature (difference) distribution, $G_{ar}=2500\text{m}^3/\text{h}$; (d) Humidity ratio (difference) distribution, $G_{ar}=2500\text{m}^3/\text{h}$

Along the wheel thickness direction, the states of the desiccant change with the air. When $G_r = G_p$, the temperature and humidity ratio variation of the desiccant can better adapt to the variation of the two streams of air. The Δt and $\Delta\omega$ fields are much more uniform when G_r = 2500m^3/h than when G_r = 800m^3/h. When G_r = 800m^3/h (F_r = 3.13), heat transfer driving force Δt varies from 0 ~ 6℃, and mass transfer driving force $\Delta\omega$ varies from 0 ~ 2g/kg. When G_r = 2500m^3/h (F_r = 1), Δt varies from 1 ~ 2℃, and $\Delta\omega$ varies within a range of about 0.5g/kg.

The average Δt and $\Delta\omega$ in the angle direction at the dehumidification side coincide with those at the regeneration side for both G_r = 800m^3/h and G_r = 2500m^3/h, as illustrated in Fig. 3-7. This is because $F_{deh} = F_{reg}$, and according to Eq. (3-23), average Δt and $\Delta\omega$ values are identical for the dehumidification side and the regeneration side.

$$hF_{deh}\Delta t_{deh} = hF_{reg}\Delta t_{reg}, \quad h_m F_{deh}\Delta\omega_{deh} = h_m F_{reg}\Delta\omega_{reg} \tag{3-23}$$

When G_r = 800m^3/h, $\Delta t_{deh} = \Delta t_{reg}$ = 2.07℃, and $\Delta\omega_{deh} = \Delta\omega_{reg}$ = 0.63g/kg. When G_r = 2500 m^3/h, $\Delta t_{deh} = \Delta t_{reg}$ = 1.72℃, and $\omega_{deh} = \Delta\omega_{reg}$ = 0.63g/kg.

(2) Case 2: variable A_r (1 ~ 5), F_r = 1 ($G_p = G_r$ = 2500m^3/h)

Fig. 3-8 shows the unmatched coefficients of Δt and $\Delta\omega$, as well as the regeneration temperature, exergy efficiency, and exergy destruction of each part, when A_r changes from 1 to 5.

As discussed in Case 1, because $G_r = G_p$, the unmatched coefficients at the dehumidification side and regeneration side are close to 1, and A_r has little effect on them. However, A_r has significant influence on the unmatched coefficient of the whole wheel. As A_r approaches 1, the unmatched coefficient of the whole wheel approaches 1. This is because when $A_r \neq 1$, according to Eq. (3-23), $\Delta t_{deh} \neq \Delta t_{reg}$ and $\Delta\omega_{deh} \neq \Delta\omega_{reg}$, which makes the Δt and $\Delta\omega$ fields of the whole wheel less uniform. Thus, as A_r increases from 1 to 5, the exergy destruction increases, the required regeneration temperature increases from 67.5℃ to 71.9℃, and the exergy efficiency decreases from 43.0% to 38.1%. Compared to F_r(Case 1), the impact of A_r is less significant.

Fig. 3-9 shows the average t_a, t_d, Δt, ω_a, ω_d, and $\Delta\omega$ in the angle direction at the dehumidification side and regeneration side as a function of the wheel thickness when A_r = 3.

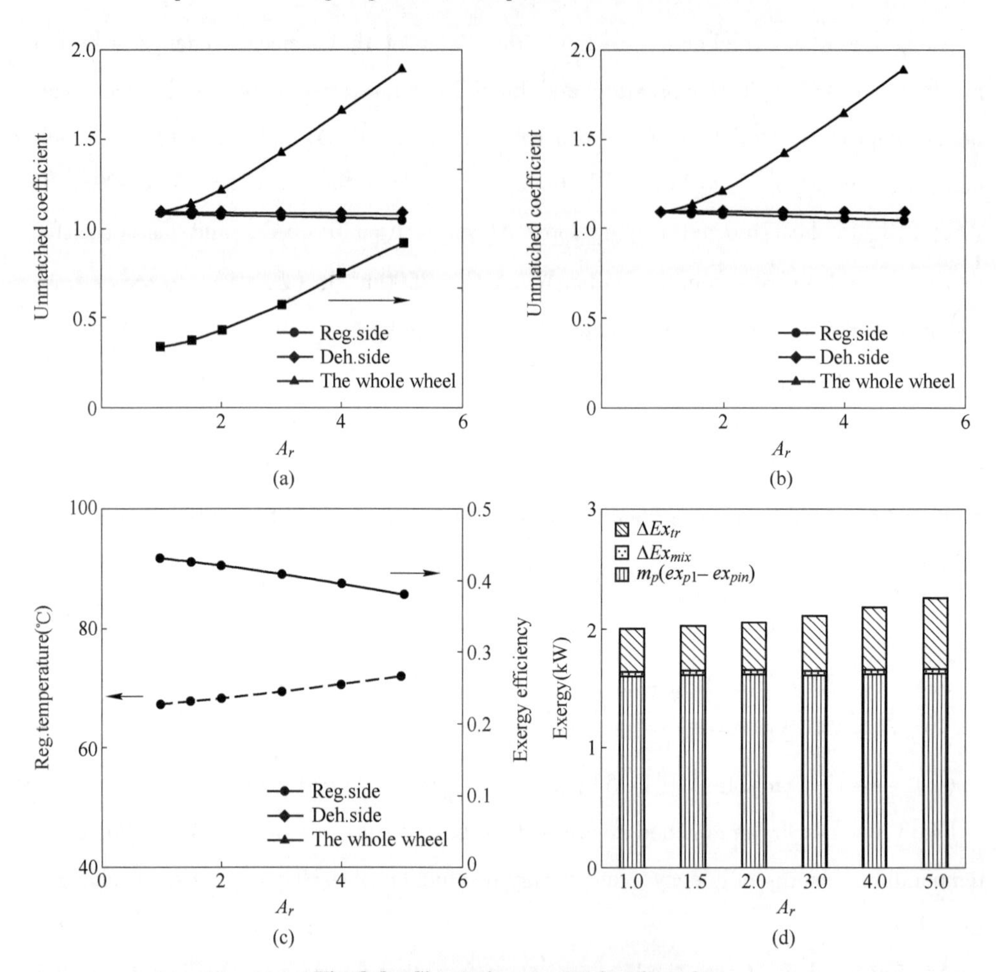

Fig. 3-8 Simulation results for Case 2

(a) Unmatched coefficient of Δt; (b) Unmatched coefficient of Δd; (c) Reg. temperature and exergy efficiency; (d) Exergy

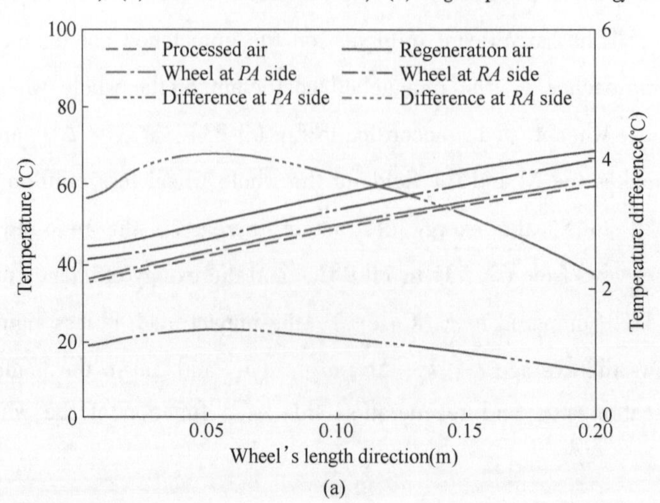

(a)

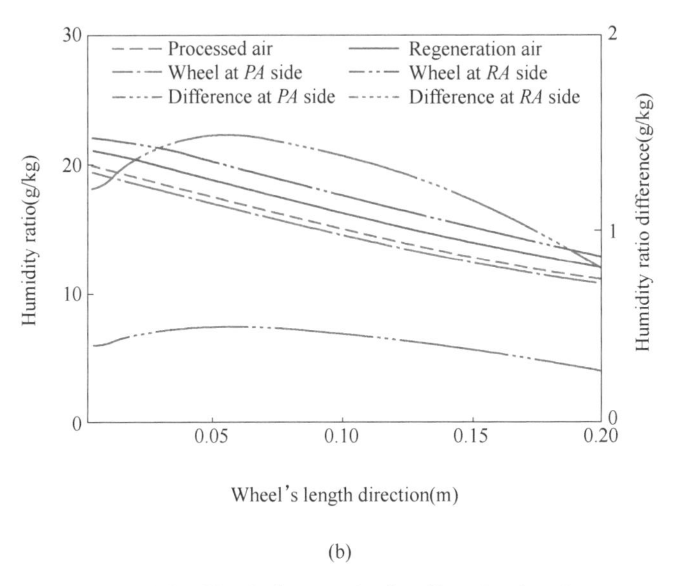

(b)

Fig. 3-9 Simulation results for Case 2: $A_r=3$

(a) Temperature (difference) distribution; (b) Humidity ratio (difference) distribution

Figs. 3-7 (c) ~ (d) illustrate the case when $A_r=1$. Because $F_r=1$, both temperature and humidity ratio differences are close to uniform for both the dehumidification side and regeneration side. However, when $A_r \neq 1$, the average Δt and $\Delta\omega$ values in the angle direction as a function of wheel thickness at the dehumidification side are different from those at the regeneration side. When $A_r=3$, the $\Delta t_{deh}=1.15$℃, $\Delta t_{reg}=3.46$℃, $\Delta\omega_{deh}=0.42$g/kg, and $\Delta\omega_{reg}=1.26$g/kg.

(3) Case 3: variable A_r (1~3), $F_r=A_r$

Fig. 3-10 shows the effects of A_r ($F_r=A_r$) on the performance of the desiccant wheel.

A larger A_r means a smaller G_r, and according to Case 1, the unmatched coefficients for the dehumidification side and regeneration side will be larger. The difference between F_{deh} and F_{reg} negatively impacts the uniformity of the whole wheel. Thus, a higher A_r leads to higher unmatched coefficients. As A_r increases from 1 to 3, the exergy destruction increases, the required regeneration temperature increases from 67.5℃ to 131.8℃, and the exergy efficiency decreases from 43.0% to 23.0%. Fig. 3-11 shows the average t_a, t_d, Δt, ω_a, ω_d, and $\Delta\omega$ in the angle direction at the dehumidification side and regeneration side as a function of the wheel thickness when $A_r=3$.

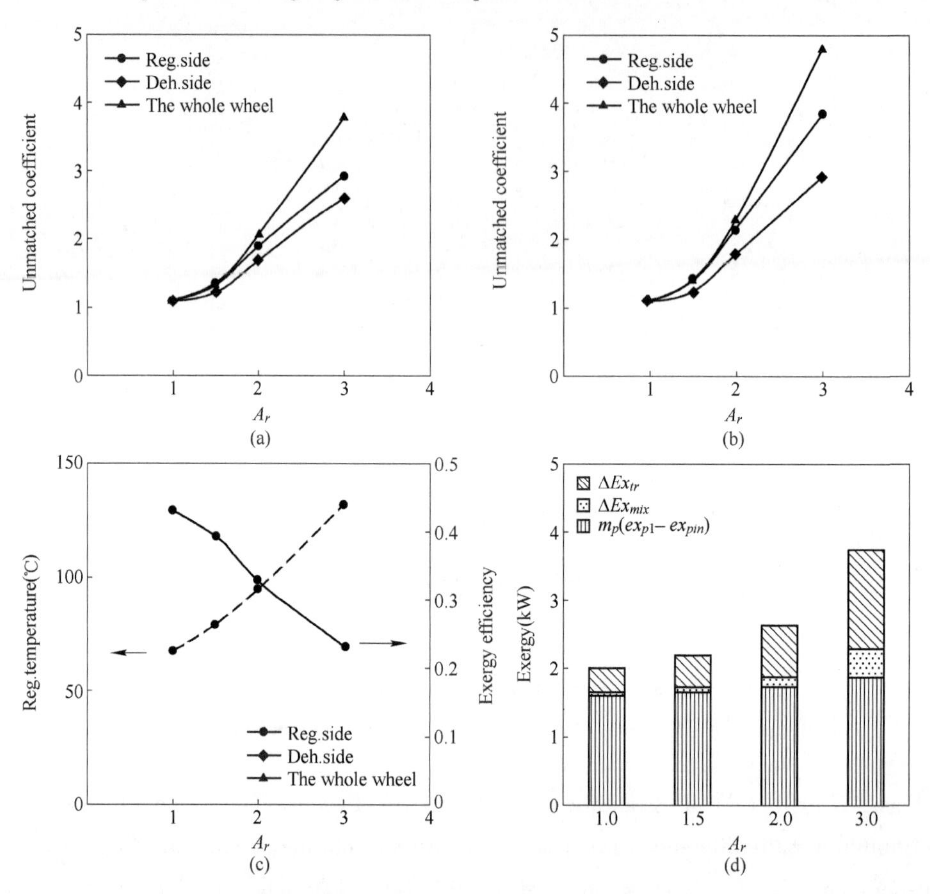

Fig. 3-10 Simulation results for Case 3

(a) Unmatched coefficient of Δt; (b) Unmatched coefficient of Δd;
(c) Reg. temperature and exergy efficiency; (d) Exergy

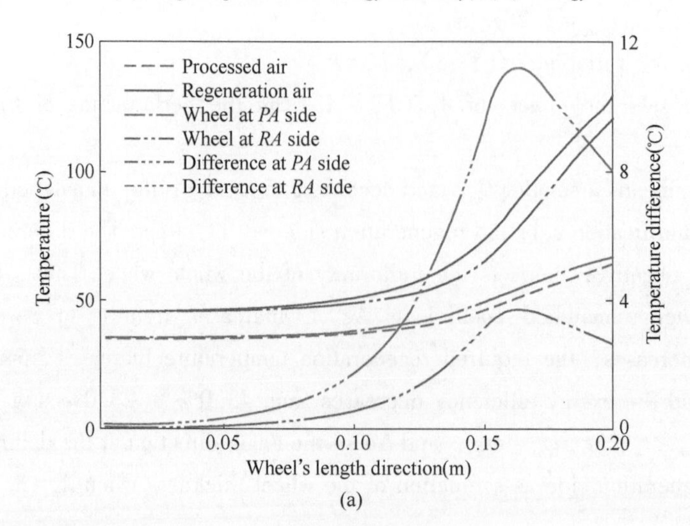

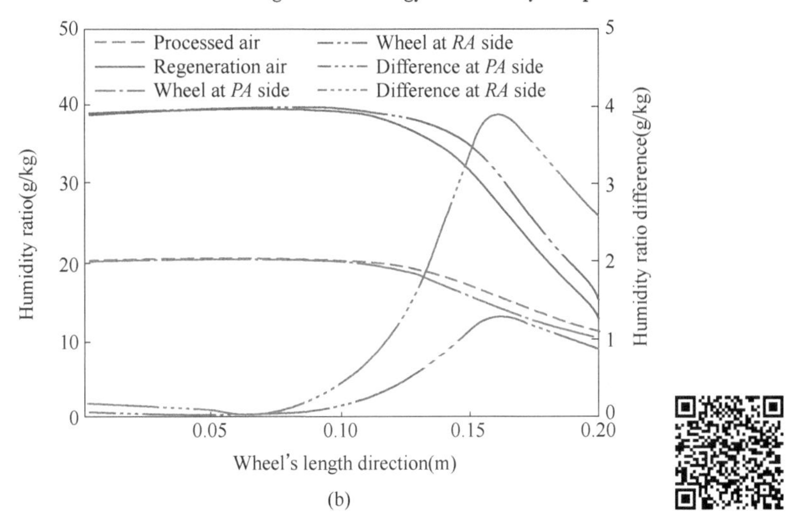

Fig. 3-11 Simulation results for Case 3: $A_r = 3$

(a) Temperature (difference) distribution; (b) Humidity ratio (difference) distribution

Figs. 3-7 (c) ~ (d) illustrate the case when $A_r = 1$. When $A_r = 3$, $\Delta t_{deh} = 1.35℃$, $\Delta t_{reg} = 4.04℃$, $\Delta\omega_{deh} = 0.42$g/kg, and $\Delta\omega_{reg} = 1.26$g/kg. For Case 2 and Case 3, assuming that $\zeta_{t,\ deh} = \zeta_{t,\ reg}$, according to Eq. (3-22), the lowest ζ_t is obtained when $A_r = 1$; when $A_r > 1$, ζ_t is larger than both $\zeta_{t,\ deh}$ and $\zeta_{t,\ reg}$. These phenomena can be observed in Fig. 3-8 and Fig. 3-11.

In conclusion, when the total heat and mass transfer is determined, the change of the unmatched coefficients influences heat and mass exergy destruction, which will influence the exergy efficiency and the regeneration temperature in the end. As discussed above, the regeneration temperature can be decreased from above 130℃ to below 70℃ when the uniformity of the heat and mass transfer driving forces improves. $A_r = 1$ and $F_r = 1$ are recommended for the smallest ΔEx, which will be adopted in the following analysis.

3.3 Decreasing thermal exergy obtained by the processed air

According to Eq. (3-5) and Eq. (3-8), the exergy provided by the regeneration air and exergy efficiency are not only influenced by exergy destruction, but also by the thermal exergy obtained by the processed air. This section mainly focuses on reducing the regeneration temperature by decreasing the thermal exergy obtained by the processed air.

When the dehumidification capacity is fixed, the chemical exergy obtained by the processed air is fixed. After being dehumidified, the thermal exergy of the processed air is increased because of the temperature increase. The obtained thermal exergy has to be provided by the regeneration air. When the thermal exergy obtained by the processed air can be decreased, the exergy provided by the regeneration air can be decreased, as can the regeneration temperature. This section analyzes the factors influencing thermal exergy, and provides corresponding solutions for reducing the thermal exergy obtained by the processed air.

3.3.1 Factors influencing thermal exergy

Eq. (3-24) shows the first-order deviation of ex_{th} to T_a. When T_a is equal to T_0, ex_{th} equals 0, and ex_{th} increases as T_a becomes larger (or smaller than T_0). The farther T_a is from T_0, the more influence dT_a will have on the change of ex_{th}, as shown in Fig. 3-2 (a). Therefore, to reduce ex_{th}, the processed air should stay in a low temperature range.

$$dex_{th}/dT_a = c_{pm}(1 - T_0/T_a) \tag{3-24}$$

As for desiccant wheel dehumidification, the processed air is handled near the isenthalpic line. Assuming that the handling process is isenthalpic, Eq. (3-25) shows that the value of the air temperature change during dehumidification will be the same when the air humidity ratio change is the same.

$$c_{pm}\Delta t_a + r\Delta\omega_a = 0 \tag{3-25}$$

As shown in Fig. 3-12, when the dehumidification capacity remains constant, if pre-cooling is adopted, the temperature will change at a lower temperature range compared to the occasions without pre-cooling. Moreover, if the dehumidification is divided into several stages and cooling is adopted between each stage, the sum of the temperature rise after each stage of the desiccant wheel is the same with the temperature rise after dehumidification in single-stage desiccant wheel system. However, the temperature will stay in an even lower temperature range for multi-stage desiccant dehumidification and cooling.

Under the assumption that the processed air is dehumidified along the isenthalpic line, Table 3-2 shows the effects of the number of stages and pre-cooling on the temperature variation range, as well as the total thermal exergy obtained by the processed air during all stages of dehumidification.

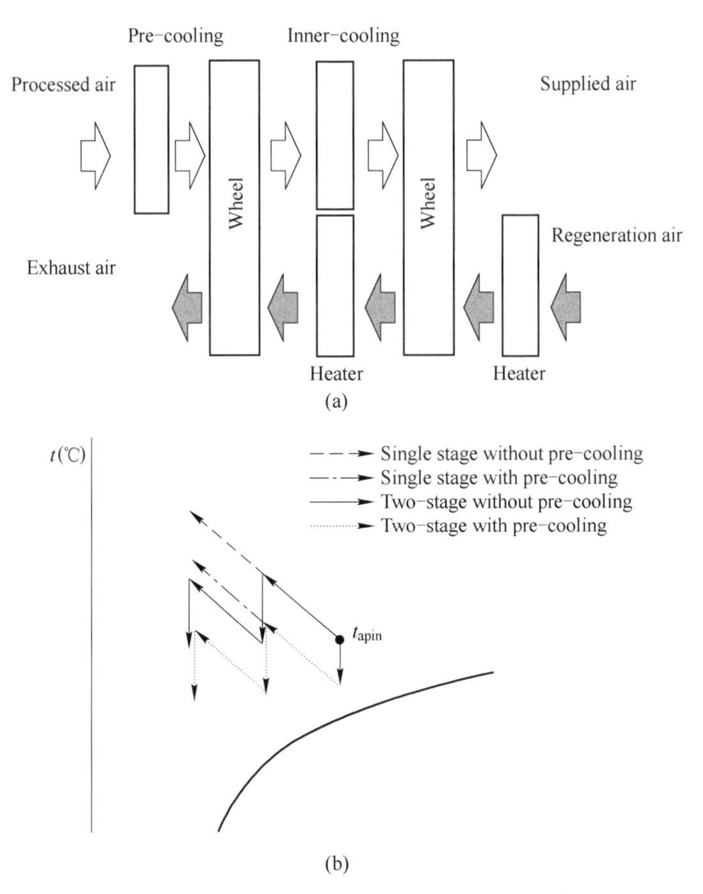

Fig. 3-12 Schematic and air handling process for single-stage and two-stage desiccant dehumidification with and without pre-cooling

(a) Two-stage desiccant wheel with pre-cooling; (b) Air-handling process

Table 3-2 Effects of number of stages and pre-cooling on temperature variation range and thermal exergy obtained by the processed air

Number of stages	Without pre-cooling (air cooled to 35℃ before dehumidification)		With pre-cooling (air cooled to 25℃ before dehumidification)	
	Temperature variation range (℃)	Obtained thermal exergy (kJ/kg)	Temperature variation range (℃)	Obtained thermal exergy (kJ/kg)
1	35.0~64.9	1.367	25.0~54.9	0.450

Continued Table 3-2

Number of stages	Without pre-cooling (air cooled to 35℃ before dehumidification)		With pre-cooling (air cooled to 25℃ before dehumidification)	
	Temperature variation range (℃)	Obtained thermal exergy (kJ/kg)	Temperature variation range (℃)	Obtained thermal exergy (kJ/kg)
2	35.0~49.9	0.704	25.0~39.9	−0.255
4	35.0~42.5	0.358	25.0~32.5	−0.625
6	35.0~40.0	0.240	25.0~30.0	−0.751
8	35.0~38.7	0.180	25.0~28.7	−0.814
12	35.0~37.5	0.121	25.0~27.5	−0.878

The humidity ratio of the processed air decreases from 20g/kg to 11g/kg. When multi-stage dehumidification is adopted, the humidity ratio change of the processed air during each stage is the same. Thus, the temperature variation range and thermal exergy change after dehumidification for each stage are identical. As shown in Table 3-2, for single-stage dehumidification, when the processed air is pre-cooled to 25℃, the temperature variation range of the processed air in the desiccant wheel is 25~54.9℃, with a total thermal exergy change during dehumidification of 0.450kJ/kg. The temperature variation range in the desiccant wheel without pre-cooling is 35~64.9℃, with a total thermal exergy change during dehumidification of 1.367kJ/kg. For multi-stage dehumidification, if the temperature is reduced to 25℃ before each stage of the desiccant wheel, the total thermal exergy change during dehumidification (i.e., in all stages) is much lower than when the temperature is only reduced to 35℃. If the number of stages is sufficiently high, when the temperature is reduced to 25℃ before each stage of the desiccant wheel, the temperature can stay in a range lower than T_0. Thus, thermal exergy change can be negative during dehumidification.

Effect of stage number on the temperature range, as well as the differences in thermal exergy change during dehumidification processes with or without pre-cooling, are illustrated in Fig. 3-13.

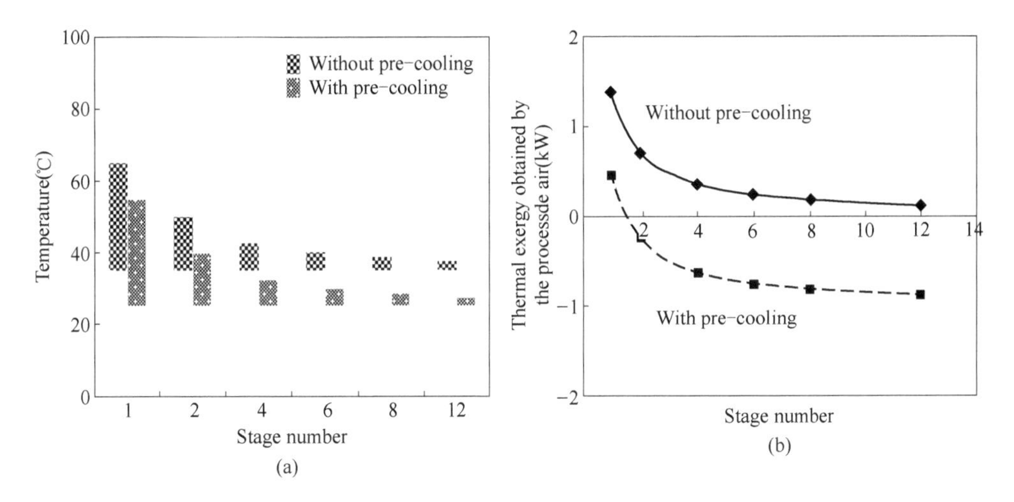

Fig. 3-13 Effects of number of stages and pre-cooling

(a) Temperature variation range; (b) Exergy obtained by processed air during dehumidification

As the number of stages increases, the total thermal exergy changes during all stages of dehumidification decreases. However, the rate of improvement slows down over time.

3.3.2 Case studies

To study the influences of pre-cooling and number of stages on the thermal exergy obtained by the processed air, as well as their influence on the regeneration temperature and the exergy efficiency of the dehumidification process, a desiccant wheel system was designed and analyzed. The wheel's structure and operating conditions are the same as those listed in Table 3-1; $A_r = 1$ and $G_p = G_r = 2500\text{m}^3/\text{h}$. The supplied air humidity ratio is fixed at 11g/kg. When the wheel is divided into several stages, the wheel thickness remains the same for all stages, and the total thickness is 200mm. Regeneration air is heated to the same temperature before entering each wheel, and the processed air is cooled down to the same temperature before entering each wheel.

Four cases (B, C, D, and E) are compared, as shown in Table 3-3. For B and D, two-stage dehumidification is adopted, and wheel thickness is 0.1m for each desiccant wheel; for C and E, four-stage dehumidification is adopted, and wheel thickness is 0.05m for each desiccant wheel. For B and C, the processed air is cooled down to 35℃ before dehumidification. For D and E, the processed air is cooled down to 25℃ before dehumidification. The air handling processes of Cases B ~ E are drawn in psychrometric charts, as shown in Fig. 3-14.

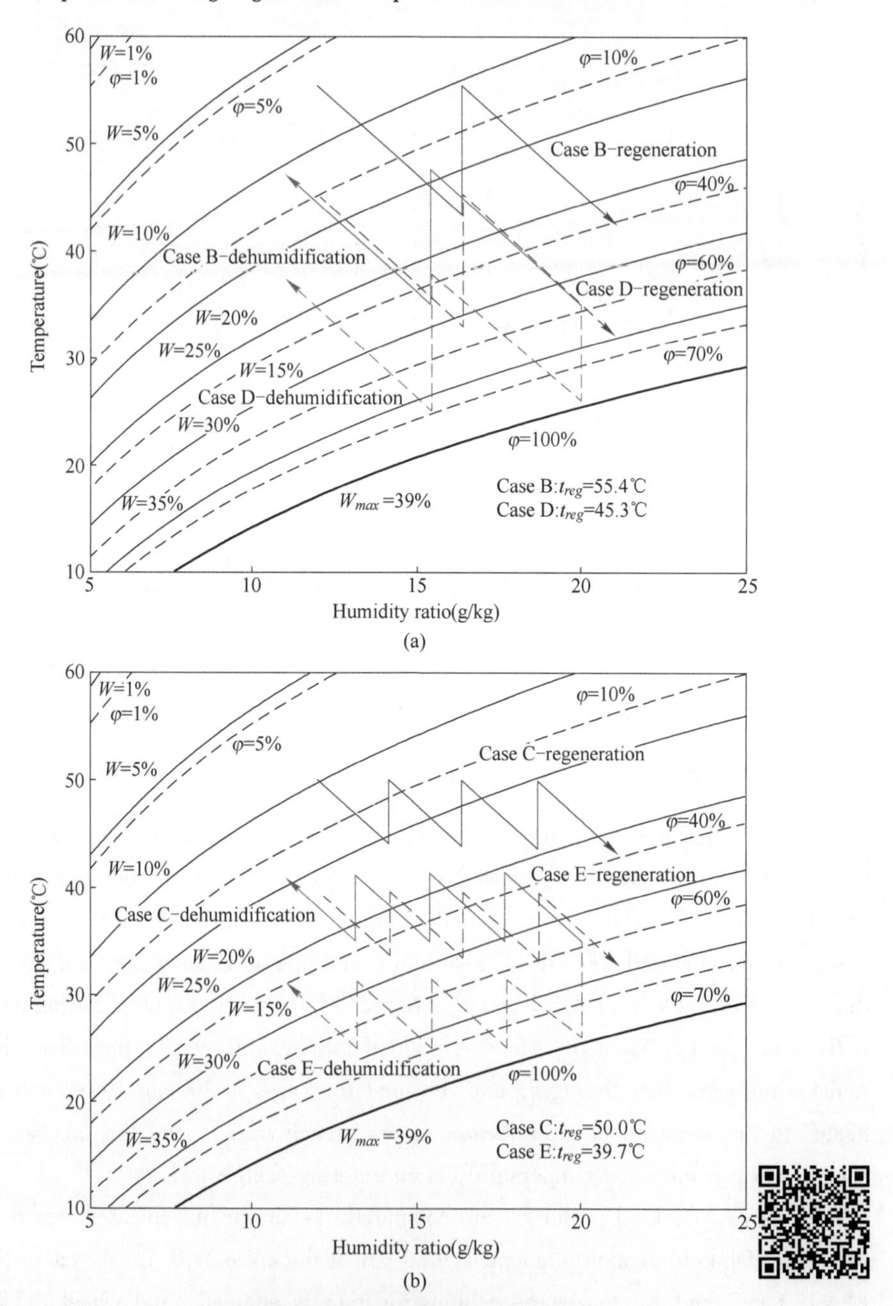

Fig. 3-14 Air handling processes

(a) Cases B and D; (b) Cases C and E

To achieve the required humidity ratio of the supplied air, the regeneration temperatures for B, C, D, and E are 55.4℃, 50.0℃, 45.3℃, and 39.7℃, respectively.

Table 3-3 Regeneration temperature and exergy information for different cases

Case	Number of stages	Pre-cooling (inlet temp. of PA to DW)	t_{reg} (℃)	Thermal exergy obtained by PA (kW)	Chemical exergy obtained by PA (kW)	Exergy destruction (kW)	Exergy provided by RA (kW)	η_{ex}
A	1	no (35.0℃)	67.5	0.75	0.86	0.39	2.00	43.0%
B	2	no (35.0℃)	55.4	0.39	0.86	0.42	1.67	51.4%
C	4	no (35.0℃)	50.0	0.20	0.86	0.48	1.54	55.9%
D	2	yes (25.0℃)	45.3	-0.25	0.86	0.44	1.05	81.5%
E	4	yes (25.0℃)	39.7	-0.46	0.86	0.50	0.89	95.9%

Cases B ~ E are compared with the most basic case (Case A), a single-stage dehumidification process without pre-cooling, illustrated in Figs. 3-3 (c) ~ (d). Cases A ~ E are compared in Fig. 3-15 and Table 3-3. For all five cases, the changes in exergy destruction are minimal. However, the thermal exergy obtained by the processed air during all stages of dehumidification decreases significantly from Case A to Case E, which benefits from the lower temperature range of the processed air during dehumidification process from Case A to Case E. Moreover, the exergy provided by the regeneration air during all stages of dehumidification decreases from 2.00kW in Case A to 0.89kW in Case E. The exergy efficiency for the desiccant wheel, i.e., the chemical exergy obtained by the processed air divided by the exergy provided by the regeneration air during all stages of dehumidification, written as Eq. (3-8), increases from 43.0% to 95.9%, and the required regeneration temperature decreases from 67.5℃ to 39.7℃.

Based on these results, it can be concluded that multi-stage dehumidification and pre-cooling are effective ways to reduce the regeneration temperature. The greatest effect of pre-cooling on the reduction of the thermal exergy obtained by the processed air was observed between Cases C and D. However, as the number of stages increases, the rate of improvement slows down.

3.4 Conclusions

This chapter primarily discussed ways to reduce the regeneration temperature of a desiccant wheel system using exergy analysis. The following conclusions are noteworthy:

(1) When the dehumidification capacity is fixed, reductions in exergy destruction and thermal exergy obtained by the processed air will lead to a decrease in the exergy

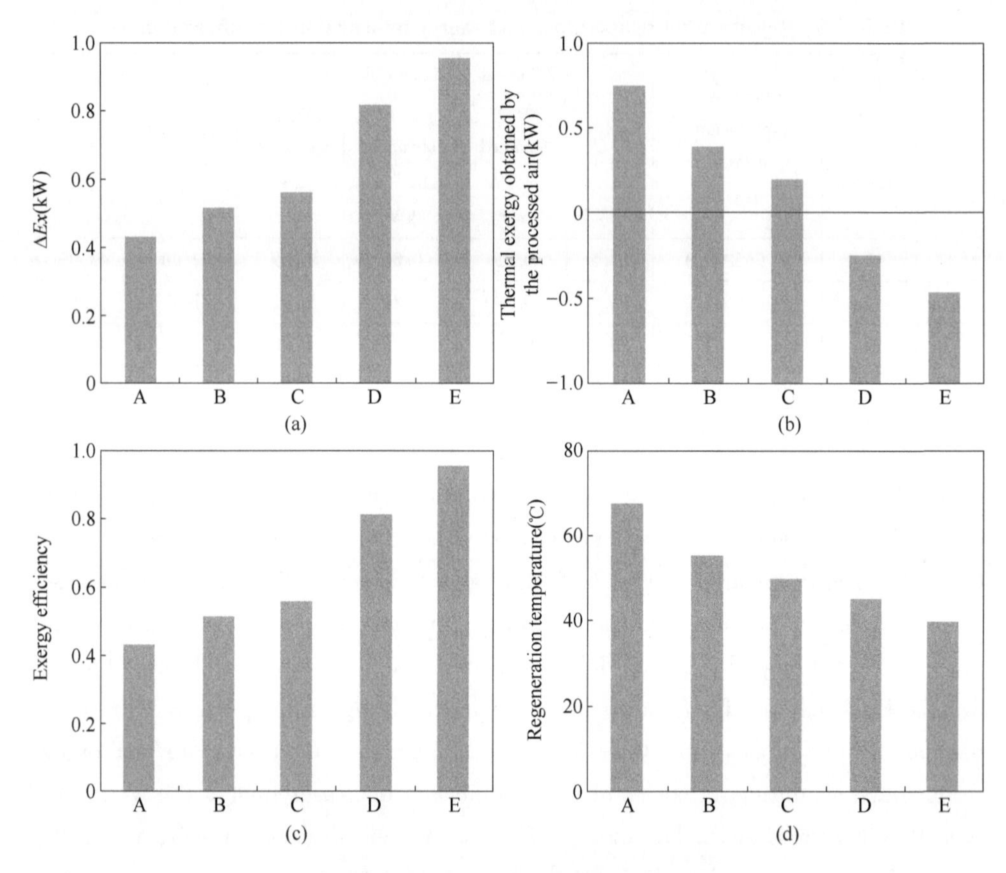

Fig. 3-15 Comparison of different cases

(a) Exergy destruction; (b) Thermal exergy obtained by processed air; (c) Exergy efficiency; (d) Regeneration temperature

provided by the regeneration air, resulting in a lower regeneration temperature.

(2) When the heat and mass transfer area and capacity are determined, heat and mass transfer exergy destruction (ΔEx_{tr}) in the wheel is influenced by the unmatched coefficients of the heat and mass transfer driving forces; A_r and F_r have significant influences on the unmatched coefficients. In order to reduce ΔEx_{tr} , it is adviced to set $A_r = F_r = 1$. The regeneration temperature could be reduced from above 130℃ to below 70℃ when the air was dehumidified from 20g/kg to 11g/kg for single-stage dehumidification.

(3) The exergy obtained by the processed air is determined by the temperature of the air during dehumidification. To reduce this part of the exergy, multi-stage dehumidification and pre-cooling are recommended. The required regeneration

temperature could be lower than 40℃ with pre-cooling and multi-stage dehumidification.

Nomenclature

A	facial area, m^2
A_r	facial area ratio of dehumidification side to regeneration side, dimensionless
c_{pm}	specific heat capacity of humid air, J/kg
ex	exergy per mass flow of dry air, kJ/kg
Ex	exergy flow rate, kW
ΔEx	exergy destruction, kW
F	heat and mass transfer area, m^2
F_r	air flow rate ratio of the processed air to the regeneration air, dimensionless
G	air volume flow rate, m^3/h
h	heat transfer coefficient, W/(m^2 · ℃)
h_m	mass transfer coefficient, kg/(m^2 · s)
k	grid number at the angle direction
L	wheel thickness
Le	Lewis number, dimensionless
M	mass removal capacity, kg/s
$\dot{m}$	mass flow rate of the air, kg/s
n	grid number along the thickness direction
NTU	number of transfer units, dimensionless
Q	cooling/heating capacity, W
R_a	gas constant for air, kJ/(mol · K)
r	heat of vaporization, J/kg
S_l	slope of saturated air line
T	temperature, K

t temperature, ℃

Δt absolute temperature difference between air and desiccant, ℃

Greek symbols

ω humidity ratio, g/kg

$\Delta\omega$ absolute ω difference between air and desiccant, g/kg

ϕ relative humidity, %

ζ unmatched coefficient

η_{ex} exergy efficiency of desiccant wheel

Subscripts

a air

th thermal

ch chemical

d desiccant

deh dehumidification

dew dew point

h heat

in inlet

m mass

mix mixture

out outlet

p processed air

r regeneration air

reg regeneration

tr transfer

References

[1] R Tu, Y Hwang, T Cao, et al. Investigation of adsorption isotherms and rotational speeds for low temperature regeneration of desiccant wheel systems [J]. Int. J. Refrig, 2018, 86: 495-509.

[2] R Tu, X H Liu, Y Jiang. Lowering the regeneration temperature of a rotary wheel dehumidification system using exergy analysis [J]. Energy Conver. and Manag, 2015, 89: 162-174.

[3] ASHRAE. ASHRAE handbook e HVAC systems and equipment [M]. Atlanta, GA: American Society of Heating, Refrigerating, and Air-Conditioning Engineers, 2008.

[4] D La, Y Li, Y J Dai, et al. Effect of irreversible processes on the thermodynamic performance of open-cycle desiccant cooling cycles [J]. Energy Convers. Manage, 2013, 67: 44-56.

[5] K Mehmet. Energy and exergy analyses of an experimental open-cycle desiccant cooling system [J]. Appl. Therm. Eng, 2003, 24 (5): 919-932.

[6] W M Worek, W Zheng. Thermodynamic properties of adsorbed water on silica gel exergy losses in adiabatic sorption processes [J]. Journal of Thermophys. and Heat Transfer, 2012, 5 (3): 435-440.

[7] Z Y Guo, S Q Zhou, Z X Li, et al. Theoretical analysis and experimental confirmation of the uniformity principle of temperature difference field in heat exchanger [J]. Int. J. Heat Mass Trans, 2002, 45 (10): 2119-2127.

[8] T Zhang, X H Liu, L Zhang, et al. Match properties of heat transfer and coupled heat and mass transfer processes in air-conditioning system [J]. Energy Convers. Manage, 2012, 59: 103-113.

[9] T Zhang, X H Liu, Y Jiang. Performance comparison of liquid desiccant air handling processes from the perspective of match properties [J]. Energy Convers. Manage, 2013, 75: 51-60.

[10] R Tu, X H Liu, Y Jiang. Performance comparison between enthalpy recovery wheels and dehumidification wheels [J]. Int J Refrig, 2013, 36 (8): 2308-2322.

Chapter 4 Effects of Adsorption Isotherms and Rotation Speed on Regeneration Temperature

Compared with single stage systems, air in two-stage systems is handled in a higher relative humidity range. Besides, for deep dehumidification application, air is treated in a lower relative humidity range as compared with the regular dehumidification application. Therefore, recommended physical parameters for desiccant materials applied for low relative humidity dehumidification and high relative humidity dehumidification should be analyzed. Effects of the adsorption isotherms on the regeneration temperature of single stage and two-stage desiccant wheel systems under humid- and mild-climate for regular and deep dehumidification applications are discussed in this chapter. Moreover, effects of rotational speed are taken into consideration. The results provide guidelines for selecting desiccant materials regarding shape factor, maximum water capacity, and rotational speed ranges for low temperature regeneration of desiccant wheel dehumidification systems.

4.1 The equilibrium isotherms of the desiccant wheel

Eq. (1-3) is a common equilibrium isotherm equation for adsorption materials. Shape factor (C) and maximum water capacity (W_{max}) are two parameters relating to adsorption properties of desiccant materials. The adsorption isotherms in Table 1-1 can be simplified as Eq. (1-3) with different C and W_{max}, as shown in Fig. 1-6. Combining the adsorption isotherm equations and the Clapeyron equation shown in Eq. (1-5), the equilibrium humidity ratio of the desiccant (ω_d), which is influenced by t_d and W/W_{max}, can be obtained, shown as Eq. (1-6).

Eq. (1-6) is adopted in the mathematical model to investigate the effects of adsorption isotherm in the name of C and W_{max} on dehumidification performances of desiccant wheel systems. The parameters used in the model is shown in Table 4-1.

Table 4-1 Parameters used in the model

f	k_d [kW/(m · ℃)]	r_s (kJ/kg)	ρ_{ad} (kg/m³)	x	c_{pad} [kJ/(kg · ℃)]	Mol (kg/kmol)
0.1765	0.00022	2.65×10^3	1129	0.7	0.92	18

Continued Table 4-1

σ	W_{max} (kg/kg)	C	ρ_d (kg/m^3)	a (m)	c_{pd} [kJ/(kg · ℃)]	d_h (m)
0.7	0.39	0.5	978	11×10^{-10}	0.912	0.0012

4.2 Air dehumidification at high and low relative humidity

In this section, dehumidification at high relative humidity and low relative humidity are discussed. Air handling processes of single stage and two-stage desiccant wheel dehumidification systems are investigated under mild- and humid-climates for regular and deep dehumidification applications, respectively.

4.2.1 System description

Single stage and two-stage desiccant wheel dehumidification systems are shown in Fig. 4-1.

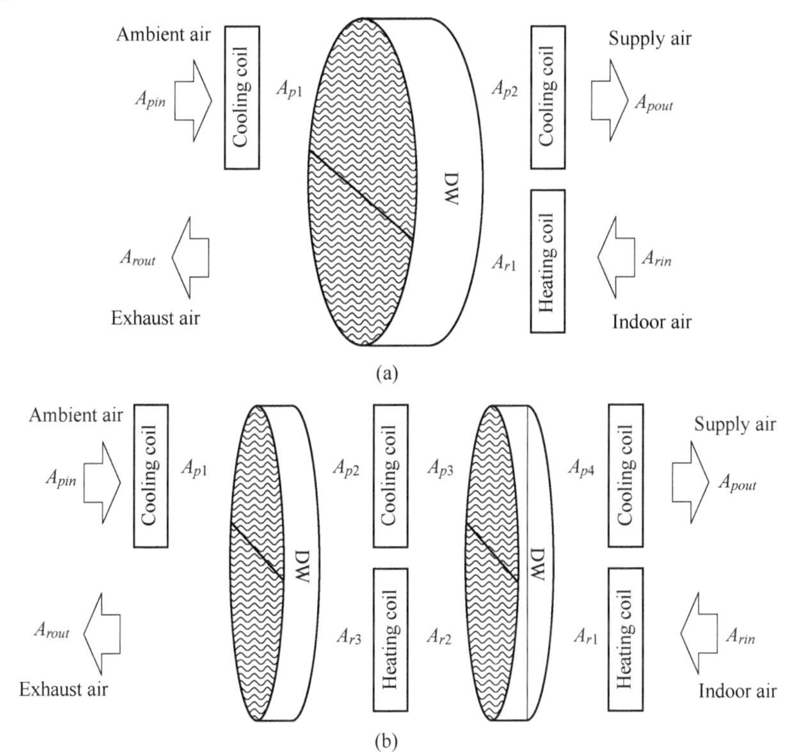

Fig. 4-1 Schematics of single stage and two-stage systems

(a) Single stage system; (b) Two-stage system

For both systems, the process air is cooled down by a cooling coil before entering the desiccant wheel. Similarly, the regeneration air is heated to the required regeneration temperature (t_{reg}) before regenerating the desiccant wheel. It should be emphasized that lower t_{reg} under the same working condition is beneficial for the adoption of low-grade heating sources such as solar energy and heat pump system, and smaller heating capacity of heat sources. Therefore, the following discussions are based on reducing t_{reg}.

The dimensions of desiccant wheels and working conditions for simulation studies are summarized in Table 4-2.

Table 4-2 Design conditions for the simulation analysis.

Working conditions	Desiccant wheel	Heat exchange coil
Process air volume flow rate: 2500m³/h BSC: 33℃, 19g/kg; WSC: 33℃, 14g/kg Regeneration air: 2500m³/h, 26℃, 12g/kg Supplied air for regular dehumidification: 2500m³/h, 25℃, 9g/kg Supplied air for deep dehumidification: 2500m³/h, 25℃, 4g/kg	Radius: 0.5m Total thickness: 0.2m Air path: 2mm high and 2mm wide *Nu*: 2.463	Temperature after the cooling coils: t_c = 25℃ Temperature after the heating coils: t_{reg}

Each desiccant wheel in the two-stage system has the samecross sectional area and half the thickness of the desiccant wheel in the single stage system. There are no physical models for cooling coils or heaters. The process air is cooled to t_c by each cooling coil. Similarly, the regeneration air is heated to the regeneration temperature (t_{reg}) by each heater. t_c is fixed at 25℃ in the following discussion. t_{reg}, which varies with working conditions, supply air humidity ratio, rotational speed, C and W_{max}, is to be determined through simulation. Beijing summer working condition (BSC: 33℃, 19g/kg), which represents humid climate, and Washington airport summer working condition (WSC: 33℃, 14g/kg), which represents mild climate, are selected. 9g/kg and 4g/kg are selected as the supply air humidity ratio (humidity ratio of A_{Pout} in Fig. 4-1) representing regular and deep dehumidification, respectively.

4.2.2 Air handling processes of different cases

Four cases, which are regular dehumidification under BSC (case 1), regular dehumidification under WSC (case 2), deep dehumidification under BSC (case 3) and deep dehumidification under WSC (case 4), are discussed for the two systems.

The air handling processes (the cooling process from A_{Pin} to A_{P1} and the heating process from A_{Rin} to A_{R1} in Fig. 4-1 are not drawn on the psychrometric charts) for the above four cases of the single stage and two-stage systems are shown in Fig. 4-2.

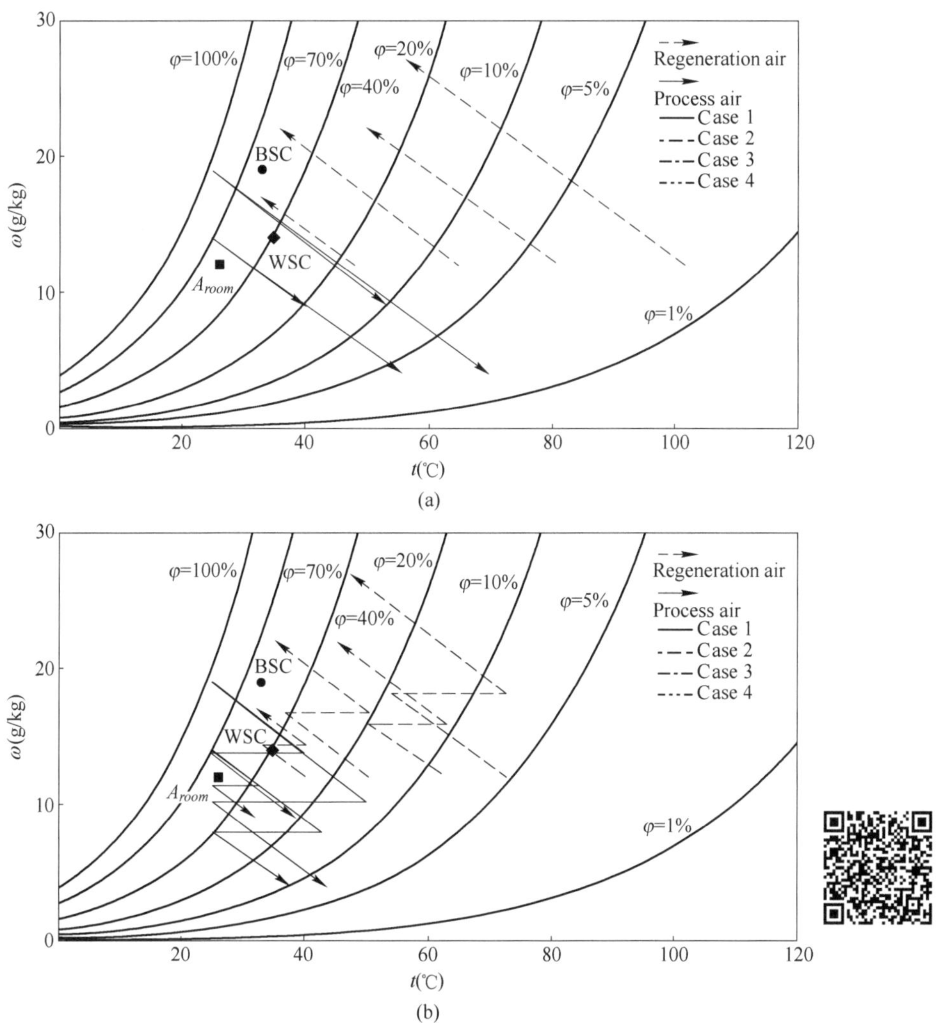

Fig. 4-2 Air handling processes for the four cases when C = 0.3, W_{max} = 0.39kg/kg and RS = 20r/h

(a) Single stage system; (b) Two-stage system

The corresponding regeneration temperature are shown in Fig. 4-3.

C and $W_{\max}$ are 0.3 and 0.39kg/kg, respectively, representing silica-gel in Fig. 1-6 (a). Rotational speed (RS) of the desiccant wheel is fixed at 20r/h. It is illustrated by Fig. 4-2 that for the same system, the air is handled at a lower relative humidity range for the deep dehumidification (case 3 and case 4) as compared with the regular dehumidification (case 1 and case 2), and the corresponding t_{reg} is higher. For the same case, the air is handled at a higher relative humidity in the two-stage system as

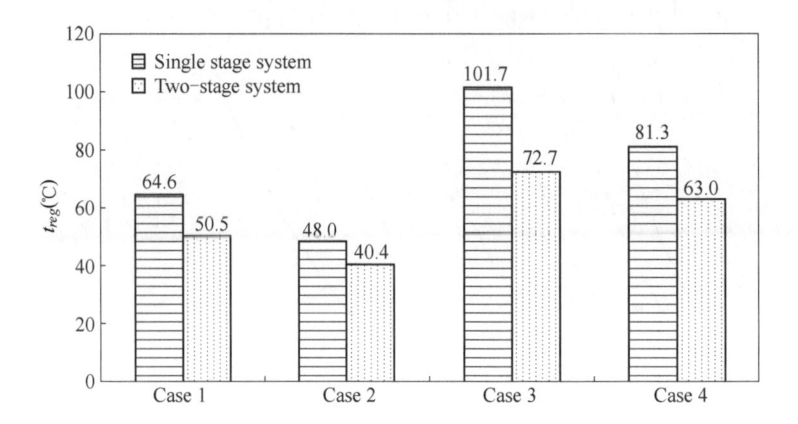

Fig. 4-3 t_{reg} of different scenarios when C=0.3, W_{max}=0.39kg/kg and RS=20r/h

compared with the single stage system, resulting in lower t_{reg}. t_{reg} varies from 40.4℃ (case 2 of the two-stage system) to 101.7℃ (case 3 of the single stage system). Fig. 4-3 shows that t_{reg} differences between the single stage and the two-stage systems range from 7.6℃ (case 2) to 29℃ (case 3).

The above discussions are based on fixed C, W_{max} and RS. In the next part, selections of C and W_{max} for high and low relative humidity dehumidification are analyzed with the aim of achieving relatively low t_{reg}. RS are taken into consideration and the following discussion is based on the optimal RS.

4.3 Effects of RS, C and W_{max} on t_{reg}

It is suggested that desiccant materials, of which C is smaller than 1, are suitable for dehumidification[5]. According to Fig. 1-6, C varies from 0.3 to 1, and W_{max} varies from 0.24kg/kg to 0.4kg/kg. In this section, cases with C equaling to 0.3, 0.5 and 1 are selected to cover adsorption materials in a wide range of C. W_{max} equaling to 0.3kg/kg and 0.8kg/kg are selected to represent adsorption material with regular and high water storage capacity. There are six combinations, which are C = 0.1 and W_{max} = 0.3kg/kg, C = 0.1 and W_{max} = 0.8kg/kg, C = 0.3 and W_{max} = 0.3kg/kg and C = 0.3 and W_{max} = 0.8kg/kg, C = 1 and W_{max} = 0.3kg/kg and C = 1 and W_{max} = 0.8kg/kg.

First, the effects of RS on t_{reg} are discussed. Next, the recommended C and W_{max} are discussed under the optimal RS (ORS). Finally, RS ranges which gives relatively low t_{reg} are analyzed.

4.3.1 Effects of RS on t_{reg}

The effects of RS on t_{reg} of single stage and two-stage systems under case 1 to case 4 are shown in Figs. 4-4～4-7.

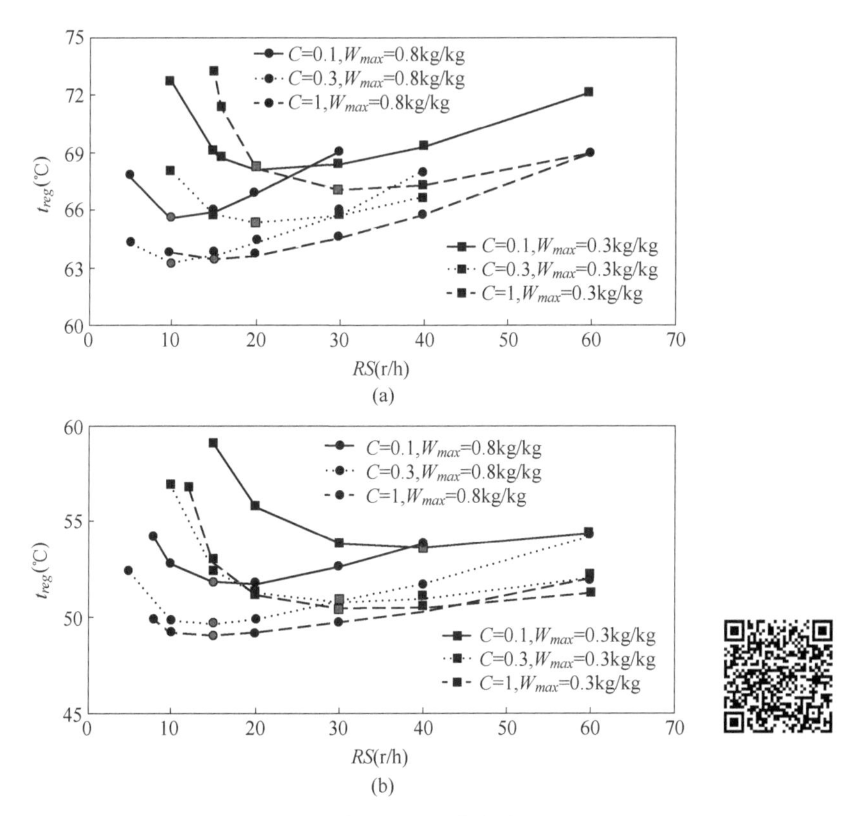

Fig. 4-4 Effects of C, W_{max} and RS on t_{reg} for case 1 of single stage and two-stage systems

(a) Single stage system; (b) Two-stage system

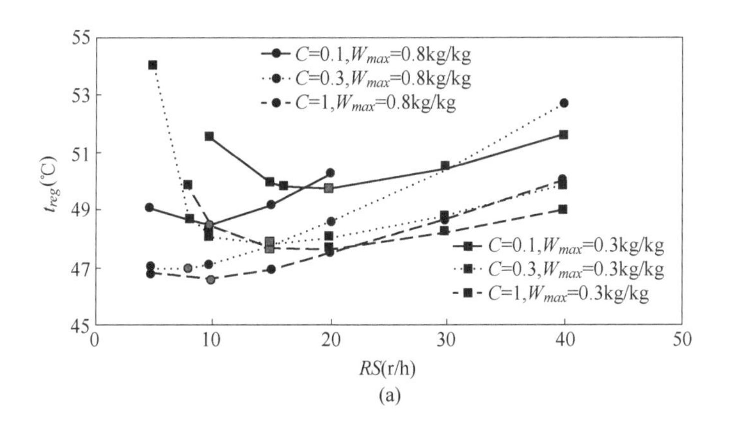

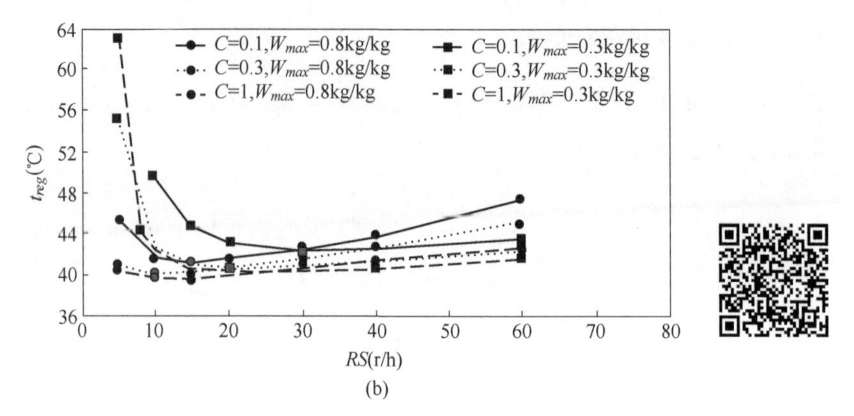

Fig. 4-5 Effects of C, W_{max} and RS on t_{reg} for case 2 of single stage and two-stage systems

(a) Single stage system; (b) Two-stage system

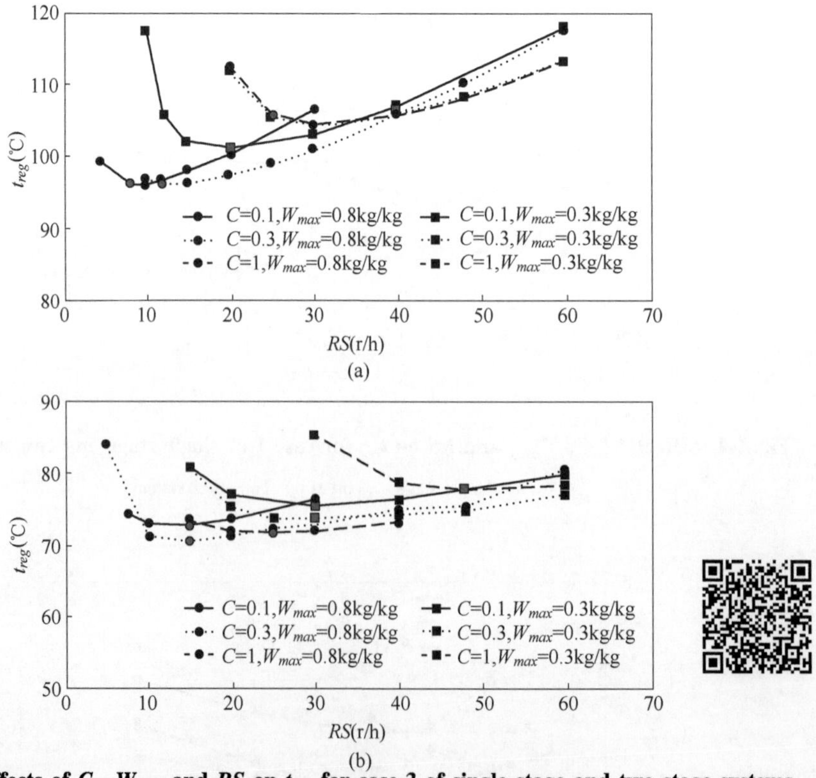

Fig. 4-6 Effects of C, W_{max} and RS on t_{reg} for case 3 of single stage and two-stage systems

(a) Single stage system (results of $C=1$ and $W_{max}=0.3$kg/kg are not shown in this figure);

(b) Two-stage system

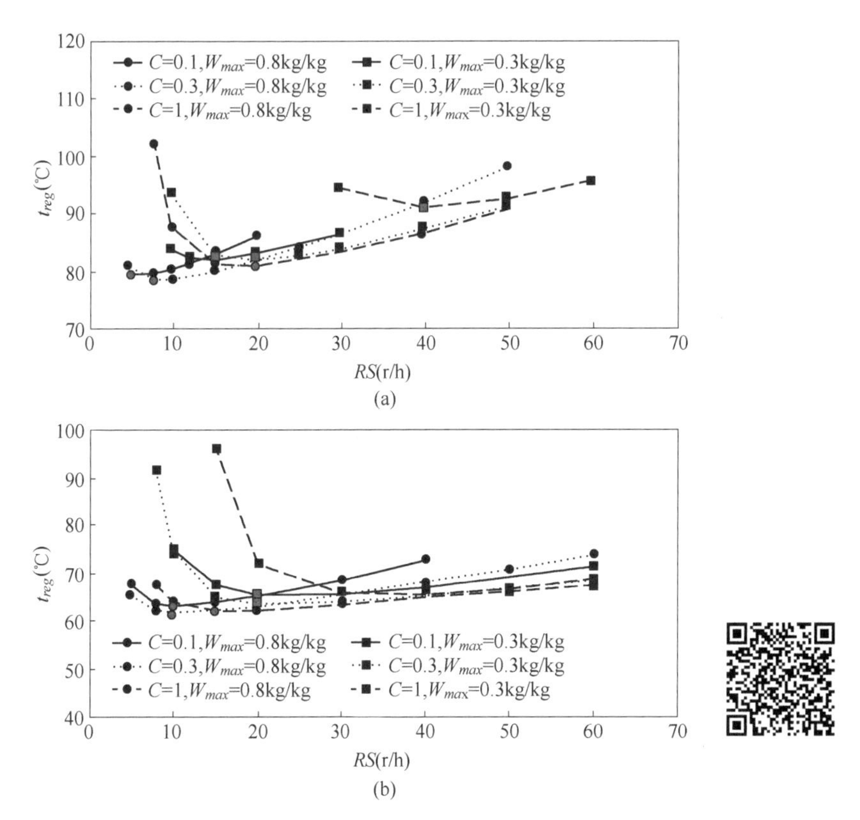

Fig. 4-7 Effects of C, W_{max} and RS on t_{reg} for case 4 of single stage and two-stage systems

(a) Single stage system; (b) Two-stage system

There are two things to be noticed. First, there are cross points among the t_{reg}-RS lines for the six groups' combinations of C and W_{max} in each Figure. In other word, the preferred combination of C and W_{max}, which results in the lowest t_{reg}, differs with RS. Taking Fig. 4-4 as an example, when $RS = 10$r/h, the combination of $C = 0.3$ and $W_{max} = 0.8$kg/kg results in the lowest t_{reg}. When $RS = 40$r/h, the combination of $C = 0.3$ and $W_{max} = 0.8$kg/kg results in the highest t_{reg} and the combination of $C = 1$ and $W_{max} = 0.8$kg/kg results in the lowest t_{reg}. Second, t_{reg} of the two-stage system may be higher than that of the single stage system if the RS is not properly selected. Taking Fig. 4-5 as an example, as for the combination of $C = 0.3$ and $W_{max} = 0.3$kg/kg for case 2, when RS is 5r/h, t_{reg} of the two-stage system (55℃) is slightly higher than that of the single stage system (54℃).

Based on the above two reasons, it is important to select the results at the ORS for discussion. It is demonstrated in Figs. 4-4 ~ 4-7 that an ORS exists for each scenario. When *RS* is away from the *ORS*, t_{reg} is increased compared with the one at the *ORS*. The *ORS*, shown as red dots in Figs. 4-4~4-7, for case 1 to case 4 are listed in Tables 4-3~4-6, respectively. The *ORS* differs with C and W_{max}, and is sensitive to W_{max} and ambient conditions. With the same supply air humidity ratio, larger W_{max} and less humid ambient conditions (WSC) lead to lower *ORS*.

This can be explained from Eq. (4-1):

$$ORS = \frac{3.6m_p}{M_{DW}} \frac{\Delta\omega}{\Delta W} \tag{4-1}$$

Mass flow rate of the process air (m_p) and the mass of desiccant wheel (M_{DW}) are the same for each scenario. For the less humid ambient condition, the dehumidification capacity ($\Delta\omega$) is lower. ΔW is the average water capacity change during dehumidification process. Desiccant material with higher W_{max} can absorb more water to reach its limit and ΔW can be higher. The above two reasons explain why *ORS* is lower at larger W_{max} and less humid ambient conditions.

It can also be observed from Figs. 4-4~4-7 that when *RS* is close to *ORS*, t_{reg} doesn't change much. The ranges of *RS* with t_{reg} variation within 1℃ around the lowest t_{reg} are listed in Tables 4-3~4-6. This is more applicable for real application. The recommended *RS* ranges of single stage and two-stage systems for the regular dehumidification (case 1 and case 2) and deep dehumidification (case 3 and case 4) are demonstrated in Figs. 4-8~4-9. Similar with the optimal rotational speed, Higher W_{max} leads to lower *RS* ranges when other conditions are the same.

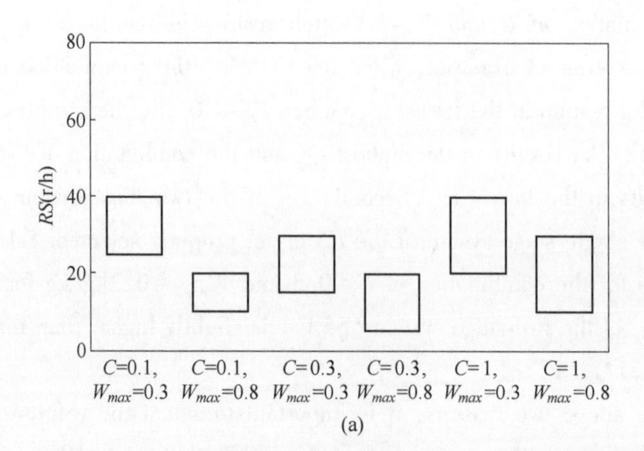

(a)

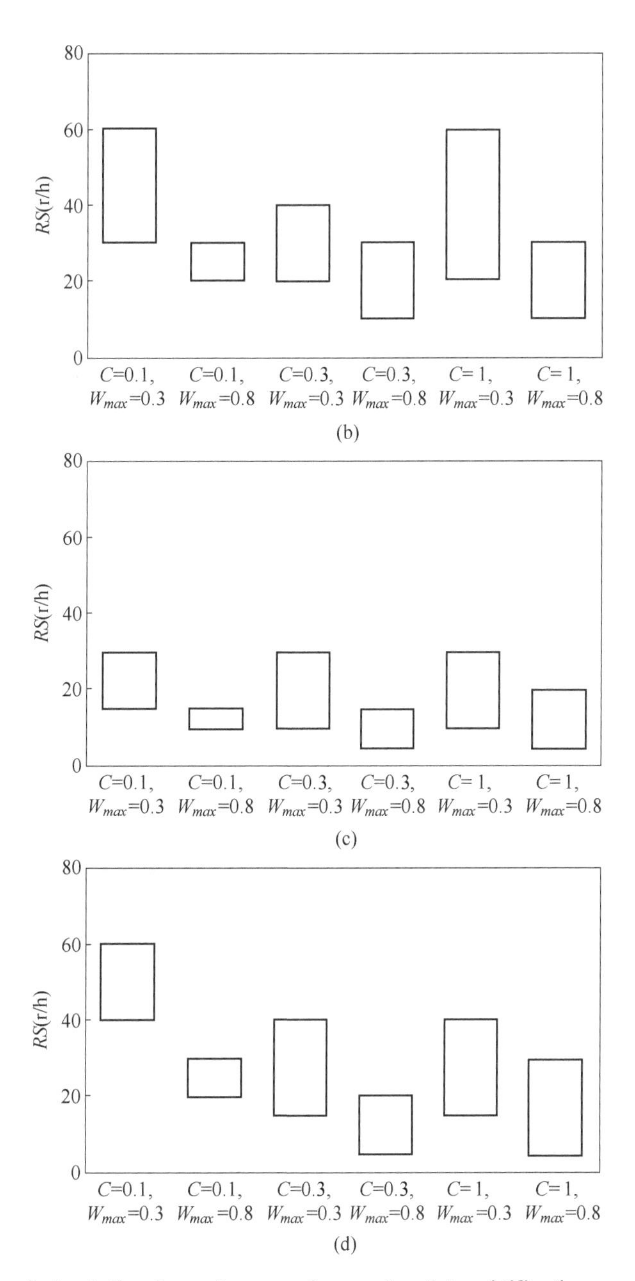

Fig. 4-8 Recommended rotational speed ranges for regular dehumidification regarding single stage and two-stage systems under different combinations of C and W_{max}

(a) Case 1: single stage system; (b) Case 1: two-stage system;

(c) Case 2: single stage system; (d) Case 2: two-stage system

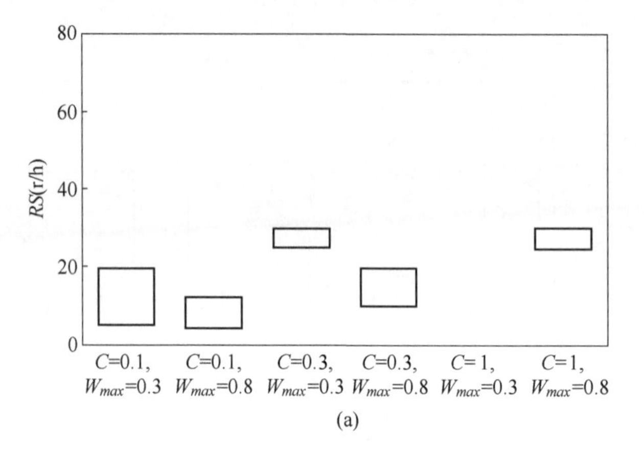
80
60
40
20
0
RS(r/h)
C=0.1, Wmax=0.3
C=0.1, Wmax=0.8
C=0.3, Wmax=0.3
C=0.3, Wmax=0.8
C=1, Wmax=0.3
C=1, Wmax=0.8
(a)

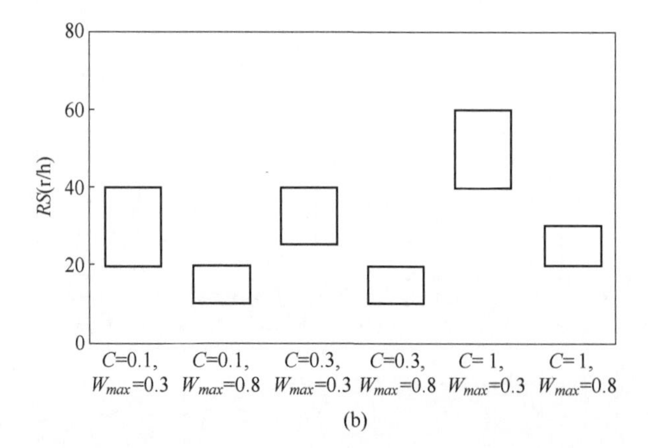
80
60
40
20
0
RS(r/h)
C=0.1, Wmax=0.3
C=0.1, Wmax=0.8
C=0.3, Wmax=0.3
C=0.3, Wmax=0.8
C=1, Wmax=0.3
C=1, Wmax=0.8
(b)

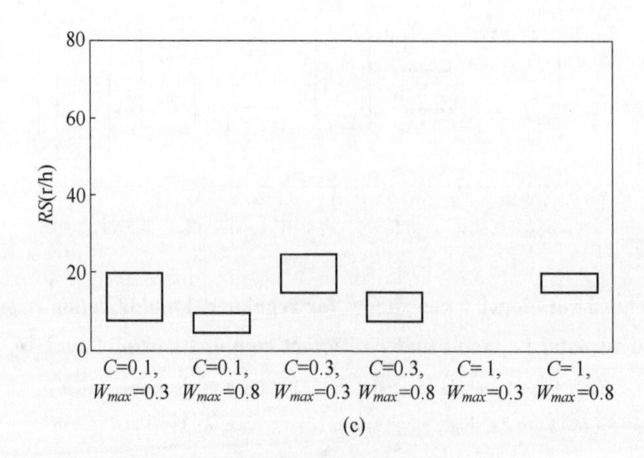
80
60
40
20
0
RS(r/h)
C=0.1, Wmax=0.3
C=0.1, Wmax=0.8
C=0.3, Wmax=0.3
C=0.3, Wmax=0.8
C=1, Wmax=0.3
C=1, Wmax=0.8
(c)

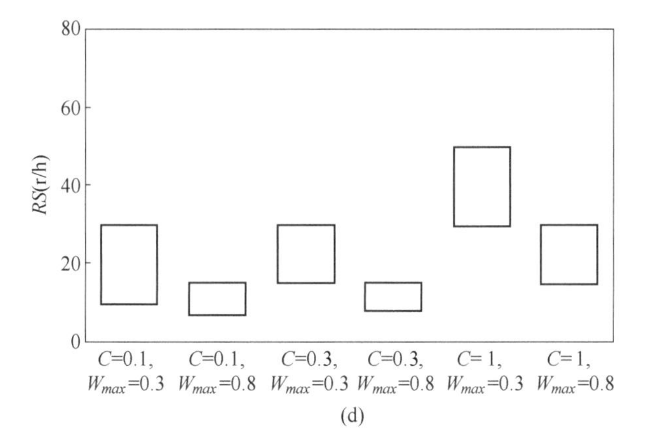

Fig. 4-9 Recommended rotation speed ranges for deep dehumidification regarding single stage and two-stage systems under different combinations of C and W_{max}

(a) Case 3: single stage system; (b) Case 3: two-stage system;
(c) Case 4: single stage system; (d) Case 4: two-stage system

4.3.2 Suggested C and W_{max} and the recommended RS Ranges

t_{reg} under ORS or recommended rotational speed ranges are listed in Tables 4-3~4-6 for different combinations of C and W_{max} for the single stage and two-stage systems under different cases.

Table 4-3 Optimal RS, recommended range of RS and corresponding t_{reg} for case 1

System	Single stage		Two-stage	
	ORS (range) (r/h)	t_{reg} (range) (℃)	ORS (range) (r/h)	t_{reg} (range) (℃)
C=0.1, W_{max}=0.3kg/kg	20 (15~40)	68.2 (68.2~69.3)	40 (30~60)	53.6 (53.6~54.4)
C=0.1, W_{max}=0.8kg/kg	10 (10~20)	65.6 (65.6~66.7)	15 (10~30)	51.8 (51.8~52.7)
C=0.3, W_{max}=0.3kg/kg	20 (15~30)	65.3 (65.3~65.7)	30 (20~40)	50.8 (50.8~51.2)
C=0.3, W_{max}=0.8kg/kg	10 (5~20)	63.3 (63.3~64.3)	15 (10~30)	49.7 (49.7~50.8)
C=1, W_{max}=0.3kg/kg	30 (20~40)	66.9 (66.9~68.1)	30 (20~60)	50.4 (50.4~51.3)
C=1, W_{max}=0.8kg/kg	15 (10~30)	63.4 (63.4~64.6)	15 (10~30)	49.1 (49.1~49.7)

Table 4-4 Optimal *RS*, recommended range of RS and corresponding t_{reg} for case 2

System	Single stage		Two-stage	
	ORS (range) (r/h)	t_{reg} (range) (℃)	*ORS* (range) (r/h)	t_{reg} (range) (℃)
C=0.1, W_{max}=0.3kg/kg	20 (15~30)	49.8 (49.8~50.5)	30 (20~60)	42.3 (42.3~43.5)
C=0.1, W_{max}=0.8kg/kg	10 (5~15)	48.5 (48.5~49.2)	15 (10~30)	41.3 (41.3~42.6)
C=0.3, W_{max}=0.3kg/kg	15 (10~30)	47.8 (47.8~48.8)	20 (15~40)	40.7 (40.7~41.2)
C=0.3, W_{max}=0.8kg/kg	8 (5~15)	47.0 (47.0~47.8)	10 (5~20)	40.0 (40.0~40.8)
C=1, W_{max}=0.3kg/kg	15 (10~30)	47.7 (47.7~48.4)	20 (15~40)	40.2 (40.2~40.7)
C=1, W_{max}=0.8kg/kg	10 (5~20)	46.6 (46.6~47.5)	10 (5~30)	39.6 (39.6~40.6)

Table 4-5 Optimal *RS*, recommended range of RS and corresponding t_{reg} for case 3

System	Single stage		Two-stage	
	ORS (range) (r/h)	t_{reg} (range) (℃)	*ORS* (range) (r/h)	t_{reg} (range) (℃)
C=0.1, W_{max}=0.3kg/kg	20 (15~20)	101.0 (101.0~101.8)	30 (20~40)	75.5 (75.5~76.8)
C=0.1, W_{max}=0.8kg/kg	8 (8~12)	95.9 (95.9~96.7)	15 (10~20)	72.7 (72.7~73.6)
C=0.3, W_{max}=0.3kg/kg	30 (25~30)	104.3 (104.3~105.3)	30 (25~40)	73.5 (73.5~74.2)
C=0.3, W_{max}=0.8kg/kg	12 (10~20)	96.0 (96.0~97.1)	15 (10~20)	70.7 (70.7~71.3)
C=1, W_{max}=0.3kg · kg			48 (40~60)	77.7 (77.7~78.6)
C=1, W_{max}=0.8kg/kg	30 (25~30)	104.4 (104.4~105.5)	25 (20~30)	71.7 (71.7~72.1)

Table 4-6 Optimal *RS*, recommended range of RS and corresponding t_{reg} for case 4

System	Single stage		Two-stage	
	ORS (range) (r/h)	t_{reg} (range) (℃)	ORS (range) (r/h)	t_{reg} (range) (℃)
C=0.1, W_{max}=0.3kg/kg	15 (12~20)	82.2 (82.2~83.2)	20 (20~30)	65.5 (65.5~65.7)
C=0.1, W_{max}=0.8kg/kg	5 (5~10)	79.5 (79.5~80.4)	10 (8~15)	63.3 (63.3~63.9)
C=0.3, W_{max}=0.3kg/kg	20 (15~25)	82.1 (82.1~82.8)	20 (15~30)	63.8 (63.8~65.1)
C=0.3, W_{max}=0.8kg/kg	8 (8~15)	78.3 (78.3~80.0)	10 (8~15)	61.7 (61.7~62.2)
C=1, W_{max}=0.3kg/kg	40	91.9	40 (30~50)	65.5 (65.5~66.3)
C=1, W_{max}=0.8kg/kg	20 (15~20)	81.8 (81.8~81.2)	15 (15~30)	62.2 (62.2~63.4)

It is common for the four cases that for both single stage and two-stage systems, t_{reg} is lower when adsorption materials with larger W_{max} is adopted. Larger W_{max} is recommended especially for dehumidification at lower relative humidity, such as the single stage system, humid ambient conditions and deep dehumidification.

C has different effects on t_{reg} between the regular dehumidification and deep dehumidification cases. For the regular dehumidification cases (case 1 and case 2), larger C ($C=1$) is recommended while smaller C ($C=0.3$) is preferred for the deep dehumidification cases (case 3 and case 4). For regular dehumidification of the single stage system operated under case 1, shown in Table 4-3, the effects of C on t_{reg} are minor when W_{max} is large (0.8kg/kg). And $C=0.3$ and $C=1$ are both preferred. For deep dehumidification of the single stage system operated under case 3, shown in Table 4-5, t_{reg} of the scenario when $C=0.1$ is slightly lower than that of $C=0.3$ when W_{max} is large (0.8kg/kg). However, the difference is so small that $C=0.1$ and $C=0.3$ are both preferred.

It is concluded that for dehumidification at high relative humidity ranges with low t_{reg}, i.e. two-stage system and less humid ambient condition for regular dehumidification, adsorption materials with C closer to 1 are recommended. For dehumidification at low relative humidity ranges with high t_{reg}, i.e. deep dehumidification, smaller C is preferred. The recommend C, W_{max} and RS ranges for different cases of the single stage and two-stage systems are listed in Table 4-7. For regular dehumidification, $C=1$ and $W_{max}=0.8$kg/kg are recommended for the single stage and two-stage systems under a wide range of working conditions (BSC and WSC). Whereas, for deep dehumidification $C=0.3$ and $W_{max}=0.8$kg/kg are preferred for the single stage and two-stage systems under a wide range of working conditions. The recommended RS ranges of the single stage and two-stage systems under BSC and WSC for regular and deep dehumidification applications are listed in Table 4-7.

Table 4-7 Recommended range of RS, C and W_{max}

Dehumidification	Stage	BSC	WSC	BSC and WSC	Single-and two-stage systems	Regular and deep dehumidification
Regular dehumidification: $C=1$, $W_{max}=0.8$kg/kg	Single stage	10~30 r/h	5~20 r/h	10~20 r/h	10~20r/h	10~15r/h
	Two-stage	10~30 r/h	5~30 r/h	10~30 r/h		
Deep dehumidification: $C=0.3$, $W_{max}=0.8$kg/kg	Single stage	10~20 r/h	8~15 r/h	10~15 r/h	10~15r/h	
	Two-stage	10~20 r/h	8~15 r/h	10~15 r/h		

The *RS* ranges of 10 ~ 15r/h works for the regular and deep dehumidification of the single stage and two-stage system under a wide range of working conditions.

4.4 Discussions

The previous analysis shows different requirements of shape factor (C) for dehumidification at low relative humidity range and high relative humidity range, or, in other words, regular dehumidification and deep dehumidification. The recommended *RS* ranges shown in Table 4-8 are proposed based on a fixed air flow rate, thickness of desiccant wheel and W_{max}.

Table 4-8 Recommended *RS* range for desiccant wheels with different thickness (C=0.3, W_{max}=0.8kg/kg)

L (m)	Working condition	*ORS* (range) (r/h)	t_{reg} (range) (℃)	Recommended *RS* range (r/h)
0.1	BSC	25 (20~40)	83.1 (83.1~84.0)	20~30
	WSC	20 (15~30)	70.2 (70.2~70.8)	
0.2	BSC	15 (10~20)	70.7 (70.7~71.3)	10~15
	WSC	10 (8~15)	61.7 (61.7~62.2)	
0.3	BSC	8 (6~15)	66.8 (66.8~67.9)	6~10
	WSC	8 (5~10)	59.3 (59.3~59.6)	

The reasons for the effects of shape factors (C) are theoretically discussed in this section. And the effects of wheel thickness and W_{max} on the recommended *RS* ranges are analyzed afterwards.

4.4.1 Theoretical analysis of the effects of *C* for different dehumidification applications

It is shown in Eq. (4-13) that ω_d is influenced by t_d, C and W/W_{max}. Fig. 4-10 shows the relations of ω_d and t_d at the same W/W_{max}, which is called the iso W/W_{max} lines, on the psychrometric chart. When C equals to 1, the iso W/W_{max} lines of desiccant material overlap the iso-relative humidity lines of the air with the same value. The iso W/W_{max} lines of the same value are greatly different among adsorption materials with different C.

The effects of C on the performances of solid desiccant dehumidification can be analyzed through $\partial\omega_d/\partial C$ and $\partial t_d/\partial C$, expressed as Eqs. (4-2) ~ (4-3), respectively:

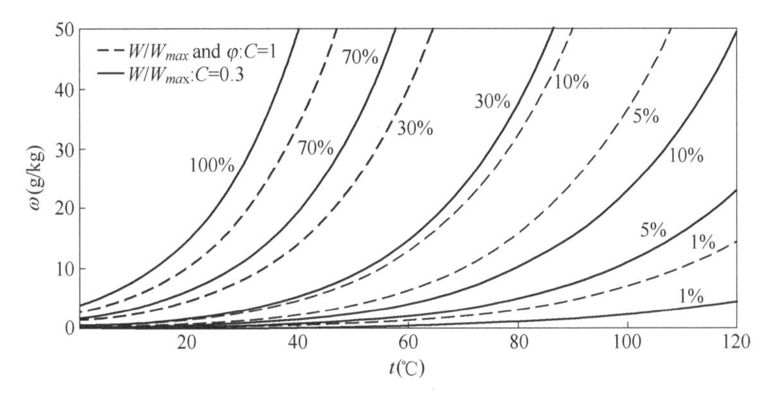

Fig. 4-10 Iso W/W_{max} lines of desiccant materials on the psychrometric chart when C equals to 1 and 0.3

$$\frac{\partial \omega_d}{\partial C}=\frac{\partial \omega_d}{\partial \phi_d}\cdot\frac{\partial \phi_d}{\partial C} \tag{4-2}$$

$$\frac{\partial t_d}{\partial C}=\frac{\partial t_d}{\partial \phi_d}\cdot\frac{\partial \phi_d}{\partial C} \tag{4-3}$$

According to Eq. (4-2), $\partial\omega_d/\partial\phi_d$ and $\partial t_d/\partial\omega_d$ can be written as Eqs. (4-4) ~ (4-5), respectively:

$$\frac{\partial \omega_d}{\partial \phi_d}=\frac{\omega_d}{\phi_d}(1+1.51\omega_d) \tag{4-4}$$

$$\frac{\partial \tau_d}{\partial \phi_d}=-\frac{(t_d+273.15)^2}{5294\phi_d} \tag{4-5}$$

According to Eq. (4-3), $\frac{\partial \phi_d}{\partial C}$ can be written as Eq. (4-6):

$$\frac{\partial \phi_d}{\partial C}=\frac{\phi_d(1-\phi_d)}{C} \tag{4-6}$$

Combining Eqs. (4-4) ~ (4-6), Eqs. (4-2) ~ (4-3) are expressed as Eqs. (4-7) ~ (4-8), respectively:

$$\frac{\partial \omega_d}{\partial C}=\frac{\omega_d(1+1.61\omega_d)(1-\phi_d)}{C} \tag{4-7}$$

$$\frac{\partial t_d}{\partial C}=\frac{(1-\phi_d)(t_d+273.15)^2}{5294C} \tag{4-8}$$

According to Eqs. (4-6) ~ (4-7), $\partial\omega_d/\partial C$ is positive and $\partial t_d/\partial C$ is negative. This means that when t_d and W/W_{max} are fixed, ω_d is higher of desiccant material with higher C. In other word, to get the same ω_d at the same W/W_{max}, temperature of desiccant material with higher C is lower, which is beneficial for the regeneration process. However, this has negative effects on the dehumidification process for higher ω_d of desiccant material with higher C at the same W/W_{max} and t_d. Therefore, when the inlet temperature and humidity ratio of the process air are fixed as discussed in this paper, to reach the same supply air humidity ratio, W/W_{max} of desiccant material with higher C should be lower. It is illustrated in Fig. 4-10 that t_d increases with the reduction of W/W_{max} at the fixed ω_d and C. Because of lower W/W_{max} ranges especially for single stage and deep dehumidification scenarios, t_{reg}, which is the inlet temperature of the regeneration air, may be increased when desiccant material with higher C is adopted.

Taking single stage system as an example, the states of desiccant materials when $C=1$ and $C=0.3$ ($W_{max}=0.8$kg/kg) under BSC for deep dehumidification application and WSC for regular dehumidification application are shown in Figs. 4-11 ~ 4-12, respectively.

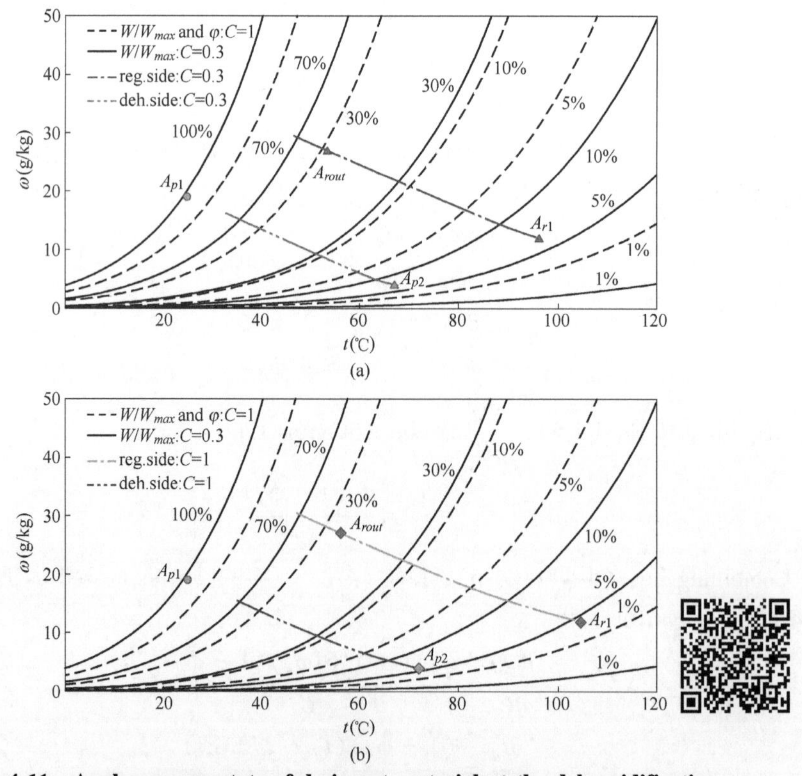

Fig. 4-11 Angle average state of desiccant material at the dehumidification side and regeneration side along the wheel thickness direction: BSC for deep dehumidification application

(a) $C=0.3$ and $W_{max}=0.8$kg/kg; (b) $C=1$ and $W_{max}=0.8$kg/kg

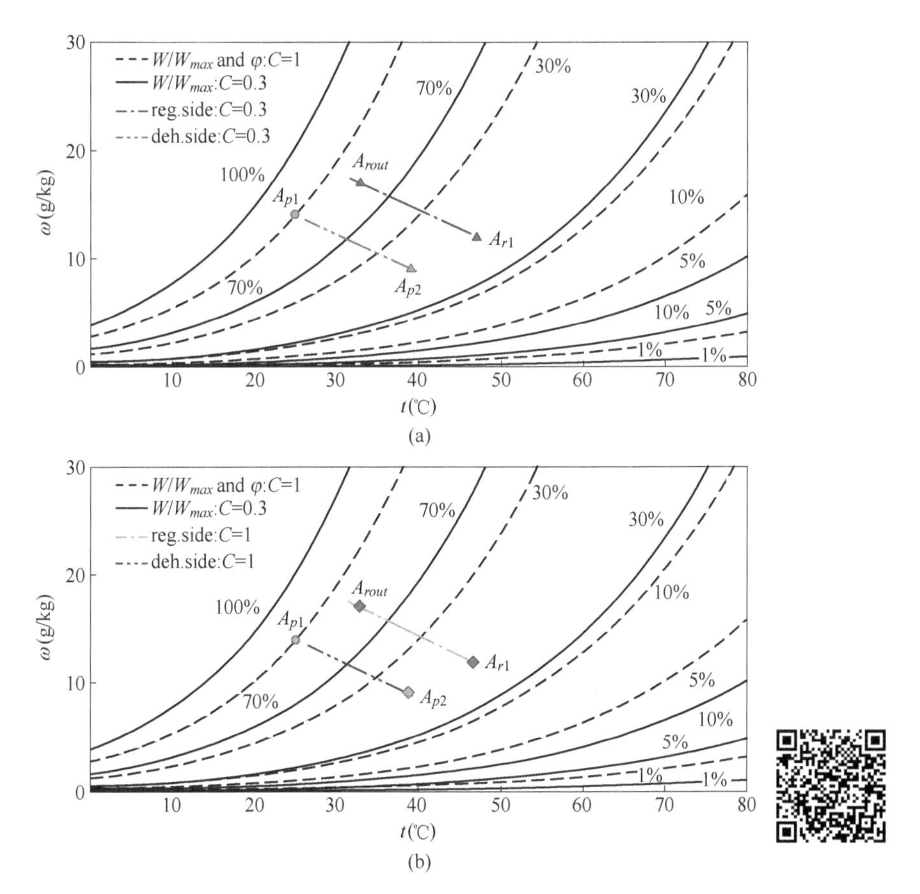

Fig. 4-12 Angle average state of desiccant material at the dehumidification side and regeneration side along the wheel thickness direction: WSC for regular dehumidification application

(a) $C=0.3$ and $W_{max}=0.8$kg/kg; (b) $C=1$ and $W_{max}=0.8$kg/kg

Fig. 4-11 is the results of BSC for deep dehumidification application. It shows that the relative humidity of supply air is close to 1%. The angle average W/W_{max} of desiccant material along the wheel thickness direction is from 6% to 78% when $C=0.3$, while it is from 1% to 35% when $C=1$. For low relative humidity dehumidification, the negative effects of low W/W_{max} of higher C (W/W_{max} is 1% when $C=1$) dominates, resulting in higher t_{reg} (104.4℃) as compared with the scenario when $C=0.3$.

Fig. 4-12 is the results of WSC for regular dehumidification application. It shows that the relative humidity of supply air is close to 15%. Similar with BSC for the deep dehumidification application, the angle average W/W_{max} of desiccant material along the wheel thickness direction when $C=0.3$, which is from 43% to 62%, is higher than those when $C = 1$, which is from 15% to 49%. However, for dehumidification at high relative humidity ranges, the advantages of lower t_d at the same W/W_{max} and ω_d of higher C ($C = 1$) outstands, especially at higher W/W_{max}, resulting in lower t_{reg} (46.6℃) as compared with the scenario when $C = 0.3$.

4.4.2 Influencing factors of the recommended *RS* ranges

The recommended *RS* ranges in the previous discussions are based on the fixed air flow rate, desiccant wheel thickness and radius, and W_{max}. From the mass balance equation shown in Eq. (4-9), *RS* can be expressed as Eq. (4-10).

$$G_a \rho_a \Delta\omega = RS \cdot \pi R^2 L \rho_{ad} \Delta W \tag{4-9}$$

$$RS\left(\frac{\Delta W}{W_{max}} \frac{1}{\Delta\omega}\right) = \frac{G_a \rho_a}{\pi R^2 L \rho_{ad} W_{max}} = \frac{u_a \rho_a}{2\rho_{ad}} \frac{1}{L W_{max}} \tag{4-10}$$

where $\Delta\omega$ is the humidity ratio differences between the inlet and outlet process air; ΔW is the average water capacity change of the desiccant material leaving and entering the dehumidification region; u_a is the air velocity and L is the wheel thickness. Normally u_a should be in an appropriate range to guarantee a satisfying heat and mass transfer coefficient, which is fixed at 1.77m/s in the discussion. $\Delta\omega$ is the task to be accomplished. Therefore, the optimal *RS* is related to L and W_{max}.

It is shown in Table 4-7 that there exist overlaps of recommended *RS* between deep dehumidification and regular dehumidification. According to Table 4-8, the universal recommended *RS* range, which is 10～15r/h when $L = 0.2$m and $W_{max} = 0.8$kg/kg, is the same with the results of two-stage system under BSC and WSC for deep dehumidification. In this subsection, the effects of L and W_{max} on the recommended *RS* ranges are discussed based on the two-stage system under BSC and WSC for deep dehumidification application. C is 0.3 for the following discussion.

L equaling to 0.1m and 0.3m are selected for discussion when $W_{max} = 0.8$kg/kg. The variations of t_{reg} with *RS* under BSC and WSC when $L = 0.1$m and 0.3m are shown in Fig. 4-13.

The optimal *RS*, recommended *RS* ranges and the corresponding t_{reg} are listed in Table 4-7. The results show that under the optimal *RS*, the increase of L is beneficial for the reduction of t_{reg}. Taking the deep dehumidification of the two-stage system as an example, when L increases from 0.1m to 0.2m, t_{reg} reduces from 83.1℃ to 70.7℃ under BSC. However, the improvement rate slows down as further increase of L, with a reduction of around 4℃ when L increased from 0.2m to 0.3m. As discussed in this section, the recommended *RS* ranges, which is for high or low relative humidity dehumidification

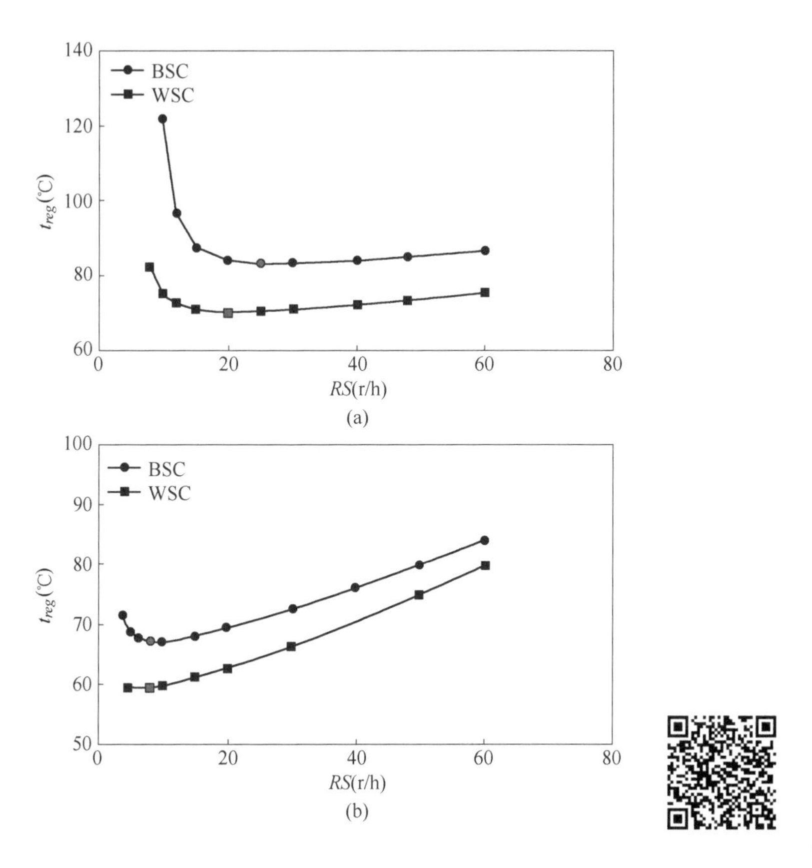

Fig. 4-13 Variation of t_{reg} with RS for different wheel thickness: deep dehumidification of the two-stage system (C=0.3 and W_{max}=0.8kg/kg)

(a) L=0.1m; (b) L=0.3m

of single and two-stage system under a wide range of working conditions, is 10~15r/h when L=0.2m. When L=0.1m, the recommended RS range is 20~30r/h, which is half the results of L=0.2m. When L=0.3m, the recommended RS range is 6~10r/h, which is around 1/3 the results of L=0.1m. Therefore, $RS \cdot L$ can be regarded as constant. According to Eq. (4-24), $(\Delta W/W_{max})/\Delta\omega$ under the optimal cases is almost the same under different L.

W_{max} equaling to 0.3kg/kg, 0.5kg/kg and 0.8kg/kg are selected for discussion and L is fixed at 0.2m. The recommended RS ranges for different W_{max} are shown in Table 4-9. It shows that larger W_{max} results in lower RS ranges and lower t_{reg}.

The results of $RS_{max} \cdot W_{max}$, $RS_{ave} \cdot W_{max}$ and $RS_{min} \cdot W_{max}$ are shown in Fig. 4-14.

Table 4-9 Recommended *RS* range for desiccant material with different W_{max} (C=0.3, L=0.2m)

W_{max} (kg/kg)	Working condition	*ORS* (range) (r/h)	t_{reg} (range) (℃)	Recommended *RS* range (r/h)
0.3	BSC	30 (25~40)	73.5 (73.5~74.2)	25~30
	WSC	20 (15~30)	63.8 (63.8~65.1)	
0.5	BSC	20 (15~30)	71.7 (71.7~72.5)	15~25
	WSC	15 (10~25)	62.4 (62.4~63.4)	
0.8	BSC	15 (10~20)	70.7 (70.7~71.3)	10~15
	WSC	10 (8~15)	61.7 (61.7~62.2)	

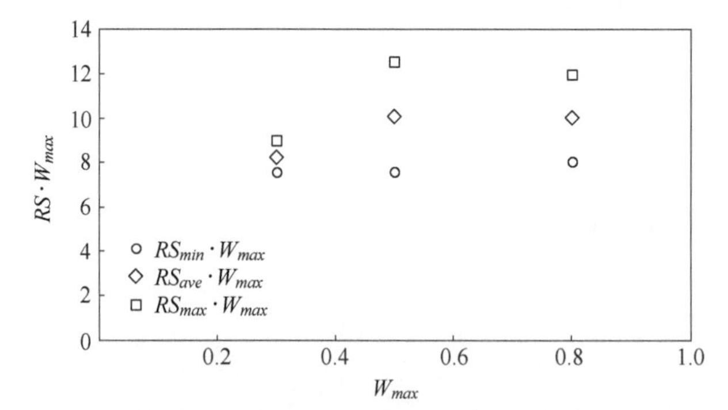

Fig. 4-14 W_{max} multiplying the maximum, average and minimum values of each *RS* ranges for different W_{max}

It shows that $RS \cdot W_{max}$ can be regarded as constant under different W_{max}. According to Eq. (4-10), $(\Delta W/W_{max})/\Delta\omega$ under the optimal cases is almost the same.

Since $(\Delta W/W_{max})/\Delta\omega$ under the optimal cases is almost the same for different L and W_{max}, $RS \cdot LW_{max}$ can be regarded as a constant. *ORS* or optimal *RS* range for desiccant wheels with different L and desiccant materials with different W_{max} can be calculated accordingly. For desiccant wheels discussed in Chapter 5, the density of adsorption material is 1129kg/m^3, and it takes 70% of the wheel's volume. The wheel's facial area is evenly divided between the process air and the regeneration air with the facial air velocity being around 1.77m/s. The recommended *RS* is 25 ~ 30r/h when W_{max} = 0.3kg/kg and L=0.2m. Based on this, recommended *RS* ranges for desiccant material with different W_{max} and desiccant wheels with different L are listed in Table 4-10.

Table 4-10 Recommended *RS* range for desiccant material with different W_{max} and *L*

Recommended *RS* ranges		Total wheel thickness in the system		
		0.1m	0.2m	0.3m
W_{max}	0.3kg/kg	50~60r/h	25~30r/h	17~20r/h
	0.4kg/kg	38~45r/h	19~23r/h	13~15r/h
	0.5kg/kg	30~36r/h	15~18r/h	10~12r/h
	0.6kg/kg	25~30r/h	13~15r/h	8~10r/h
	0.8kg/kg	19~23r/h	9~11r/h	6~8r/h

4.4.3 Case studies

In this subsection, performances of the system shown in Fig. 4-1 (a), which is used to process ambient air (2500m^3/h, 33℃, 19g/kg) and regenerated by the indoor air (2500m^3/h, 26℃, 12g/kg), operated under the above recommended parameters are discussed. The supply air humidity ratio is 9g/kg. The structure of the desiccant wheel is the same as Table 4-2 except that the thickness is 0.15m. According to the above analysis, the recommended parameters are: $C = 0.3$, $W_{max} = 0.8$kg/kg, $RS = 15$r/h. Another group of parameters of $C = 0.1$, $W_{max} = 0.3$kg/kg and $RS = 10$r/h is selected to evaluate performance improvement.

Under the recommended parameters, namely $C = 0.3$, $W_{max} = 0.8$kg/kg, $RS = 15$r/h, t_{reg} is 65.7℃. When C is changed to 0.1, t_{reg} is increased to 68.0℃. When C is changed to 0.1 and W_{max} is changed to 0.3kg/kg, t_{reg} is increased to 73.6℃. When C is changed to 0.1 and W_{max} is changed to 0.3kg/kg and RS is changed to 10r/h, t_{reg} is increased to 82.6℃. The increase of t_{reg} will increase heating capacity of heating sources, affect adoption of low-grade heat sources and reduce system performances.

4.5 Conclusions

This chapter aims to suggest the proper adsorption isotherms from the aspects of shape factor (C) and maximum water capacity (W_{max}), as well as rotational speed (RS) ranges, for high and low relative humidity dehumidification applications to achieve low temperature regeneration (t_{reg}). The performances of single stage and two-stage desiccant wheel systems under mild (WSC: 33℃ and 14g/kg) and humid (BSC: 33℃ and 19g/kg) working conditions for regular (supply air: 9g/kg) and deep dehumidification (supply air: 4g/kg) applications were analyzed through simulation. The mathematical model was validated by the experiment results. The main conclusions are as follows:

(1) Compared with the regular dehumidification, the air is handled at a lower relative humidity range for the deep dehumidification and the corresponding t_{reg} is higher. For the same working condition and supply air humidity ratio, the air is handled at a higher relative humidity for the two-stage system as compared with the single stage system, resulting in lower t_{reg}.

(2) The proper C and W_{max} are suggested under the optimal rotational speed. Large W_{max} is preferred to reduce t_{reg} especially for deep dehumidification of the single-stage system under humid climate. For dehumidification at high relative humidity ratio ranges with low t_{reg}, adsorption materials with C equaling to 1 are preferred. $C=1$ and $W_{max}=0.8$kg/kg is recommended for the single stage and two-stage systems for regular dehumidification. Whereas, for dehumidification at low relative humidity ranges with high t_{reg}, smaller C is preferred. $C=0.3$ and $W_{max}=0.8$kg/kg is preferred for the single stage and two-stage systems for deep dehumidification.

(3) The reasons for the different requirements of C for low and high relative humidity dehumidification are explained theoretically. As compared with lower C, the pros of higher C is the reduction of t_{reg} because of lower t_d at the same W/W_{max} and ω_d. However, the cons of higher C is that W/W_{max} range should be lower to meet the dehumidification requirement, which may lead to the increase of t_{reg}. The advantages of higher C dominate for high relative humidity dehumidification and higher C is preferred. The disadvantages of higher C dominate for low relative humidity dehumidification and lower C is preferred.

(4) When W_{max} equals to 0.8kg/kg and wheel thickness is 0.2m, the recommended *RS* range that works for the regular and deep dehumidification of the single stage and two-stage system under a wide range of working conditions (BSC and WSC) is 10 ~ 15r/h. Further analyses showed that the *ORS* or optimal *RS* range is mainly influenced by the thickness of the desiccant wheel and the maximum water capacity of the desiccant material. $RS \cdot L \cdot W_{max}$ can be regarded as a constant. *ORS* or optimal *RS* range for wheels with different L and material with different W_{max} can be calculated accordingly.

Nomenclature

A	area
a	pore radius, m

BSC	Beijing Summer Condition
WSC	Washington airport Summer Condition
C	shape factor of desiccant material
COP	Coefficient of Performance
c_p	specific heat, kJ/(kg · K)
d_h	hydraulic diameter, m
D_A	ordinary diffusion coefficient, m^2/s
D_S	surface diffusion coefficient, m^2/s
f	area ratio
G	volume flow rate, m^3/s
h	heat transfer coefficient, kW/(m^2 · K)
h_m	mass transfer coefficient, kg/(m^2 · s)
h_v	heat of vaporization, kJ/kg
k	thermal conductivity, kW/(m · K)
L	wheel thickness, m
Le	Lewis number
m	mass flow rate, kg/s
M	mass of desiccant wheel, kg
Nu	Nusselt number
ORS	Optimal Rotational Speed
Pa	standard atmospheric pressure, Pa
P_{vs}	saturated vapor pressure, Pa
P	process air
R	regeneration air
r_s	adsorption or desorption heat, kJ/kg

RS	Rotational Speed
t	Celsius temperature, ℃
u	velocity, m/s
W	adsorption capacity of desiccant materials, $kg_{water}/kg_{dry\ adsorbent}$
x	volume ratio of adsorption material
x^*	mass ratio of adsorption material
Z	thickness, m

Greek symbols

ω	humidity ratio, g/kg
φ	relative humidity ratio
τ	time
ρ	density, kg/m^3
σ	porosity
ξ	tortuosity factor, dimensionless

Subscripts

a	air
ave	average
ad	adsorption material
c	cold
d	desiccant material
p	process
reg	regeneration
in	inlet
out	outlet
max	maximum

min	minimum
w	water
DW	desiccant wheel

References

[1] L Z Zhang. Total Heat Recovery: Heat and Moisture Recovery from Ventilation Air [M]. New York: Nova Science Publisher, 2008.

[2] R Tu, X H Liu, Y Jiang. Performance comparison between enthalpy recovery wheels and dehumidification wheels [J]. Int. J. Refrig, 2013, 36 (8): 2308-2322.

[3] R Tu, X H Liu, Y Jiang. Performance analysis of a two-stage desiccant cooling system [J]. Appl. Energy, 2014, 13: 1562-1574.

[4] P Majumdar. Heat and Mass transfer in composite desiccant pore structures for dehumidification [J]. Sol. Energy, 1998, 62 (1): 1-10.

[5] R Tu, Y Hwang, T Cao, et al. Investigation of adsorption isotherms and rotational speeds for low temperature regeneration of desiccant wheel systems [J]. Int. J. Refrig, 2018, 86: 495-509.

Chapter 5 Irreversible Processes of Dehumidification Systems with Single Stage Desiccant Wheel

In this chapter, the exergy destructions of heating and cooling sources, as well as that of heat and mass transfer devices, were all considered. First, a reversible desiccant dehumidification and cooling system was designed, and the changes in *COP* and exergy efficiency when each device became irreversible were examined. The results can provide guidance for system design optimization. Based on these results, a ventilation cycle is investigated from the perspective of exergy destruction, and the performance improvement potentials of this system were explored. Finally, several system improvement methods were proposed and examined through simulation.

5.1 Performance analysis of a reversible *DDCS*

The typical *DDCS*, as shown in Fig. 5-1, includes a desiccant wheel used for air dehumidification (A_{pin} to A_{p1}), a heater used to heat the regeneration air (A_{r1} to A_{r2}), and a cooler used to cool down the dried air (A_{p2} to A_{pout}).

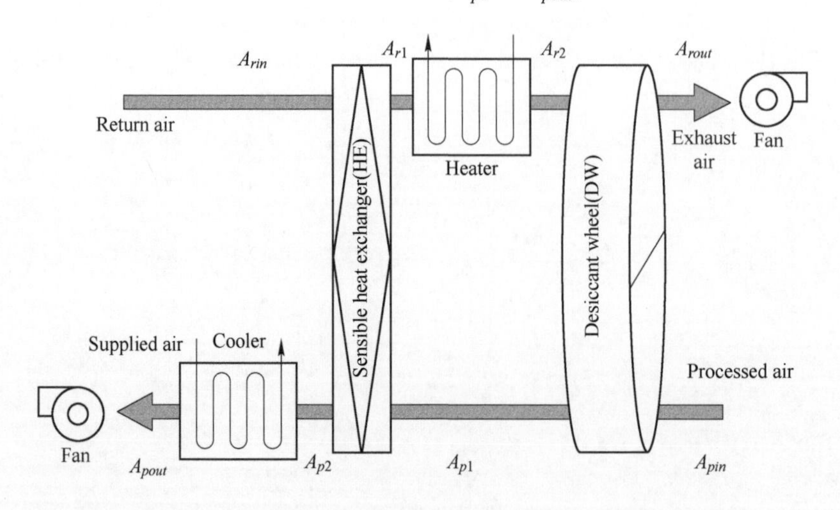

Fig. 5-1 Schematic of a desiccant dehumidification and cooling system (*DDCS*)

In addition, a sensible heat exchanger is usually adopted to recover the sensible heat between the processed air (A_{p1} to A_{p2}) and the regeneration air (A_{rin} to A_{r1}), reducing the required cooling and heat capacity in the *DDCS*. The reversible *DDCS* introduced in this section is illustrated in Fig. 5-1. This reversible *DDCS* has no exergy destruction and requires the minimum power input to handle the processed air to the required state. The influence of the irreversible process of each main device on the performance of the *DDCS* will be discussed in the following subsections.

5.1.1 Introduction of the reversible *DDCS*

For the reversible *DDCS*, there is no exergy destruction. The schematic of the air handling process for the reversible *DDCS* is shown in Fig. 5-2. The following assumptions are made: (1) the ideal desiccant wheel with no exergy destruction is adopted, and the air is treated along the isenthalpic line; (2) in order to achieve the minimum cooling capacity requirement of the air ($\dot{Q}_{cs,\ ideal}$: to cool the processed air from A_{p2} to A_{pout}) and the minimum heat capacity requirement of the air ($\dot{Q}_{hs,\ ideal}$: to heat the regeneration air from A_{r1} to A_{r2}), the area of the sensible heat exchanger is infinite, and the heat exchange efficiency is 100%; and (3) infinite-stage ideal refrigerators and heat pumps that obey the reverse Carnot cycle are utilized as coolers (cooling the processed air from A_{p2} to A_{pout} in Fig. 5-2) and heaters (heating the regeneration air from A_{r1} to A_{r2} in Fig. 5-2), respectively.

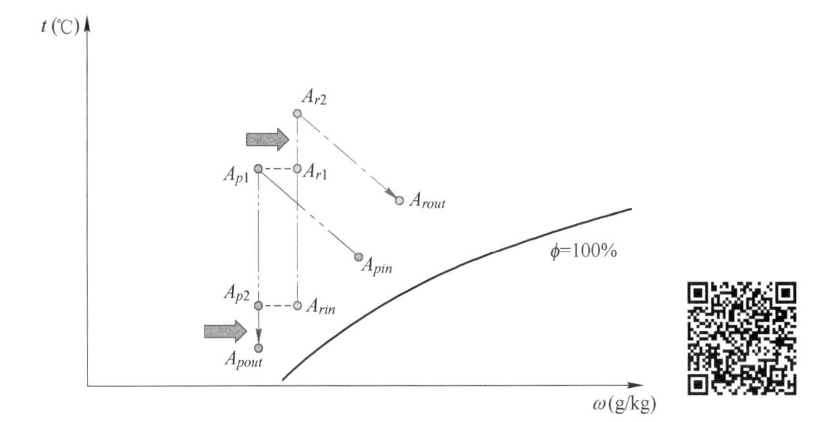

Fig. 5-2 Air handling process of the reversible *DDCS*

The preconditions of the ideal desiccant wheel are as follows: the air mass flow rates of the processed air and the regeneration air are the same; the desiccant wheel is evenly divided between the two streams of air; and the heat and mass transfer area is infinite. For the ideal desiccant wheel, there is no exergy destruction; thus, the exergy provided by the regeneration air is equal to the exergy obtained by the processed air. Based on the exergy analysis results, energy and mass conservation equations, and the assumption of the isenthalpic air handling process, the ideal regeneration temperature (t_{r2} in Fig. 5-2) can be calculated. However, the processed air inlet (A_{pin} in Fig. 5-2) and outlet humidity ratio and the regeneration air inlet humidity ratio (ω_{p1} and ω_{r2}, respectively, in Fig. 5-2) must be determined first. Such an ideal regeneration temperature is the lowest value that the actual desiccant wheel system can achieve. Fig. 5-3 shows the calculation logic of the ideal regeneration temperature (t_{r2} in Fig. 5-2).

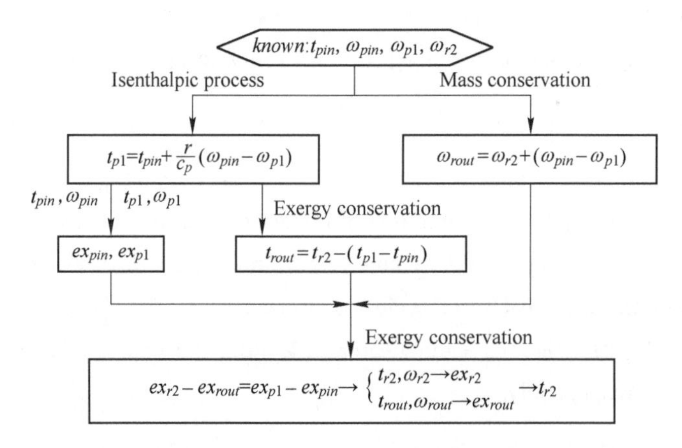

Fig. 5-3 Calculation logic of the ideal regeneration temperature (t_{r2} in Fig. 5-2) of the reversible *DDCS*

For the infinite-stage reverse refrigerators in Fig. 5-4 (a), the evaporating temperature equals to the processed air temperature along the processed air flow direction, and the condensing temperature equals to the ambient air temperature. For the infinite-stage reverse heat pumps, as seen in Fig. 5-4 (b), the condensing temperature equals to the regeneration air temperature along the air flow direction, and the evaporating temperature equals to the ambient air temperature. Thus, there is no exergy destruction for the refrigerators and heat pumps, and the minimum required power of the refrigerators ($\dot{W}_{cs,ideal}$: power needed to provide $\dot{Q}_{cs,ideal}$) and heat pumps ($\dot{W}_{hs,ideal}$: power needed to provide $\dot{Q}_{hs,ideal}$) can be achieved.

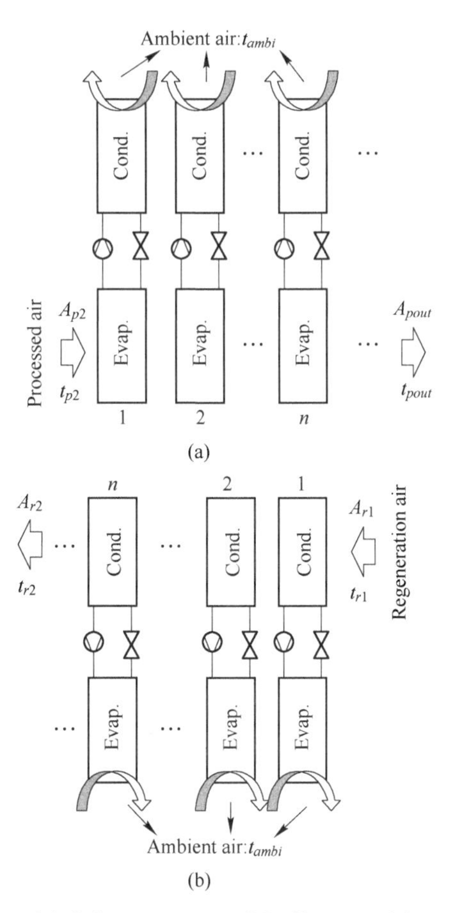

Fig. 5-4 Schematics of infinite-stage reversible Carnot refrigerator and heat pump

(a) An infinite-stage reversible Carnot refrigerator used as the cooler;

(b) An infinite-stage reversible Carnot heat pump used as the heater

5.1.2 Performance analysis of the ideal *DDCS*

$\dot{Q}_{cs,\ ideal}$ and $\dot{Q}_{hs,\ ideal}$ can be calculated by Eq. (5-1) and Eq. (5-2), respectively:

$$\dot{Q}_{cs,\ ideal} = \dot{m}_p (i_{p2} - i_{pout}) \tag{5-1}$$

$$\dot{Q}_{hs,\ ideal} = \dot{m}_r (i_{r2} - i_{r1}) \tag{5-2}$$

where $\dot{m}$ is the air mass flow rate in kg/s; i is the specific enthalpy of moisture air in kJ/kg; and $\dot{Q}$ has the unit of kW. The subscripts p and r stand for the processed air and the regeneration air, respectively; the numbers in the air temperature subscripts can be seen in Fig. 5-1.

The minimum power needed by the ideal cooling and heating sources can be calculated by Eq. (5-3) and Eq. (5-4), respectively:

$$\dot{W}_{cs,\ ideal} = \dot{m}_p c_{pa}\left[(T_{pout} - T_{p2}) - T_0 \ln \frac{T_{pout}}{T_{p2}}\right] \tag{5-3}$$

$$\dot{W}_{hs,\ ideal} = \dot{m}_r c_{pa}\left[(T_{r2} - T_{r1}) - T_0 \ln \frac{T_{r2}}{T_{r1}}\right] \tag{5-4}$$

where T is the temperature in Kelvin; c_{pa} is the specific heat capacity of the air in kJ/(kg · ℃); and T_0 is the temperature in Kelvin of the reference state. $\dot{W}$ has the unit of kW. COP_{ideal} and $\eta_{ex,\ ideal}$ of the reversible *DDCS* can be calculated by Eq. (5-5) and Eq. (5-6), respectively:

$$COP_{ideal} = \frac{\dot{Q}_p}{\dot{W}_{ideal}} = \frac{\dot{m}_p(i_{pin} - i_{pout})}{\dot{W}_{cs,\ ideal} + \dot{W}_{hs,\ ideal}} \tag{5-5}$$

$$\eta_{ex,\ ideal} = \frac{\dot{Ex}_p}{\dot{W}_{ideal}} = \frac{\dot{m}_p(ex_{pout} - ex_{pin})}{\dot{W}_{cs,\ ideal} + \dot{W}_{hs,\ ideal}} \tag{5-6}$$

where ex is the specific exergy of moisture air in kJ/kg; and $\dot{Ex}$ has the unit of kW.

The specific exergy of moisture air can be calculated using Eq. (5-7)[1]. The first term on the right side is the thermal exergy due to the temperature difference between the air (T_a) and the reference state (T_0), and the second term on the right side is the chemical exergy due to the humidity ratio difference between the air (ω_a) and the reference state (ω_0).

$$ex = c_{pa}T_0\left(\frac{T_a}{T_0} - 1 - \ln\frac{T_a}{T_0}\right) + R_a T_0\left[(1 + 1.608 \times 10^{-3}\omega_a)\ln\frac{1 + 1.608 \times 10^{-3}\omega_0}{1 + 1.608 \times 10^{-3}\omega_a} + 1.608 \times 10^{-3}\omega_a \ln\frac{\omega_a}{\omega_0}\right] \tag{5-7}$$

where ω has the unit of g/kg.

Fig. 5-5 shows the effects of the supplied humidity ratio on the ideal regeneration temperature, COP_{ideal}, and $\eta_{ex,\ ideal}$ under Beijing summer conditions (33℃, 19g/kg); the indoor exhaust air (26℃, 12g/kg) is used for regeneration, and the supplied air temperature is fixed at 24℃.

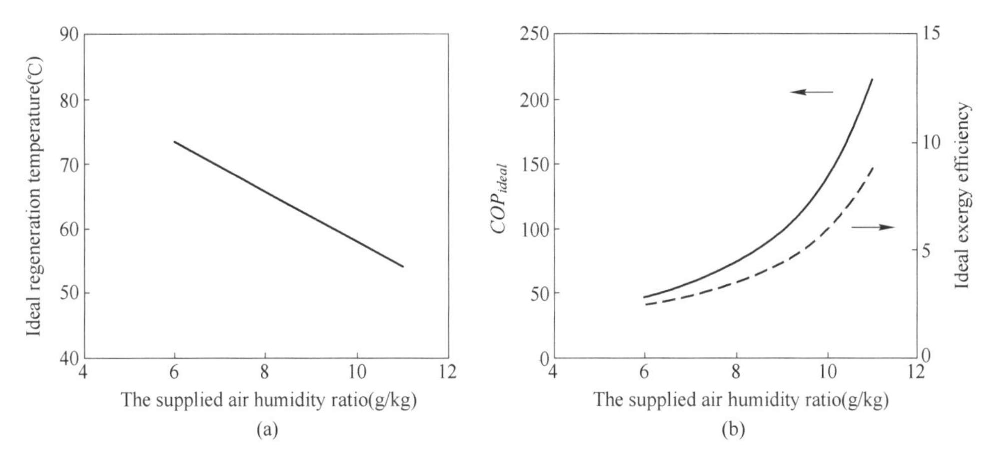

Fig. 5-5 Effects of supplied air humidity ratio on ideal performances

(a) Ideal regeneration temperature; (b) Ideal *COP* and ideal exergy efficiency

The saturated condition at the ambient air temperature (33℃) is chosen as the reference state[2, 3]. With the adoption of heat recovery, $\eta_{ex,\ ideal}$ can be higher than 1. It can be seen that COP_{ideal} and $\eta_{ex,\ ideal}$ are much higher than those of the actual *DDCS*[2, 4]. There are several reasons for this: in the ideal system, there is no exergy destruction; the heat exchanger can completely recover the heat between the two streams of air; and the heating and cooling sources are ideal. Thus, the power required by the heat and cooling sources is much lower than in the actual system. The decrease in performance for the actual *DDCS* is due to the irreversible processes of desiccant wheels, heat exchangers, and heat/cooling sources.

5.1.3 Effects of non-ideal processes on the performance of the *DDCS*

The performance of the actual *DDCS* was obtained through numerical simulation. There are several main devices of the *DDCS*, including desiccant wheels, heat exchangers, heat pumps (or refrigerators), etc. The mathematical model of desiccant wheels utilized in this chapter was established in previous research and validated by experimental results[3,5,6]. This model considers the resistance of both gas and solid sides. For the solid side, the ordinary diffusion of the vapor, surface diffusion of the absorbed water, and heat conductivity of the desiccant material in the wheel thickness direction are all taken into account. The ε-*NTU* method[1] was used to simulate the heat exchangers (e.g., air-to-air, air-to-water, and air-to-refrigerant heat exchangers).

For the actual *DDCS*, the refrigerators and heat pumps are usually single-stage devices (as seen in Fig. 5-6), not Carnot cycle-based devices. Thermodynamic perfectness η_{HP} can be utilized to describe the performance difference between an actual refrigerator and an ideal Carnot cycle refrigerator working at the same evaporating and condensing temperatures, as shown in Eq. (5-8). The value of η_{HP} is lower than 1.

$$\eta_{HP} = \frac{COP_{actual}}{COP_{Carnot}}, \text{ where } COP_{Carnot} = \frac{T_{evap}}{T_{cond} - T_{evap}} \tag{5-8}$$

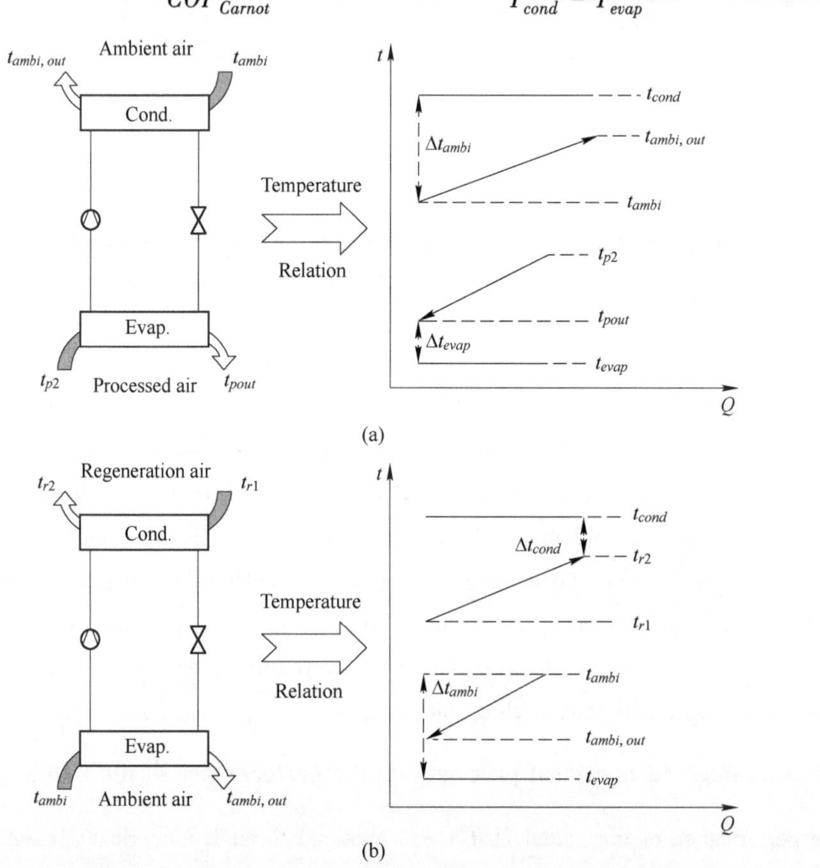

Fig. 5-6 Schematics and temperature relations of single-stage refrigerator and heat pump

(a) A single-stage refrigerator; (b) A single-stage heat pump

Temperature differences exist for the evaporators and condensers, as shown in Fig. 5-6. Based on Eq. (5-8), the power input of actual single-stage refrigerators and heat pumps can be calculated with Eq. (5-9) and Eq. (5-10), respectively:

$$\dot{W}_{cs} = \frac{\dot{Q}_{cs}}{COP_{actual}} = \dot{m}_p(i_{p2} - i_{pout})\frac{t_{ambi} - t_{pout} + \Delta t_{evap} + \Delta t_{ambi}}{(t_{pout} - \Delta t_{evap} + 273.15)\,\eta_{HP}} \tag{5-9}$$

$$\dot{W}_{hs} = \frac{\dot{Q}_{hs}}{COP_{actual}} = \dot{m}_r(i_{r2} - i_{r1}) \frac{t_{r2} - t_{ambi} + \Delta t_{cond} + \Delta t_{ambi}}{(t_{r2} + \Delta t_{cond} + 273.15)\eta_{HP}} \tag{5-10}$$

where t is the temperature in Celsius. The numbers in the air temperature subscripts in Eqs. (5-9) ~ (5-10) are shown in Figs. 5-1 and 5-6.

When an electrical heater is used instead of a heat pump, as shown in Fig. 5-6 (b), the power input of the electrical heater can be calculated using Eq. (5-11).

$$\dot{W}_{hs} = \dot{m}_r(i_{r2} - i_{r1}) \tag{5-11}$$

In the following discussion, electrical heaters and heat pumps are chosen as the typical heating sources, and refrigerators are chosen as cooling sources.

COP and η_{ex} of the actual system are defined in Eqs. (5-12) ~ (5-13), respectively:

$$COP = \frac{\dot{Q}_p}{\dot{W}} = \frac{\dot{m}_p(i_{pin} - i_{pout})}{\dot{W}_{cs} + \dot{W}_{hs}} \tag{5-12}$$

$$\eta_{ex} = \frac{\dot{E}x_p}{\dot{W}} = \frac{\dot{m}_p(ex_{pout} - ex_{pin})}{\dot{W}_{cs} + \dot{W}_{hs}} \tag{5-13}$$

It can be seen that to achieve the same $\dot{Q}_p$ and $\dot{E}x_p$ of the processed air, the lower the power input ($\dot{W}_{cs} + \dot{W}_{hs}$) is, the higher COP and η_{ex} will be. Apparently, based on Eq. (5-10) and Eq. (5-11), electrical heating systems perform worse than heat pump systems.

Considering the irreversible heating and cooling sources, Eq. (5-14) and Eq. (5-15) describe the exergy destruction of the heating source and that of the cooling source, respectively (i. e., $\Delta\dot{E}x_{hs}$ and $\Delta\dot{E}x_{cs}$, respectively), which refer to the exergy provided by the heating or cooling sources, respectively, minus the exergy obtained by the air.

$$\Delta\dot{E}x_{hs} = \dot{W}_{hs} - \dot{m}_r(ex_{r2} - ex_{r1}) \tag{5-14}$$

$$\Delta\dot{E}x_{cs} = \dot{W}_{cs} - \dot{m}_p(ex_{pout} - ex_{p2}) \tag{5-15}$$

For the heat pump and refrigerator, $\Delta\dot{E}x_{hs}$ and $\Delta\dot{E}x_{cs}$ are produced in the process of heat transfer from the high-temperature side to the low-temperature side. For the electrical heater, $\Delta\dot{E}x_{hs}$ is produced during the electricity-heat conversion process. For the ideal cooling and heating sources shown in Fig. 5-4, $\Delta\dot{E}x_{cs} = \Delta\dot{E}x_{hs} = 0$.

Thus, exergy efficiency in the actual *DDCS* can be rewritten as Eq. (5-16). The exergy destruction of the desiccant wheel ($\Delta \dot{E}x_{DW}$), the sensible heat exchanger ($\Delta \dot{E}x_{HE}$), the cooling source ($\Delta \dot{E}x_{cs}$), and the heat source ($\Delta \dot{E}x_{hs}$) are all considered.

$$\eta_{ex} = \frac{\dot{m}_p(ex_{pout} - ex_{pin})}{\dot{m}_p(ex_{pout} - ex_{pin}) + \dot{m}_r(ex_{rout} - ex_{rin}) + \Delta \dot{E}x_{DW} + \Delta \dot{E}x_{HE} + \Delta \dot{E}x_{CS} + \Delta \dot{E}x_{hs}} \tag{5-16}$$

It can be seen from Eq. (5-13) and Eq. (5-16) that under the fixed Ex_t, the exergy destruction of both the heat and mass transfer devices and that of the cooling and heating sources have to be reduced in order to decrease ($\dot{W}_{cs} + \dot{W}_{hs}$) and improve the system performance (η_{ex} and *COP*) in the *DDCS*.

In order to investigate the performance of *DDCS*s that change from being reversible to being irreversible, systems A to F (described in Table 5-1), which become irreversible step by step and approach the actual system, were designed.

Table 5-1 Description of systems A to F

System	Description
A	Ideal system, with no exergy destruction; $\eta_{ex,\ cs} = \eta_{ex,\ hs} = 1$
B	Based on A, the ideal desiccant wheel is replaced by an actual desiccant wheel with the parameters listed in Table 5-2
C	Based on B, a single-stage refrigerator and heat pump with $\Delta t_{evap} = \Delta t_{cond} = 2$ and $\Delta t_{ambi} = 7$ are adopted to replace the infinite-stage reversible Carnot refrigerator and heat pump
D	Based on C, η_{HP} of the refrigerator and that of the heat pump are both equal to 0.5
E	Based on D, the heat exchange efficiency of the heat recovery exchanger becomes 0.7
F	Based on E, an electrical heater is adopted to replace the single-stage heat pump

Under the working conditions and wheel structure listed in Table 5-2, the required regeneration temperatures for the ideal desiccant wheel and the real desiccant wheel are 57.8℃ and 67.5℃, respectively. Fig. 5-7 illustrates *COP* and exergy efficiency of systems A to F.

For the reversible process (system A), *COP* and exergy efficiency are 137.9% and 502.0%, respectively. Based on system A, when the desiccant wheel becomes irreversible (system B), *COP* and exergy efficiency are reduced to 35.4% and

128.9%, respectively. Based on system B, when the heat pump system becomes irreversible (systems C and D), *COP* and exergy efficiency of system D are reduced to 11.9% and 43.4%, respectively. Based on system D, when the heat recovery exchanger becomes irreversible (system E), *COP* and exergy efficiency are reduced to 5.0% and 18.3%, respectively. Based on system E, when electrical heating is used (system F), *COP* and exergy efficiency are reduced to 1.5% and 5.6%, respectively.

Table 5-2 Working conditions and the actual desiccant wheel

Working conditions:
Processed air inlet: 33℃, 19g/kg, 0.8kg/s; Regeneration air: 26℃, 12g/kg, 0.8kg/s; Supplied air: <24℃, 10g/kg
Parameters of the actual desiccant wheel:
Wheel radius: 0.5m; Material: silica gel; Air channel structure: sinusoidal shape, 2mm high, and 2mm wide; Facial area is evenly divided between the two streams of air

It can be seen from Fig. 5-7 that as the system becomes irreversible, the power inputs of the cooling and heating sources increase considerably. This is due to two main reasons. First, the exergy destruction of both the heating and cooling sources increases as the system becomes irreversible. Second, increases in both cooling demand (1.63kW for systems A ~ D and 9.51kW for systems E ~ F) and heating demand (2.03kW for system A, 7.83kW for systems B ~ D, and 15.86kW for systems E ~ F) are due to the higher regeneration temperature of actual wheels (i. e., systems B to F) and the lower efficiency of actual heat recovery exchangers (i. e., systems E to F). According to Eqs. (5-10), (5-11) and (5-14), *COP* and η_{ex} both decrease.

For real systems, the desiccant wheel is typically 0.1 ~ 0.2m thick, the facial velocity of the air is usually 2 ~ 3m/s, and the efficiency of the heat recovery exchanger is usually lower than 70%[3]. Thus, the contributions to overall system improvement from desiccant wheels and heat recovery exchangers are limited. Therefore, the performance improvement of cooling and heating sources will have considerable influence on the overall performance improvement of the *DDCS*. In reality, systems A to D are difficult to realize. The performance of system E, with *COP* and exergy efficiency being 5.0 and 18.3% under the designed working conditions in Table 5-2, respectively, represents the relatively high standards that actual systems can achieve.

In the nextsection, the performance of an actual *DDCS* based on the ventilation cycle will be examined, and performance improvement methods will be discussed.

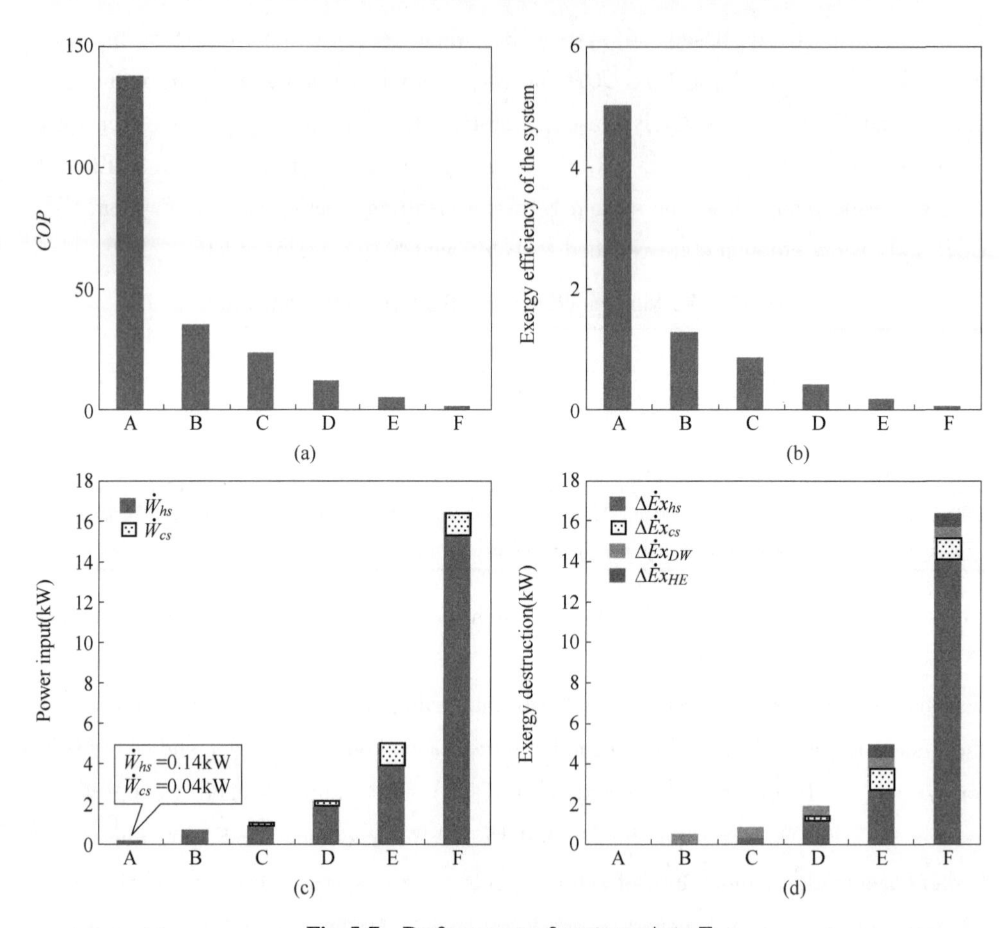

Fig. 5-7 Performance of systems A to F

(a) System *COP*; (b) η_{ex}; (c) Power input of heating source and cooling source; (d) Exergy destruction

5.2 Performance analysis of the ventilation cycle

5.2.1 System description and performance index

The schematic of the ventilation cycle is shown in Fig. 5-8, which includes a desiccant wheel (*DW*), a sensible heat recovery exchanger (*HE*), a heater used as the heating source, two direct evaporative coolers (EC_1 and EC_2) used as cooling sources, and two fans.

The ambient air is used as the processed air and the indoor exhaust air is used for regeneration. The processed air is first dehumidified and heated by *DW* (from point A_{pin} to A_{p1}), then cooled down by the regeneration air in HE (from point A_{p1} to A_{p2}), and finally cooled down and humidified by EC_1 (from point A_{p2} to A_{pout}) to the supplied

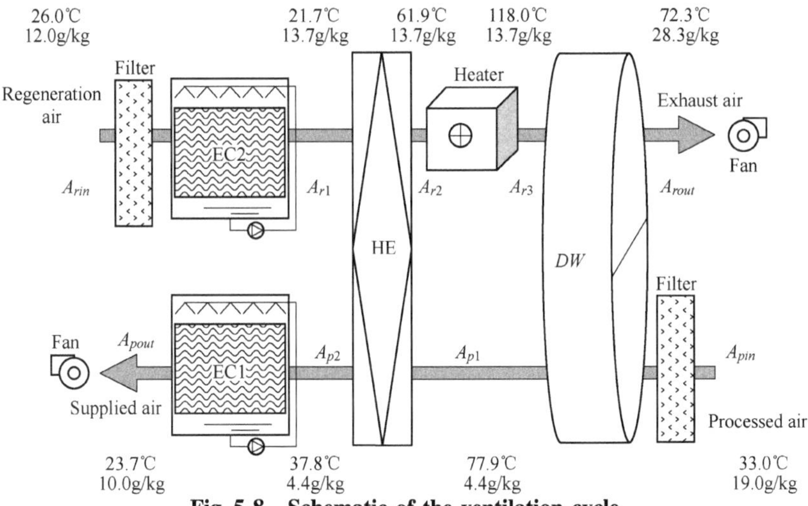

Fig. 5-8 Schematic of the ventilation cycle

state. The regeneration air is first cooled down and humidified by EC_2 (from point A_{rin} to A_{r1}), then heated by HE (from point A_{r1} to A_{r2}) and the heater (from point A_{r2} to A_{r3}) to the required regeneration temperature, and finally regenerated by DW (from point A_{r3} to A_{rout}). Fig. 5-9 shows the air handling process in a psychrometric chart, where W is the water content ratio of the desiccant material (silica gel) and ϕ is the relative humidity ratio of the humid air.

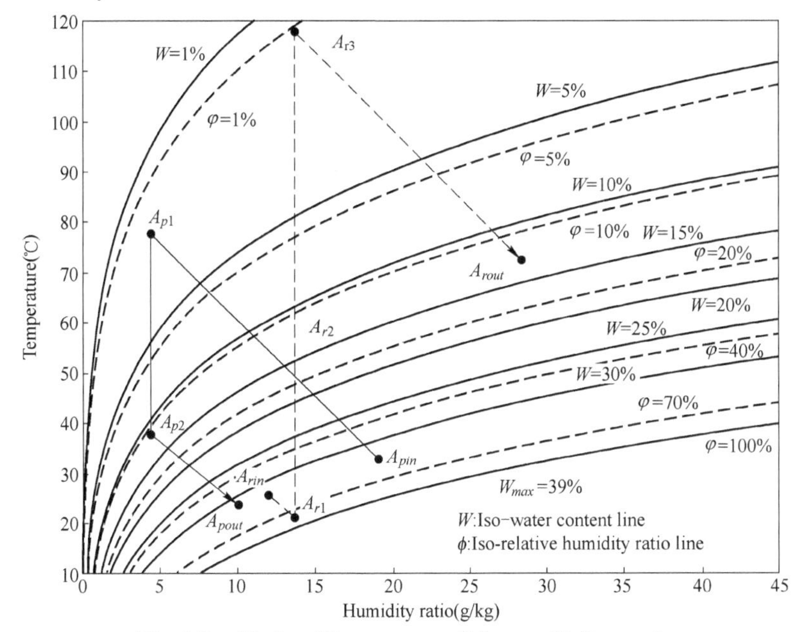

Fig. 5-9 Air handling process of the ventilation cycle

It should be noted that the processed air is cooled down by the evaporative cooler inthe ventilation cycle, so there is no need for an extra cooling source. However, the "free" cooling of EC_1 involves as acrifice: an increase in the humidity ratio of the air. As shown in Fig. 5-9, the humidity ratio of A_{p1} is much lower than that of A_{pout}, and the desiccant wheel exhibits over-dehumidification.

The over-dehumidification ratio (θ) of the ventilation cycle is defined in Eq. (5-17), and refers to the humidity ratio difference between A_{pin} and A_{p1} divided by the humidity ratio difference between A_{pin} and A_{pout}, as shown in Fig. 5-8 and Fig. 5-9. The value of θ is greater than or equal to 1.

$$\theta = \frac{\omega_{pin} - \omega_{p1}}{\omega_{pin} - \omega_{pout}} \tag{5-17}$$

An electrical heater is usually adopted as the regeneration heating source in the ventilation cycle. According to the definitions of COP and exergy efficiency, as shown in Eqs. (5-12), (5-13) and (5-16), $\dot{W}_{cs} = 0$, $\dot{W}_{hs} = \dot{m}_r(i_{r3} - i_{r2})$ and $\Delta \dot{E}x_{cs} = 0$ for such a ventilation cycle. The heat and mass transfer exergy destruction of the two direct evaporative coolers ($\Delta \dot{E}x_{EC_1}$ and $\Delta \dot{E}x_{EC_2}$) should be added into the denominator of Eq. (5-16). Similarly, when the processed air inlet and outlet conditions are fixed, the exergy destruction of the heat and mass transfer devices and that of the heater both significantly influence the exergy efficiency and COP.

Considering the power input of the two fans ($\dot{W}_{fan}$), EER of the $DDCS$ can be calculated as Eq. (5-18):

$$EER = \frac{\dot{Q}_P}{\dot{W}} = \frac{\dot{m}_p(i_{pin} - i_{pout})}{\dot{W}_{cs} + \dot{W}_{hs} + \dot{W}_{fan}} \tag{5-18}$$

5.2.2 Performance analysis of the ventilation cycle

The performance of the ventilation cycle was simulated under the working conditions listed in Table 5-2. The wheel thickness is 0.2m and NTU of the heat recovery exchanger is 2.5; the structures of EC_1 and EC_2 are identical; the water flow rate in both EC_1 and EC_2 is 0.19kg/s. When the air facial velocity is around 2m/s, the pressure drop of the desiccant wheel with the thickness being 0.2m is approximately 150Pa, while that of the heat recovery exchanger with a heat exchange efficiency around of 70% is approximately 125Pa[7], and that of the direct evaporative cooler is around 30Pa[8]. The pressure drops of the electrical heater and the filter are assumed to be 50Pa and 100Pa, respectively; the pressure drops of the processed air duct and the regeneration air duct are assumed to be 300Pa and 150Pa. If the excess pressure is 50Pa, the required pressure head for the processed air side and regeneration air side are 755Pa and 655Pa, respectively. If the efficiency of the fans is 50%, the power consumption of the processed air fan is 1.15kW, and that of the regeneration air fan is 0.91kW.

This case is analyzed under the optimal rotation speed (i.e. the same outlet humidity ratio of the processed air is realized by the lowest regeneration temperature) of the desiccant wheel, which is around 20r/h. To achieve the required supplied air state (24℃ and 10g/kg), *NTU* of each *EC* equals 1.17, and the required regeneration temperature is 118℃. The air states after each device are shown in Fig. 5-8 and Fig. 5-9. The results show that, the required regeneration temperature (118℃) is much higher than that of system E (67.5℃). This is due to two factors. First, the humidity ratio of the processed air after the wheel is as low as 4.4g/kg, with the over-dehumidification ratio (θ) being 1.62. Second, the humidity ratio of the regeneration air increases from 12.0g/kg to 13.7g/kg after EC_2. Due to these two reasons, the required regeneration temperature and power input of the heater increase considerably from 67.5℃ and 15.9kW, respectively (system E), to 118℃ and 45.0kW, respectively (ventilation cycle). *COP* and exergy efficiency of this ventilation cycle system are 0.57 and 2.1%, respectively, which are much lower than those of system E (COP = 5.0, η_{ex} = 18.3%). *EER* of the ventilation cycle is 0.55.

The exergy flow diagram of this case is shown in Fig. 5-10. It can be seen that only 2.1% of the exergy provided by the heater is used to realize the cooling and dehumidification of the processed air. 84.5% of the exergy provided by the heater is consumed in the electricity-heat conversion process; the remaining 13.4% is consumed by the exergy destruction of the heat and mass transfer devices (11.4%) and by the activity required to increase the exergy of the regeneration air (2.0%). Therefore, it is important to reduce the exergy destruction of the heat source and that of the heat and mass transfer devices in order to improve system performance. The exergy destruction of EC_1 and that of EC_2 are relatively small compared to that of the desiccant wheel and the heat exchanger. However, over-dehumidification caused by EC_1 increases the dehumidification capacity of the desiccant wheel and the heat transfer capacity of the heat exchanger. This results in a larger exergy destruction of both the desiccant wheel and heat exchanger. Meanwhile, the heat capacity of the heater increases with the increased regeneration temperature, which results in greater exergy needed from the heater and greater exergy destruction of the heat source.

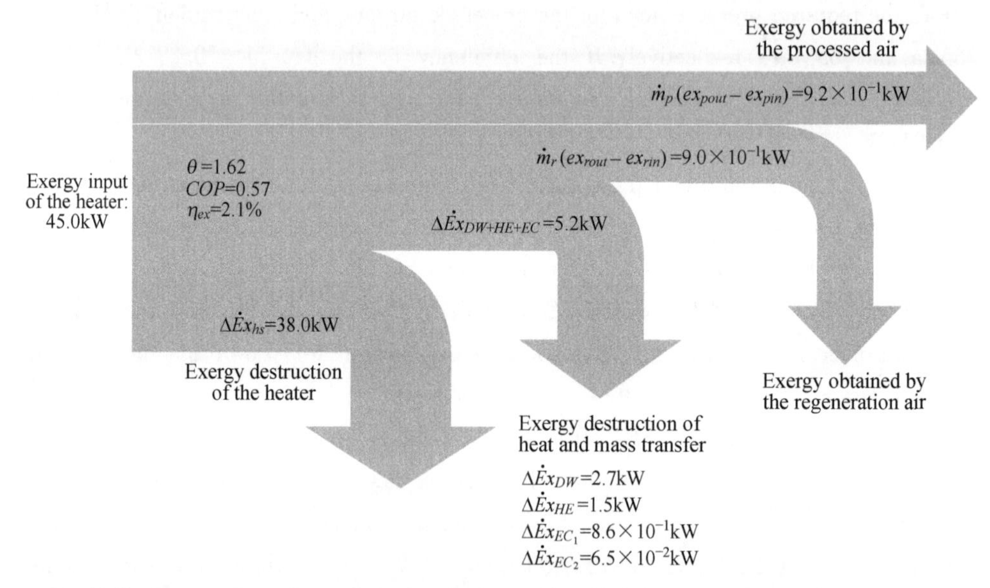

Fig. 5-10　Exergy flow chart of the ventilation cycle under the working conditions listed in Table 5-2 (wheel thickness equals 0. 2m and NTU_{HE} equals 2. 5)

In conclusion, the performance of the ventilation cycle is limited for two main reasons: over-dehumidification and adoption of the electrical heater with low exergy efficiency.

5. 2. 3　Effects of wheel thickness and heat recovery efficiency

The influences of wheel thickness (L_{DW}) and the heat transfer ability of the heat recovery device (NTU_{HE}) on the performance of the ventilation cycle are investigated in this subsection. The cases in which L_{DW} is 0. 2m and 0. 4m and NTU_{HE} is 2. 5, 4 and 9 (with a heat exchange efficiency of 71. 4%, 80% and 90%, respectively) are analyzed. The required supplied air state is fixed at 24℃ and 10g/kg.

As shown in Fig. 5-11, COP and exergy efficiency of the ventilation cycle both improve with the increase of NTU_{HE} and L_{DW}, but the influence of NTU_{HE} is more significant than that of L_{DW}. However, θ is still larger than 1, and over-dehumidification still exists.

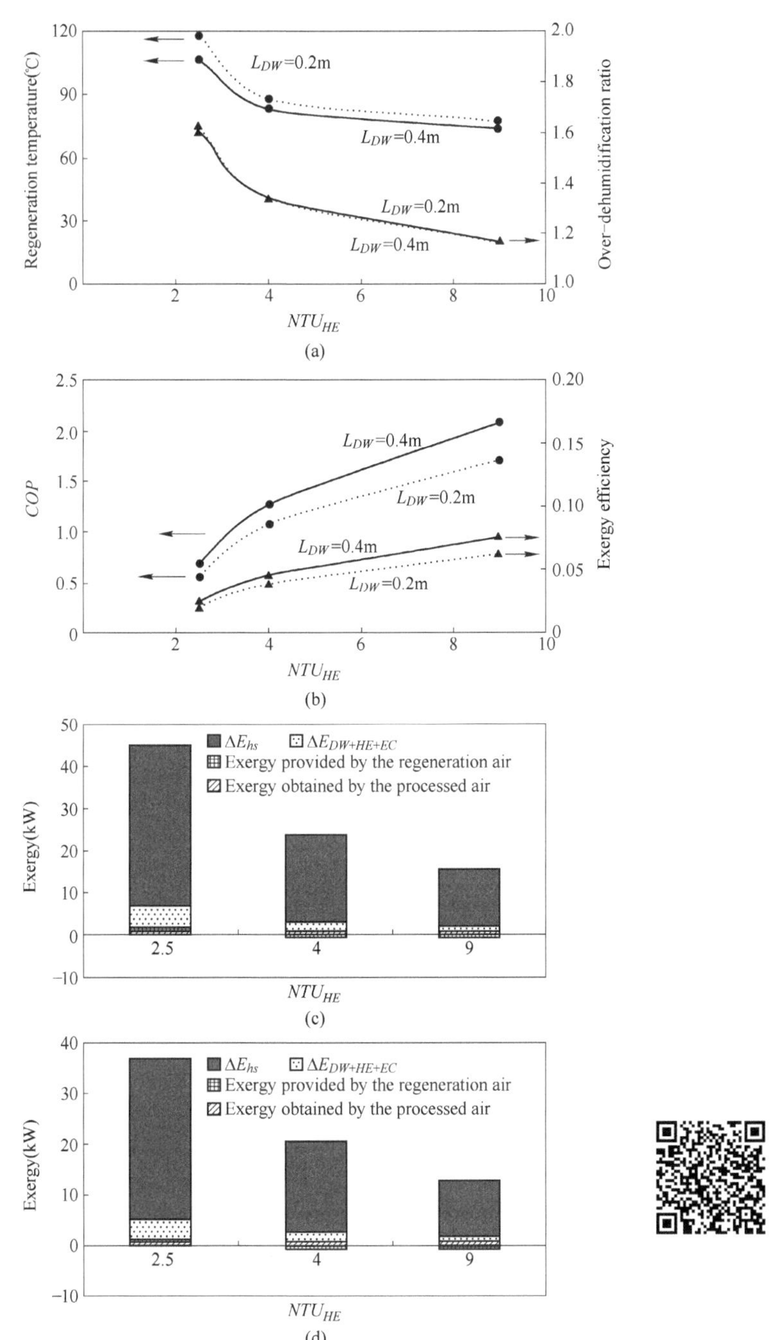

Fig. 5-11 Effects of wheel thickness (L_{DW}) and NTU_{HE} on ventilation cycle's performances

(a) Regeneration temperature and over-dehumidification ratio; (b) *COP* and exergy efficiency;
(c) Exergy results when L_{DW} = 0.2m; (d) Exergy analysis results when L_{DW} = 0.4m

Increasing L_{DW} and NTU_{HE} can increase system performance. However, the size of both the desiccant wheel and the heat recovery exchanger will increase considerably, resulting in a larger system size and higher fan power consumption. The total power consumption of the two fans increases from around 2.0kW (needed total pressure head = 1410Pa) to around 3.0kW (needed total pressure head = 2060Pa) when L_{DW} and NTU_{HE} increase from 0.2m and 2.5, respectively, to 0.4m and 9, respectively. The increase of the power consumption of the fans is smaller than the decrease of the power consumption of the heater. Therefore, *EER* increases from 0.55 (L_{DW} = 0.2m and NTU_{HE} = 2.5) to 1.68 (L_{DW} = 0.4m and NTU_{HE} = 9). In summary, increasing NTU_{HE} and L_{DW} can increase the system's performance. However, this solution has inherent problems (e.g., a larger system volume and greater fan power consumption), and θ is still greater than 1. The next section will focus on improving the system by changing its structure.

5.3 Performance improvement of the actual system

This section focuses on performance improvement methods from the perspectives of avoiding over-dehumidification and using alternative heat sources with low exergy destruction (e.g., heat pump systems) in place of the electrical heater.

5.3.1 Avoiding over-dehumidification

A sensible heat exchanger is used in place of the evaporative cooler (EC_1) to avoid over-dehumidification, as shown in Fig. 5-12. An electrical heater is still used to heat the regeneration air in this case. The air-water sensible heat exchanger and *EC* are collectively called the indirect evaporative cooler (*IEC*). The cold water comes from *EC* at the regeneration air side. After cooling the processed air, water returns back to *EC*. This system is referred to as the Improved System.

The performance of the Improved System was simulated under the working conditions listed in Table 5-2. The wheel thickness is 0.2m; the mass flow rates of the air and the water are 0.8kg/s and 0.19kg/s, respectively; and the two sensible heat exchangers have the same *NTU* and efficiency of 2.5% and 71.4%, respectively. To achieve the required supplied air state (24℃ and 10g/kg), the required regeneration temperature of the Improved System is 73.5℃, and *NTU* of *EC* is equal to 1.17. The air states of each point in the Improved System are shown in Fig. 5-12. It can be seen from the figure that over-dehumidification is avoided, and the regeneration temperature is greatly reduced from 118℃ (ventilation cycle) to 73.5℃.

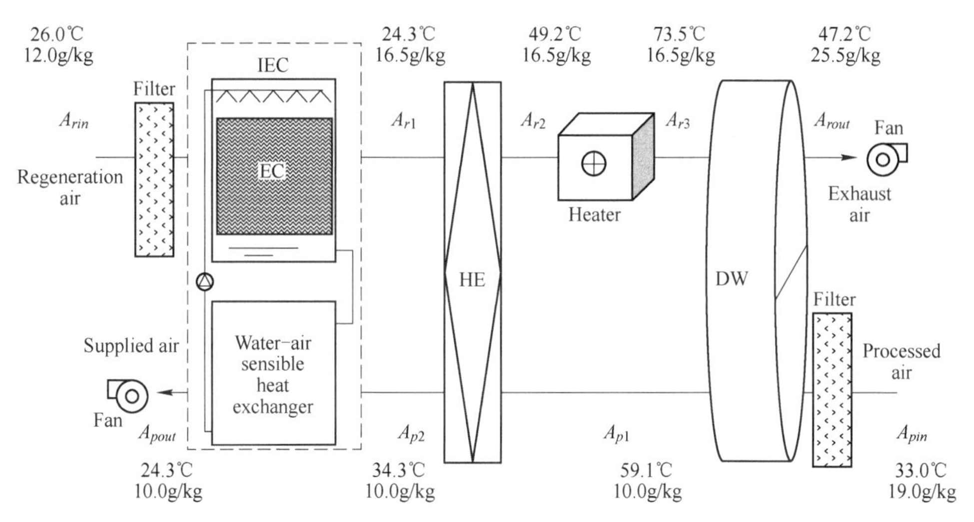

Fig. 5-12 Avoiding over-dehumidification with an indirect evaporative cooler (Improved System)

The exergy flow chart of the Improved System is illustrated in Fig. 5-13.

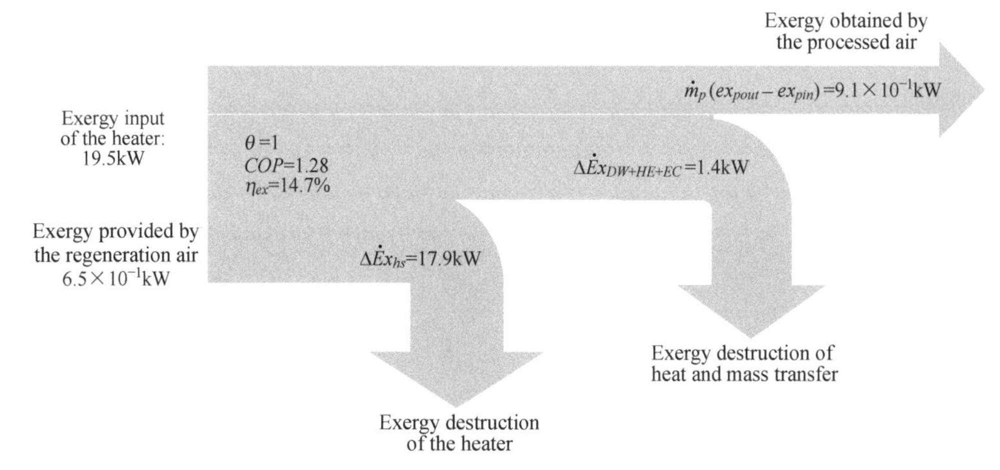

Fig. 5-13 Exergy flow chart of the Improved System

The results show that the exergy destruction of all the heat and mass transfer devices is greatly reduced from 5.2kW (Fig. 5-10) to 1.4kW. Simultaneously, the exhaust air temperature is reduced because of the lower regeneration temperature. And instead of taking exergy away from the system, as in the case illustrated in Fig. 5-10, the regeneration air provides exergy to the Improved System. The exergy destruction of the heat source is reduced from 38.0kW (Fig. 5-10) to 17.9kW, and the total exergy required by the heater is dramatically reduced from 45.0kW (Fig. 5-10) to

19.5kW. Compared to the ventilation cycle when NTU_{HE} is 2.5, COP of the Improved System increases from 0.57 to 1.28, and exergy efficiency of the Improved System increases from 2.1% to 4.7%. The pressure drop of the water-air heat exchanger is approximately 50Pa[9]. The power consumption of the fan at the processed air side is 1.08kW (needed pressure head = 775Pa), and that at the regeneration air side is 0.91kW (needed pressure head = 655Pa). EER of the Improved System is 1.16.

Therefore, over-dehumidification is avoided in the Improved System, and the regeneration temperature and exergy destruction can be decreased considerably. In addition, COP and exergy efficiency can be increased to 1.28 and 4.7%, respectively; these values are closer to those of system E compared with the ventilation cycle.

5.3.2 Adopting a heat pump system in place of the electrical heater

The results from Subsection 5.3.1 show that avoiding over-dehumidification can improve system performance. However, when an electrical heater is used, the system's performance is still low because of the high exergy destruction of the heat source. This subsection discusses the performance of the *DDCS* when a heat pump system is used, which serves as both a cooling and a heating source simultaneously.

The system structure is shown in Fig. 5-14. There is one desiccant wheel (*DW*), one heat recovery exchanger (*HE*), one evaporator to cool down the processed air after dehumidification, and two parallel condensers of the same size to heat the regeneration air before and after regeneration. Condenser 2 is mainly used to dissipate the extra heat

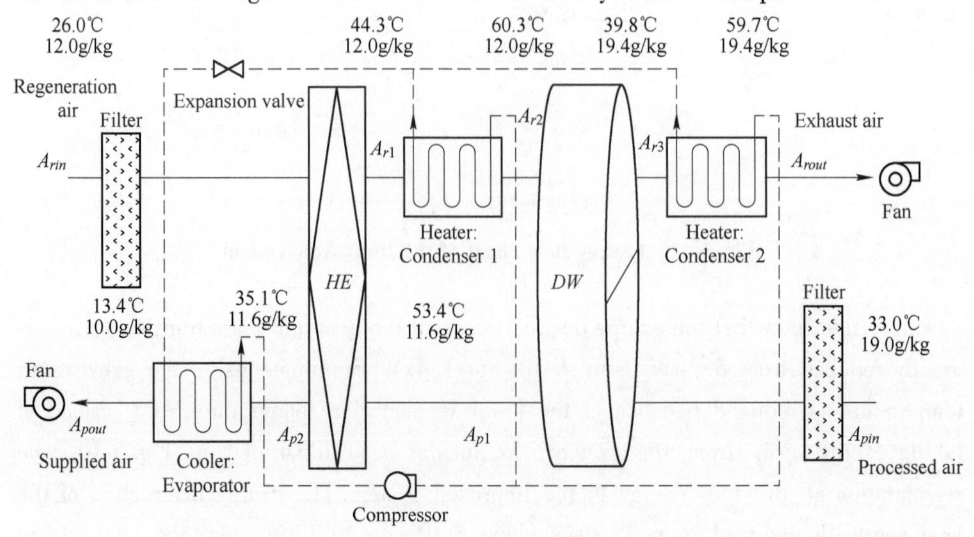

Fig. 5-14 Schematic of the heat pump-driven desiccant cooling system without pre-cooling

of condensation. The evaporator and condensers are linked with the same expansion valve and compressor.

COP and exergy efficiency of this system can be calculated by Eqs. (5-10), (5-11) and (5-14), where $\dot{W} = \dot{W}_{HP}$. Because the heat pump is used as both the heat source and the cooling source, $\Delta\dot{E}x_{hs}$ and $\Delta\dot{E}x_{cs}$ are replaced by the exergy destruction of the heat pump system ($\Delta\dot{E}x_{HP}$), which is the input power of the compressor minus the exergy obtained by the air at the evaporators and condensers.

System performance was investigated under the same working conditions listed in Table 5-2.

The humidity ratio and temperature of the supplied air are 10g/kg and lower than 24℃, respectively. The wheel is 0.2m thick; *NTU* of the evaporator is 4; *NTU* of each condenser is 2; and *NTU* of *HE* is 2. The thermodynamic perfectness of the compressor is 0.5. The air handling process is shown in Fig. 5-14 and Fig. 5-15, and the exergy flow chart is shown in Fig. 5-16.

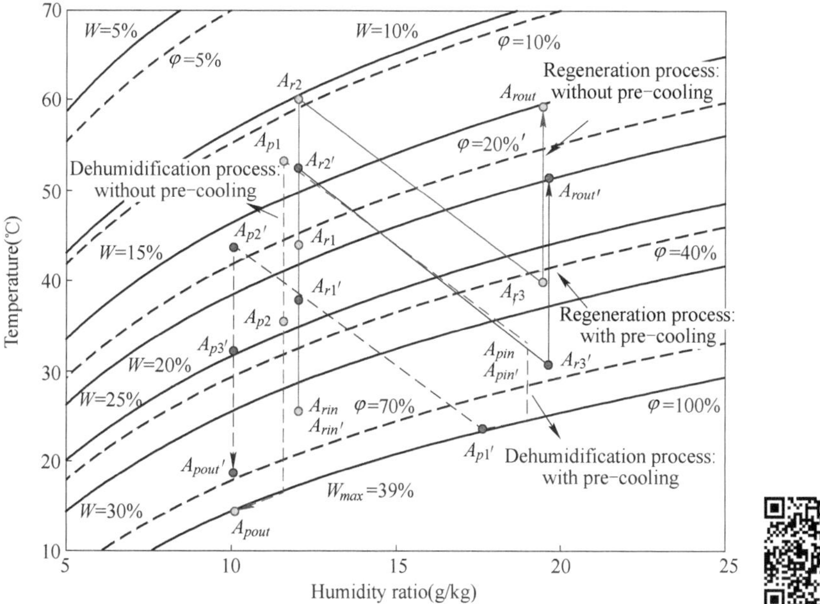

Fig. 5-15 Air handling process of the heat pump-driven desiccant cooling system

The results show that the adoption of a heat pump system can greatly increase the performance of the *DDCS*, and the regeneration temperature decreases to 60.3℃. The

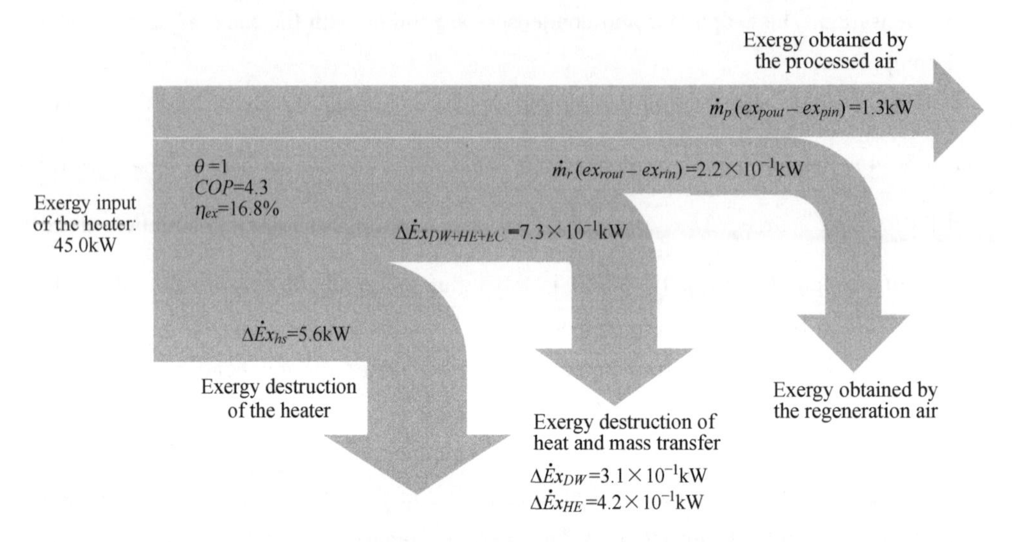

Fig. 5-16 Exergy flow chart of the heat pump-driven desiccant cooling system without pre-cooling

required condensing and evaporating temperatures are 62.8℃ and 9.2℃, respectively. The outlet processed air temperature is around 14℃. The power input of the compressor is 7.9kW, and *COP* and exergy efficiency are 4.34 and 16.8%, respectively. Based on the exergy analysis results, it is clear that the reduction of the power input of the compressor ($\dot{W}_{HP}$) benefits from the reduction in the exergy destruction of the desiccant wheel ($\Delta\dot{E}x_{DW}$) and that of the heat exchanger ($\Delta\dot{E}x_{HE}$) and heat pump system ($\Delta\dot{E}x_{HP}$).

Fig. 5-17 shows the schematic of an other heat pump-driven system, in which another evaporator is utilized before dehumidification to achieve pre-cooling. Fig. 5-15 and Fig. 5-17 show the air handling process under the same working conditions and wheel structure listed in Table 5-2; the *NTUs* of each evaporator, condenser, and *HE* are all 2. It can be seen that because of pre-cooling, the regeneration temperature decreases to 52.7℃. The required condensing and evaporating temperatures are 55.0℃ and 16.6℃, respectively. The outlet processed air temperature is around 18.7℃. Compared to the heat pump system without pre-cooling, the power input of the compressor and exergy destruction ($\Delta\dot{E}x_{DW}+\Delta\dot{E}x_{HE}+\Delta\dot{E}x_{HP}$) are reduced to 6.0kW and 5.2kW, respectively. *COP* and exergy efficiency of this system increase to 5.01 and 18.0%, respectively; these values are almost the same as those of system E (i.e., 5.0 and 18.3%).

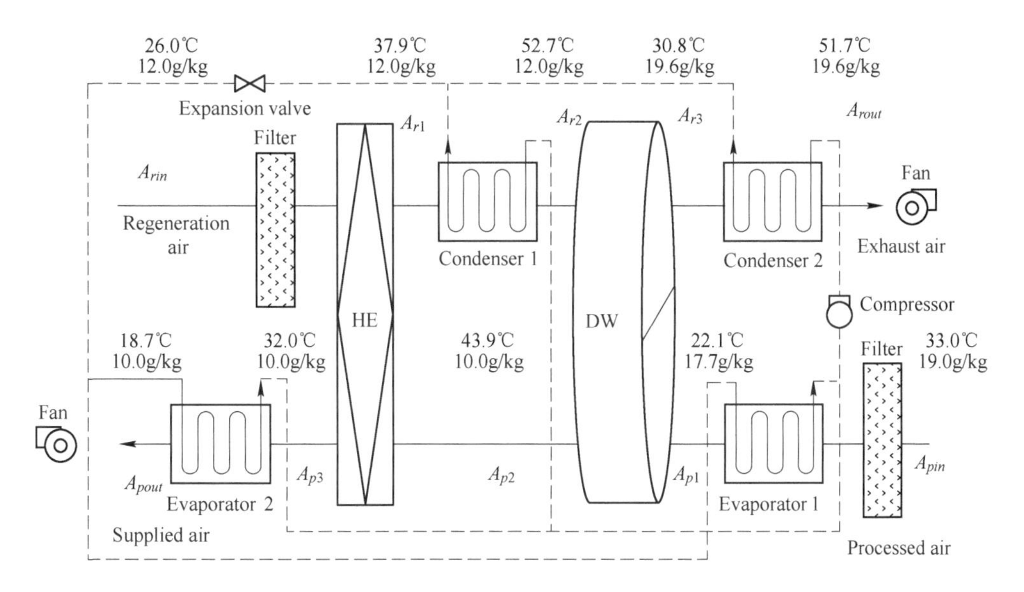

Fig. 5-17 Schematic of the heat pump-driven desiccant cooling system with pre-cooling

The pressure drop of the evaporators is the same with that of the condensers, which is 80Pa. The total pressure drop and total fan power consumption are the same for the two systems shown in Fig. 5-14 and Fig. 5-17 because of the fixed total *NTU* of the evaporators and that of the condensers. The power consumption of the processed air fan and regeneration air fan are 1.12kW (needed pressure head = 805Pa) and 0.91kW (needed pressure head = 655Pa). *EERs* of the systems shown in Fig. 5-14 and Fig. 5-17 are 3.45 and 3.74, respectively.

5.3.3 Performance comparison

The performances of the ventilation cycle, Improved System, and heat pump-driven system (with and without pre-cooling) are summarized in Table 5-3.

The results are all based on the same working conditions and wheel structure listed in Table 5-2. The pressure drops and the fan power consumptions are listed in Table 5-4.

It can be seen from Table 5-3 that when over-dehumidification is avoided and a heat pump system is used in place of the electrical heating system, the heat source temperature and exergy destruction of the heat and mass transfer devices, as well as those of the heating and cooling sources, can all be greatly reduced. This leads to a reduction of the power input. *COP* and exergy efficiency can increase from 0.57 and 2.1%, respectively, in the original ventilation cycle to 5.01 and 18.0%, respectively, in the heat pump-driven system with pre-cooling. These latter values are similar to those of system E. It can be seen from Table 5-4 that, the change of the power consumption of the fans in different systems is little compared with that of the heater or compressor. The heat pump-driven system with pre-cooling has the highest *EER*, which is 3.74.

The performances of the four systems under *ARI* Summer condition (i.e., the processed air inlet state is 35℃ and 14.3g/kg) are calculated, and the results are listed in Table 5-5.

Table 5-3 Comparison of different desiccant wheel systems under the working conditions listed in Table 5-2

System	θ	Required heat source temperature (℃)	$\dot{Q}_p$ (kW)	Exergy obtained by processed air (kW)	Exergy destruction (kW) $\Delta\dot{E}x_{DW+HE+EC}$	Power input (kW) $\Delta\dot{E}x_{hs}+\Delta\dot{E}x_{cs}$	*COP*		η_{ex}	*EER*
Ventilation cycle (Fig. 5-8)	1.62	118.0	25.4	9.2×10^{-1}	5.2	38.0	45.0	0.57	2.1%	0.55
Improved System (Fig. 5-12)	1	73.5	25.0	9.1×10^{-1}	1.4	17.9	19.5	1.28	4.7%	1.16
Heat pump-driven system without pre-cooling (Fig. 5-14)	1	62.8	34.4	1.3	7.3×10^{-1}	5.6	7.9	4.34	16.8%	3.45
Heat pump-driven system with pre-cooling (Fig. 5-17)	1	55.0	30.0	1.1	5.6×10^{-1}	4.6	6.0	5.01	18.0%	3.74

Table 5-4 Pressure drops and fan power consumption of different desiccant wheel systems (the excess pressure is 50 Pa and the efficiency of fans is 50%)

System	Pressuredrop (Pa)									Power of fans (kW)	
	DW	Filter	Air-air HE	Air-water HE	Evaporators/condensers	Heater	EC_1/EC_2	Air duct: Processed air	Air duct: Regeneration air	Processed air	Regeneration air
Ventilation cycle (Fig. 5-8)	150	100	125	-	-/-	50	30/30	300	150	1.05	9.1×10^{-1}
Improved System (Fig. 5-12)	150	100	125	50	-/-	50	-/30	300	150	1.08	9.1×10^{-1}

Continued Table 5-4

System	Pressuredrop (Pa)									Power of fans (kW)	
	DW	Filter	Air-air HE	Air-water HE	Evaporators/condensers	Heater	EC_1/EC_2	Air duct: Processed air	Air duct: Regeneration air	Processed air	Regeneration air
Heat pump-driven system without pre-cooling (Fig. 5-14)	150	100	125	-	80/80	-	-/30	300	150	1.12	9.1×10^{-1}
Heat pump-driven system with pre-cooling (Fig. 5-17)	150	100	125	-	80/80	-	-/30	300	150	1.12	9.1×10^{-1}

Table 5-5 Comparison of different desiccant wheel systems under ARI summer conditions

System	θ	Required heat source temperature (℃)	$\dot{Q}_p$(kW)	Exergy obtained by processed air(kW)	Exergy destruction (kW): $\Delta\dot{E}x_{DW+HE+EC}$	Exergy destruction (kW): $\Delta\dot{E}x_{hs}+\Delta\dot{E}x_{cs}$	Power input (kW)	*COP*	η_{ex}	*EER*
Ventilation cycle (Fig. 5-8)	1.82	76.0	17.3	6.8×10^{-1}	1.65	20.6	22.4	0.77	3.0%	0.71
Improved System (Fig. 5-12)	1	58.5	17.3	6.8×10^{-1}	6.3×10^{-1}	12.9	13.5	1.28	5.0%	1.12
Heat pump-driven system without pre-cooling (Fig. 5-14)	1	56.2	26.6	1.17	3.9×10^{-1}	3.5	4.9	5.40	23.7%	3.82
Heat pump-driven system with pre-cooling (Fig. 5-17)	1	43.8	19.0	7.3×10^{-1}	2.0×10^{-1}	1.9	2.3	8.17	31.4%	4.34

The fan power consumptions are the same with those of the Beijing summer condition (33℃, 19g/kg). The saturated condition at 35℃ is chosen as the reference state during the exergy analysis. Similar with the cases under the Beijing summer working conditions, the exergy destruction of the heat and mass transfer devices as well as heat/cooling sources is the least for the heat pump driven system with pre-cooling, with *COP*, exergy efficiency and *EER* being 8.17, 31.4% and 4.34, respectively.

5.4 Conclusions

This chapter focuses on the performance of desiccant wheel dehumidification and cooling systems (*DDCSs*). The influences of the irreversible processes of the heat and mass transfer devices and the heating and cooling sources on the performance of *DDCSs* are examined theoretically. The performance of the ventilation cycle is also analyzed, and several performance improvement methods are proposed. The main conclusions can be summarized as follows:

(1) For the reversible *DDCS*, under the designed working conditions listed in Table 5-2, *COP* and exergy efficiency are 137.9% and 502.0%, respectively. As the system becomes irreversible gradually from system A to system F, performance is reduced. *COP* and exergy efficiency of system E, which are 5.0 and 18.3%, respectively, represent the relatively high standards that actual systems can achieve.

(2) The ventilation cycle was researched under the same working conditions. *COP* and exergy efficiency are 0.57% and 2.1%, respectively. The exergy destruction of the electrical heater and that of the heat and mass transfer devices are both quite high. The performance of the actual system can be improved through avoiding over-dehumidification and adopting a heat source with lower exergy destruction.

(3) When the evaporative cooler at the processed air side is replaced by a sensible heat exchanger, over-dehumidification can be avoided, and exergy destruction of heat and mass transfer devices, as well as that of the heat source, can be reduced. When an indirect evaporating cooler is used, under the same working conditions, *COP* and exergy efficiency are 1.28% and 4.7%, respectively, which are much higher than those of the original ventilation cycle.

(4) When the electrical heating system is replaced by the heat pump system, exergy destruction of the heat source can be greatly reduced, especially when pre-cooling is adopted. Under the same working conditions, *COP* and exergy efficiency increase to 5.0% and 18.0%, respectively, in the system with pre-cooling; these performance values are close to those of system E. These results apply to the *ARI* summer condition, with *COP*, exergy efficiency and *EER* being 8.17%, 31.4% and 4.34%, respectively, for the heat pump system with pre-cooling.

Nomenclature

COP coefficient of performance

c_{pa}	specific heat at constant pressure of dry air, kJ/(kg · ℃)
ex	specific exergy of moisture air, kJ/kg
EER	energy efficiency ratio
$\dot{Ex}$	exergy rate, kW
$\Delta\dot{Ex}$	exergy destruction rate, kW
i	specific enthalpy of moisture air, kJ/kg
$\dot{m}$	mass flow rate, kg/s
NTU	number of heat transfer units
$\dot{Q}$	heat/cooling rate, kW
R_a	ideal gas constant of dry air, kJ/(mol · K)
t	Celsius temperature, ℃
T	Kelvin temperature, K
Δt	Celsius temperature difference, ℃
$\dot{W}$	input power, kW
r	specific vaporization heat, kW/kg

Greek symbols

ω	humidity ratio, g/kg
η_{ex}	exergy efficiency
η_{HP}	thermodynamic perfectness of the compressor
θ	over-dehumidification ratio
φ	relative humidity ratio
ε	heat transfer efficiency

Subscripts

a	air

ambi	ambient air
cond	condenser
cs	cooling source
DW	desiccant wheel
EC	direct evaporative cooler
evap	evaporator
ex	exergy
hs	heat source
HE	heat exchanger
HP	heat pump system
IEC	indirect evaporative cooler
in	inlet
out	outlet
p	processed air
r	regeneration air
0	reference state

References

[1] A Bejan. Advanced engineering thermodynamics [M]. 3rd ed. Hoboken (NJ): John Wiley & Sons, 2006.

[2] D La, Y Li Y, Y J Dai, et al. Effect of irreversible processes on the thermodynamic performance of open-cycle desiccant cooling cycles [J]. Energy Conver. Manage, 2013, 67: 44-56.

[3] R, Tu. Analysis of heat and mass transfer of solid desiccant dehumidification and air handling process optimization [D]. Tsinghua University, 2014.

[4] E Hürdogan, O Buyükalac, A Hepbasli, T Yilmaz. Exergetic modeling and experimental performance assessment of a novel desiccant cooling system [J]. Energy Build, 2011, 43: 1489-1498.

[5] R Tu, X H Liu, Y Jiang. Performance analysis of a two-stage desiccant cooling system [J]. Appl. Energy, 2014, 13: 1562-1574.

[6] R Tu, X H Liu, Y Jiang. Performance comparison between enthalpy recovery wheels and dehumidification wheels [J]. Int. J. Refrig, 2013, 36: 2308-2322.

[7] H Zhang, J L Li. Two-stage desiccant cooling system and its energy consumption analysis [J]. J HVAC, 1998, 28 (6): 2-5.

[8] X W Liu, X Huang, Z X Wu. Study of the filler performance in direct evaporative cooler [J]. Fluid Mach, 2010, 38 (4): 53-57.

[9] X W Du. Measurement method and experimental research of performance of heat transfer and pressure drop for fin-tube heat exchanger [D]. Shanghai Jiaotong University, 2008.

Chapter 6 Performance Influencing Factors of Single-Stage Desiccant Wheel Dehumidification Systems

This chapter aims at finding the inherent performance influencing factors for the existing ventilation systems through exergy destruction analysis. The influencing factors for exergy destructions are described and the influences of exergy destructions on t_{reg} and *COP* are discussed. Based on the exergy destruction analysis of a basic ventilation system, approaches to enhance the performances of desiccant wheel cooling systems are proposed. Two improved ventilation systems are then analyzed based on the basic ventilation cycle, which can realize low t_{reg} and have higher *COP*. This chapter will provide guidelines for efficient system design and performances evaluation.

6.1 Performance of the ventilation system

6.1.1 System description of BVS

Ventilation systems with desiccant wheel cooling are typical air handling processes for cooling systems, which are used to dehumidify the ambient air and regenerated by the indoor air. Fig. 6-1 shows the schematic and air handling processes of a basic ventilation system (BVS). For BVS, a direct evaporative cooler, which is a spray tower (*ST*), is used as the cooling device[1-3]. According to Fig. 6-1, the process air (PA) is firstly dried in a near isentropic process in a desiccant wheel (DW) (A_{pin} to A_{p1}). Then, it is cooled by a sensible heat recovery unit (HR) by the regeneration air (RA) (A_{p1} to A_{p2}). Lastly, it is cooled to the supply temperature by ST_p under an isenthalpic process (A_{p2} to A_{pout}). For RA, it is firstly cooled by ST_R under an isenthalpic process (A_{rin} to A_{r1}) before cooling PA in HR (A_{r1} to A_{r2}). After being heated in the heater (H) to the required regeneration temperature (A_{r2} to A_{r3}), RA is used to regenerate DW (A_{r3} to A_{rout}). It can be seen from the psychrometic chart in Fig. 6-1 (b) that the humidity ratio of PA after DW is lower than that of SA. Thus, there exists over dehumidification for DW.

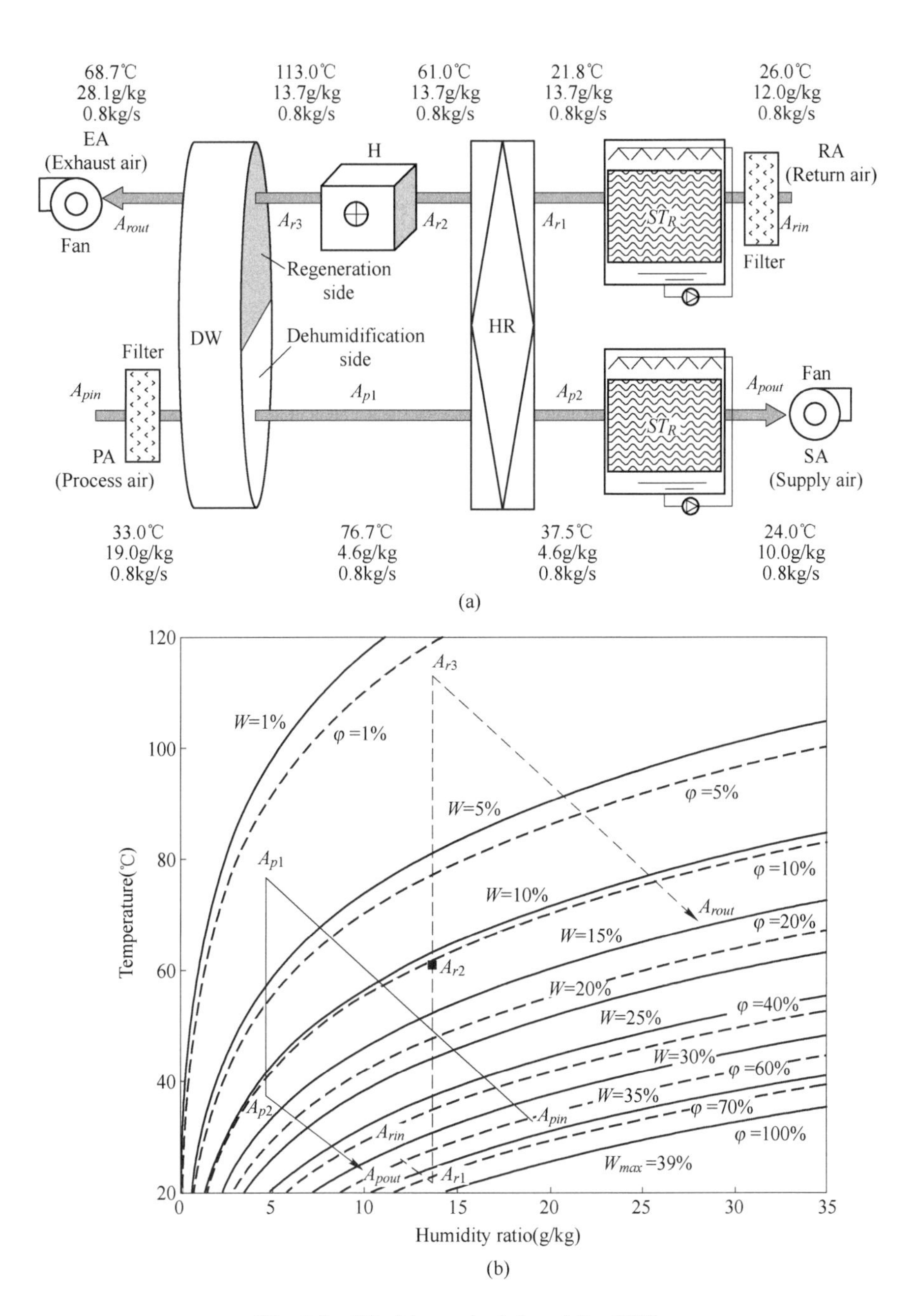

Fig. 6-1 Working principles of the BVS

(a) Schematic of BVS; (b) Air handling process on psychrometric chart

Over dehumidification ratio (*ODR*) of DW for BVS is defined as Eq. (6-1):

$$ODR = \frac{\omega_{pin} - \omega_{p1}}{\omega_{pin} - \omega_{pout}} \tag{6-1}$$

where ω_{pin}, ω_{p1} and ω_{pout} represent the humidity ratios of the air entering DW, leaving DW and entering the space, respectively, according to Fig. 6-1 (a).

The desiccant wheel, which rotates between PA and RA to realize continuous dehumidification process, is the key component for desiccant wheel cooling systems. The one-dimension double-diffusion mathematic model, which is introduced in Chapter 2, is programmed to simulate the complex transient heat and mass transfer processes between the air and the solid desiccant in the honeycomb structure of DW. Model validations with experiment results have been introduced in Chapter 2. The main parameters used in Eqs. (3-1) ~ (3-13) are listed in Table 6-1.

Table 6-1 Values of the main parameters in the simulation model of the desiccant wheel

ξ	k_s [kW/(m · K)]	r_s (kJ/kg)	ρ_{ad} (kg/m^3)	x	c_{pad} [kJ/(kg · K)]	*Mol* (kg/kmol)	*Nu*
2.8	0.22×10^{-3}	2.65×10^3	1129	0.7	0.92	18	2.46
σ	W_{max} (kg/kg)	D_0 (m^2/s)	ρ_{sub} (kg · m^3)	a (m)	c_{psub} [kJ/(kg · K)]	C	*Le*
0.7	0.39	1.6×10^{-6}	625	11×10^{-10}	0.88	0.3	1

Efficiency-*NTU* method is adopted to simulate the performances of sensible heat exchangers, i. e. HR and HE, and direct evaporative cooler, i. e. *ST* in BVS. For HR and HE in the present study, heat capacity flow rates, i. e. $\dot{m}c_p$, are identical for the two heat transfer fluids. Heat transfer efficiency for HR and HE can be expressed as Eq. (6-2). For direct evaporative cooler, the air handling process can be viewed as isenthalpic. Heat or mass transfer efficiency can be expressed as Eq. (6-3). Eq. (6-3) is also used to calculate the heat transfer efficiency of evaporators and condensers.

$$\theta = \frac{NTU}{NTU + 1} \tag{6-2}$$

$$\theta = 1 - e^{-NTU} \tag{6-3}$$

The model for *ST* in IVS1, which is based on the heat and mass transfer processes between the air and the spray water, can be referred to [4].

In the following discussions, these models are adopted to simulate heat and mass transfer processes in DW, heat exchangers and *ST*.

6.1.2 Performances of BVS

Performances of BVS are analyzed in this subsection based on simulation results. The air flow rates of PA and RA are designed the same, and the wheel is evenly divided. An electrical heater is used as the heating source. *COP* of BVS is defined as Eq. (6-4)[1-3], which is the total heat removed by the supply air from the air conditioning space divided by the power consumption of the electrical heater (P_H):

$$COP = \frac{\dot{m}_p(i_{space} - i_{pout})}{P_H} = \frac{\dot{m}_p(i_{space} - i_{pout})}{\dot{m}_r c_{pa}(t_{r3} - t_{r2})} \tag{6-4}$$

It is obvious that the lower t_{r3}(t_{reg} for BVS) is, the higher *COP* will be.

Performance analysis of BVS is based on the same ω_{pout} and fixed geometrical dimensions of each component. ω_{pout} is set to meet the latent heat removal demand of a residential or commercial building. The indoor air is set to be 26℃ and 12g/kg, and ω_{pout} is designed as 10g/kg[5]. The geometrical dimensions are designed under Beijing summer condition. The air mass flow rates of PA and RA are designed to be 0.8kg/s. Thickness and radius of *DW* are 0.2m and 0.5m, respectively. The air side *NTU* of HR (NTU_{HR}), shown in Eq. (6-5)[6], is 2.5.

$$NTU = \frac{hF}{\dot{m}_a c_{pa}} \tag{6-5}$$

The air side *NTU* for ST (NTU_{ST}), as shown in Eq. (6-5), is designed when the supply air temperature is 24℃, which is calculated to be 1.13 for BVS.

Information of air and geometrical dimensions for each component are listed in Table 6-2.

Table 6-2 Working conditions and the geometric information of each component in all the systems

Working conditions	DW for all systems	ST, HR, HE, evaporator and condenser
Beijing summer condition: 33℃, 19g/kg; ARI summer condition: 35℃, 14.3g/kg; Return air: 26℃, 12g/kg; Supply air: 10g/kg; PA and RA: 0.8kg/s	Radius: 0.5m; Thickness: 0.2m; Rotation speed=20r/h; Material: silica gel; Air channel structure: sinusoidal shape, 2mm high, and 2mm wide	NTU_{HR} in all systems: 2.5; NTU_{HE} in IVS1: 2.5; NTU_E and NTU_C in IVS2: 2.5; For HE, $(\dot{m}\cdot c_p)_a/(\dot{m}\cdot c_p)_w = 1$; NTU_{ST}: 1.13 for BVS and 1.30 for IVS1

Based on Table 6-2, the performances of BVS under Beijing summer condition are calculated. The air states after each component are shown in Fig. 6-2. The humidity ratio of *PA* after *DW* is 4. 6g/kg and *ODR* is 1. 60. The regeneration temperature (t_{reg}) is 113. 0℃. And *COP* for BVS is 0. 13.

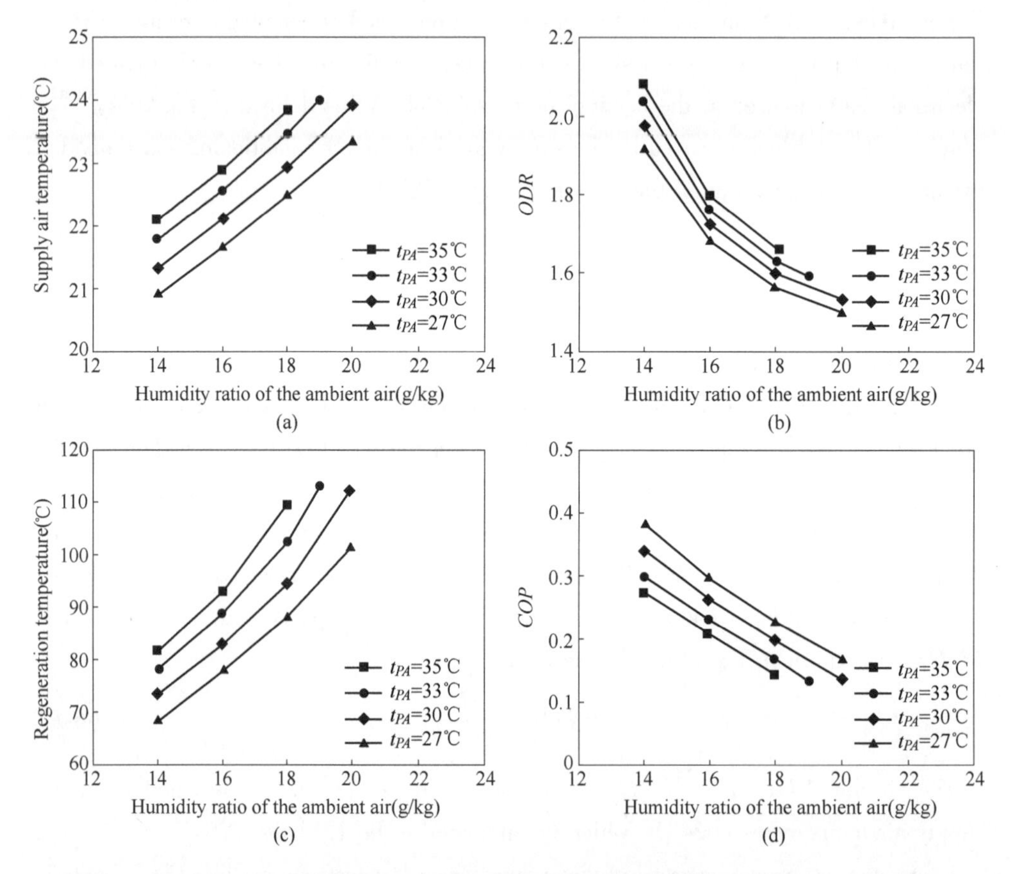

Fig. 6-2 Influences of the ambient air temperature and humidity ratio on the performance of the BVS when the process air humidity ratio is 10g/kg

(a) Supply air temperature; (b) *ODR*; (c) Regeneration temperature; (d) *COP*

Fig. 6-2 shows performances of BVS when the temperature of the ambient air changes from 35℃ to 27℃ and the humidity ratio of the ambient air changes from 20g/kg to 14g/kg. The results show that with fixed ω_{pout} (10g/kg), t_{pout} is lower than 24℃ for all cases. However, *ODR* is larger than 1. 5 and is over 2. 0 for some cases. t_{reg} of BVS is relatively high and *COP* is lower than 0. 4.

As a summary, the adoption of ST_p leads to over dehumidification for DW in BVS. t_{reg} is relatively high, which, combined with the adoption of the electrical heater, leads to low *COP* for BVS. In the next part, exergy destruction analysis is applied on ventilation systems in order to find inherent performance influencing factors for desiccant wheel cooling systems and to provide guidelines for efficient system design.

6.2 Exergetic analysis of the ventilation system

6.2.1 Performance influencing factors for the ventilation system

System boundary for exergy analysis of ventilation systems is shown in Fig. 6-3.

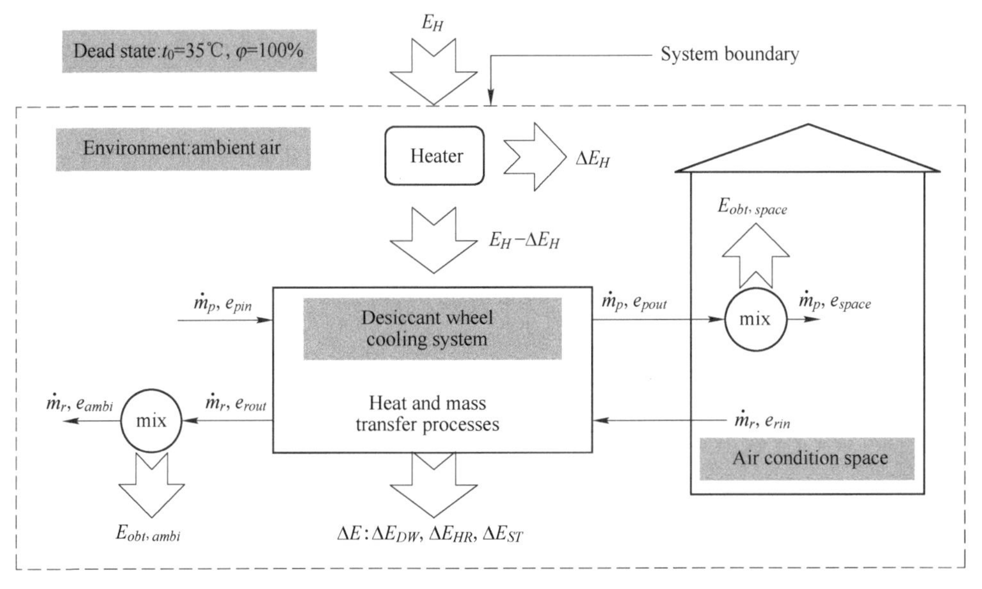

Fig. 6-3 System boundary of the desiccant wheel cooling system for exergetic analysis

Based on exergy balance equations, exergy provided by the heater (E_H) is written as Eq. (6-6):

$$\begin{aligned} E_H &= \Delta E_H + \dot{m}_r(e_{Hout} + e_{Hin}) \\ &= \Delta E_H + \Delta E + E_{obt,\ ambi} + E_{obt,\ space} - \dot{m}_p(e_{pin} - e_{space}) - \dot{m}_r(e_{rin} - e_{ambi}) \end{aligned} \quad (6\text{-}6)$$

where ΔE_H is the exergy destruction of the heater; ΔE is the total exergy destruction of heat and mass transfer in $DW(\Delta E_{DW})$, $HR(\Delta E_{HR})$ and $ST(\Delta E_{ST})$; $E_{obt,\ ambi}$ and $E_{obt,\ room}$ are exergies obtained by the ambient environment from the exhausted air and by the air conditioning space from the supply air, which can be explicitly expressed in Eqs. (6-7) ~ (6-8), respectively:

$$E_{obt,\ ambi} = \dot{m}_r(e_{rout} - e_{ambi}) \quad (6\text{-}7)$$

$$E_{obt,\ space} = \dot{m}_p(e_{pout} - e_{space}) \quad (6\text{-}8)$$

For ventilation systems discussed in this paper, $\dot{m}_p = \dot{m}_r$, $e_{rin} = e_{space}$ and $e_{pin} = e_{ambi}$.

Thus, Eq. (6-6) can be written as Eq. (6-9):

$$E_H = \Delta E_H + \Delta E + E_{obt,\ ambi} + E_{obt,\ space} \tag{6-9}$$

Exergy efficiency (η_e) of the ventilation system is defined as $E_{obt,\ space}$ divided by E_H. Combining with Eq. (6-9), η_e can be expressed as Eq. (6-10):

$$\eta_e = \frac{E_{obt,\ space}}{E_H} = \frac{E_{obt,\ space}}{\Delta E_H + \Delta E + E_{obt,\ ambi} + E_{obt,\ space}} = \frac{\dot{m}_p(e_{pout} - e_{space})}{\Delta E_H + \dot{m}_r(e_{Hout} - e_{Hin})} \tag{6-10}$$

Specific exergy of the air (e_a) is the function of temperature (T_a in Kelvin) and humidity ratio (ω_a in g/kg), shown as Eq. (6-11)[7]:

$$e_a = c_{pa}T_0\left(\frac{T_a}{T_0} - 1 - \ln\frac{T_a}{T_0}\right) + R_aT_0\left[(1 + 1.608 \times 10^{-3}\omega_a)\ln\frac{1 + 1.608 \times 10^{-3}\omega_0}{1 + 1.608 \times 10^{-3}\omega_a} + 1.608 \times 10^{-3}\omega_a\ln\frac{\omega_a}{\omega_0}\right] \tag{6-11}$$

When the supply air state is constant, $E_{obt,\ space}$ is fixed. According to Eq. (6-10), η_e is directly influenced by ΔEx_H , ΔE and $E_{obt,\ ambi}$. Higher η_e can be realized with lower ΔE_H , ΔE and $E_{obt,\ ambi}$. Eq. (6-10) also shows that e_{Hout} is influenced by ΔE and $E_{obt,\ ambi}$. With lower ΔE and $E_{obt,\ ambi}$, e_{Hout} can be decreased, leading to the increase of η_e . According to Eq. (6-11), when ω_a is fixed, the reduction of e_{Hout} means t_{Hout} or t_{reg} can be reduced. Therefore, under the same $E_{obt,\ space}$, lower ΔE and $E_{obt,\ ambi}$ lead to lower e_{Hout} , which eventually lead to lower t_{reg} and higher η_e . Thus, according to Eqs. (6-4) and (6-10) ~ (6-11), η_e, t_{reg} and *COP* are related to each other. In order to realize low temperature regeneration and high *COP* and η_e of desiccant wheel cooling systems, ΔE_H, ΔE and $E_{obt,\ ambi}$ should be decreased.

ΔE and ΔE_H are two main performance influencing factors for desiccant wheel cooling systems. The electrical heater, which represents the least efficient heat source, can be replaced by low-grade and high efficient heating methods, such as heat pumps or solar energy. When t_{reg} is reduced. ΔE is caused by profiles of temperature and humidity ratio differences between the two heat and mass transfer media, which can be written as Eq. (6-12)[8].

$$\Delta E_{htr} \approx \psi\frac{Q^2}{hF}\zeta_t,\ \Delta Ex_{mtr} \approx \vartheta\frac{M^2}{h_mF}\zeta_\omega \tag{6-12}$$

where ψ and ϑ , which are functions of T_0, dry bulb and dew point temperature of the air and the desiccant, can be assumed to be constant[8]; ζ_t and ζ_ω are unmatched coefficients[9-11], which present the uniformity of the temperature differences ($\Delta t = T_2 - T_1$) field and the humidity ratio differences ($\Delta\omega = \omega_2 - \omega_1$) field, respectively, between the two heat and mass transfer media. To reduce ΔE , unmatched coefficients of the component and ODR (relating to Q and M) should be reduced.

6.2.2 Exergy analysis of BVS

BVS is investigated in this subsection based on subsection 7.2.1. The dead state (T_0, ω_0) is chosen as the saturated state at 35℃ for the following discussions.

E_H for the electrical heater equals to the power consumption of the heater (P_H). Therefore, ΔE_H for the electrical heater in BVS is explicitly written as Eq. (6-13):

$$\Delta E_H = E_H - \dot{m}_r(e_{r3} - e_{r2}) = \dot{m}_r[c_{pa}(t_{r3} - t_{r2}) - (e_{r3} - e_{r2})] \tag{6-13}$$

The results of exergy analysis for BVS are shown in Fig. 6-4.

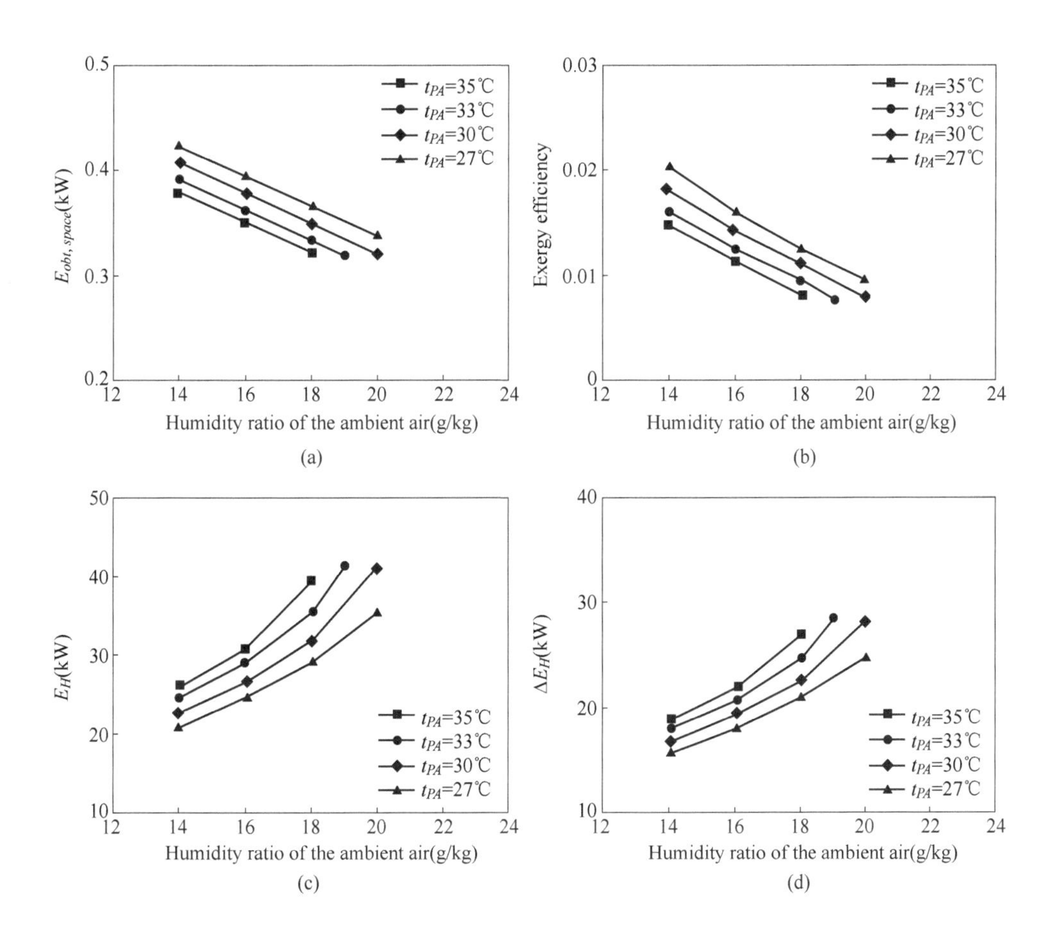

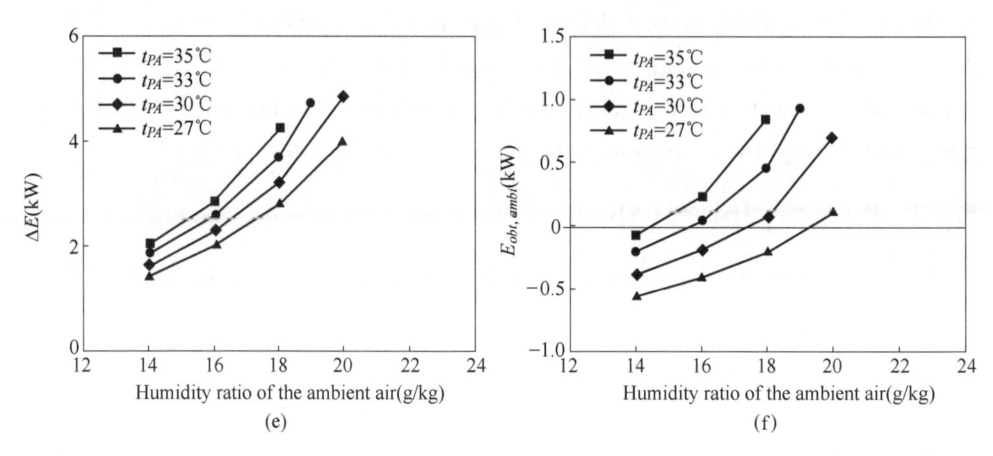

Fig. 6-4 Exergetic performance of BVS under various working conditions

(a) Exergy obtained by the air conditioning space; (b) Exergy efficiency; (c) Exergy provided by the electrical heater; (d) Exergy destruction of the electrical heater; (e) Exergy destruction of heat and mass transfer components; (f) Exergy obtained by the ambient environment

Fig. 6-4 (a) shows that $E_{obt,space}$ stays almost the same when humidity ratio of the supply air is fixed at 10g/kg. The differences are caused by the supply air temperature, referring to Fig. 6-2 (a). For all the cases, exergy efficiencies (η_e) are pretty low, which are in the range of 0.77% and 2.04% as shown in Fig. 6-4 (b). This is due to the large amount of exergy provided by the heater (E_H), shown in Fig. 6-4 (c). Exergy destruction of the electrical heater (ΔE_H) accounts for a significant part of E_H, shown in Fig. 6-4 (d). The rest of E_H is mostly consumed by the exergy destructions of heat and mass transfer components (ΔE), i. e. DW, HR and ST, shown in Fig. 6-4 (e). ΔE_{DW}, ΔE_{HR}, and ΔE_{ST} under different working conditions are shown in Fig. 6-5.

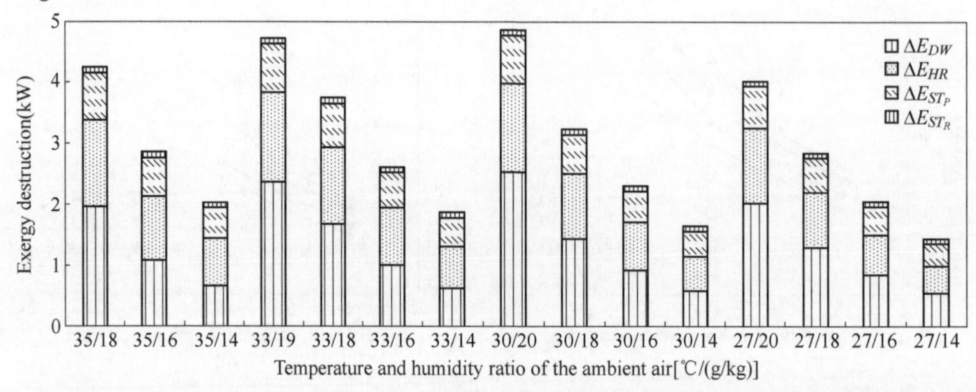

Fig. 6-5 Exergy destruction of each component in BVS under various working conditions

It can be seen that exergy destructions at DW (ΔE_{DW}) and HR (ΔE_{HR}) are responsible for around 70% ~ 80% of the total heat and mass transfer exergy destruction (ΔE). $E_{obt,\ ambi}$, which is negative for the cases when t_{reg} is low as shown in Fig. 6-4 (f), can be neglected as compared with ΔE . The exergy flow charts of BVS under Beijing summer condition and ARI summer condition are shown in Fig. 6-6.

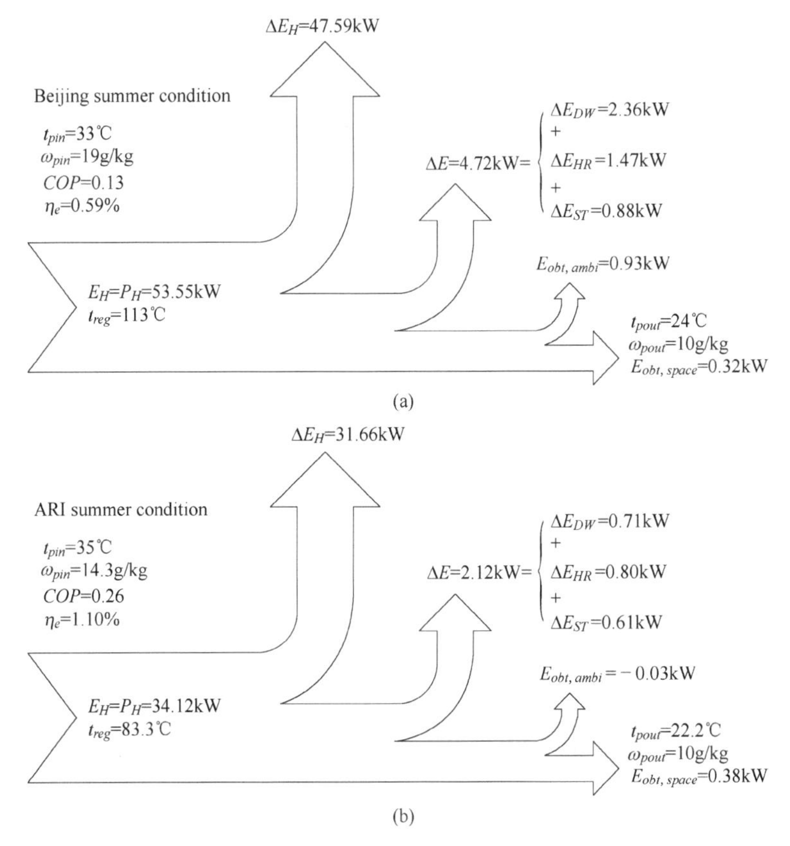

Fig. 6-6 Exergy flow chart of BVS under Beijing and ARI summer conditions

(a) Beijing summer condition; (b) ARI summer condition

The above analysis shows that the key factors influencing the performance, i. e. *COP* and η_e , of BVS are large exergy destructions at the electrical heater (ΔE_H) and at the heat and mass transfer components, i. e. DW (ΔE_{DW}) and HR (ΔE_{HR}) . It should be noticed that ΔE_{ST} is not significant compared with ΔE_{DW} and ΔE_{HR} . Does this mean the impact of ΔE_{ST} can be neglect?

Eq. (6-12) is applied to analyze the exergy destructions of sensible heat transfer components, i. e. HR, and coupled heat and mass transfer components, i. e. DW and *ST*, in BVS. The results are listed in Table 6-3.

Table 6-3 Analysis based on Eq. (6-12) for BVS, IVS1 and IVS2

City		Beijing summer condition					*ARI* summer condition				
System		BVS		IVS1	IVS2		BVS		IVS1		IVS2
COP/improving rate		0.134/0		0.281/1.1	2.485/17.6		0.263/0		0.438/0.67		4.634/16.6
η_e /improving rate		0.59%/0		1.25%/1.1	10.76%/17.1		1.10%/0		1.88%/0.71		20.63%/17.7
—		DW	HR	ST_R	ST_P	HE	DW	HR	ST_R	ST_F	HE
$hF = NTU \cdot \dot{m} \cdot c_p$ (kW/K)	BVS	11.4	2.0	14.4	14.4	—	11.4	2.0	14.4	14.4	—
	IVS1	11.4	2.0	16.6	—	2.0	11.4	2.0	16.6	—	2.0
	IVS2	11.4	2.0	—	—	—	11.4	2.0	—	—	—
Q(kW)	BVS	35.10	31.29	3.36	10.85	—	21.31	22.64	3.36	8.79	—
	IVS1	21.05	20.14	1.54	—	8.12	11.04	14.45	1.98	—	6.19
	IVS2	19.97	18.27	—	—	—	9.95	12.27	—	—	—
M(kg/s)	BVS	11.48×10^{-3}	—	1.35×10^{-3}	4.30×10^{-3}	—	6.93×10^{-3}	—	1.35×10^{-3}	3.50×10^{-3}	—
	IVS1	7.19×10^{-3}	—	3.68×10^{-3}	—		3.45×10^{-3}	—	3.13×10^{-3}	—	
	IVS2	7.19×10^{-3}	—	—	—	—	3.45×10^{-3}	—	—	—	—
ζ_t / ζ_ω	BVS	1.770/1.960	1/–	1.104/1.104	1.106/1.104	–/–	1.396/1.542	1/–	1.104/1.104	1.105/1.104	–/–
	IVS1	1.178/1.229	1/–	1.543/1.357	–/–	1/–	1.158/1.317	1/–	1.608/1.272	–/–	1/–
	IVS2	1.106/1.111	1/–	–/–	–/–	–/–	1.077/1.093	1/–	–/–	–/–	–/–
ΔE (kW)	BVS	2.36	1.47	0.69×10^{-1}	0.82	—	0.71	0.80	0.69×10^{-1}	0.54	—
	IVS1	0.51	0.62	0.21	—	0.05	0.17	0.34	0.12	—	0.06
	IVS2	0.42	0.52	—	—	—	0.11	0.24	—	—	—

Although ΔE_{ST} is not significant as compared with ΔE_{DW} and ΔE_{HR} , ST_p in BVS causes over dehumidification for DW. Under Beijing summer condition and ARI summer condition, the humidity ratios after DW are 4.6g/kg and 5.6g/kg, respectively. When ω_{pout} is 10g/kg, *ODR* are 1.60 and 2.02, respectively. Thus, it can be seen from Table 6-3 that the heat and mass transfer capacities of DW, which are 11.48×10^{-3}kg/s and 35.10kW under Beijing summer condition, and 6.93×10^{-3}kg/s and 21.31kW under ARI summer condition, are larger than required. Similarly, the temperature of PA after DW is higher because of over dehumidification at DW. This enlarges the temperature differences between RA after ST_R and PA after DW, which consequently increases the heat transfer capacity at HR under fixed NTU_{HR}. As a summary, over dehumidification caused by ST_R enlarges heat and mass transfer capacities at DW and HR, leading to higher ΔE_{DW} and ΔE_{HR} .

Unmatched coefficients of DW and HR are the other influencing factors for ΔE_{DW} and ΔE_{HR} . As for sensible heat exchangers, when heat flow capacities ($\dot{m}\cdot c_p$) of the two heat transfer media are the same, temperature differences fields are uniform[12]. In this paper, heat flow capacities of the two heat transfer media for HR are designed the same. Thus, ζ_t for HR equals to 1. Previous research on DW[8] shows that under the same working conditions, evenly divided DW with the mass flow rates of the two streams of air being the same, which is the case for DW in this paper, has the lowest ζ_t and ζ_ω. ζ_t and ζ_ω of DW are listed in Table 6-3. It can be seen that ζ_t and ζ_ω of DW under ARI summer condition are smaller than those under Beijing summer condition. Under the same heat and mass transfer area, will the fields be more uniform when heat and mass transfer capacities are smaller? This question is discussed in the next part when over dehumidification is eliminated.

Exergy analysis shows that the main performance limitation factors of BVS are the over dehumidification at DW caused by direct evaporate cooling and the adoption of low efficient electrical heater. In the next part, improved systems are suggested and their performances are discussed with BVS being the baseline.

6.3 Improved systems for the ventilation system

6.3.1 Avoiding over dehumidification of DW

As mentioned in subsection 7.2.2, DW in BVS has the drawback of over dehumidification, which is due to the isenthalpic cooling process of PA in ST_p. For the first improved ventilation system (IVS1), ST_p is replaced by a water-to-air sensible heat exchanger (HE). Electrical heater is still adopted in IVS1. Fig. 6-7 shows the schematic and air handling processes of IVS1.

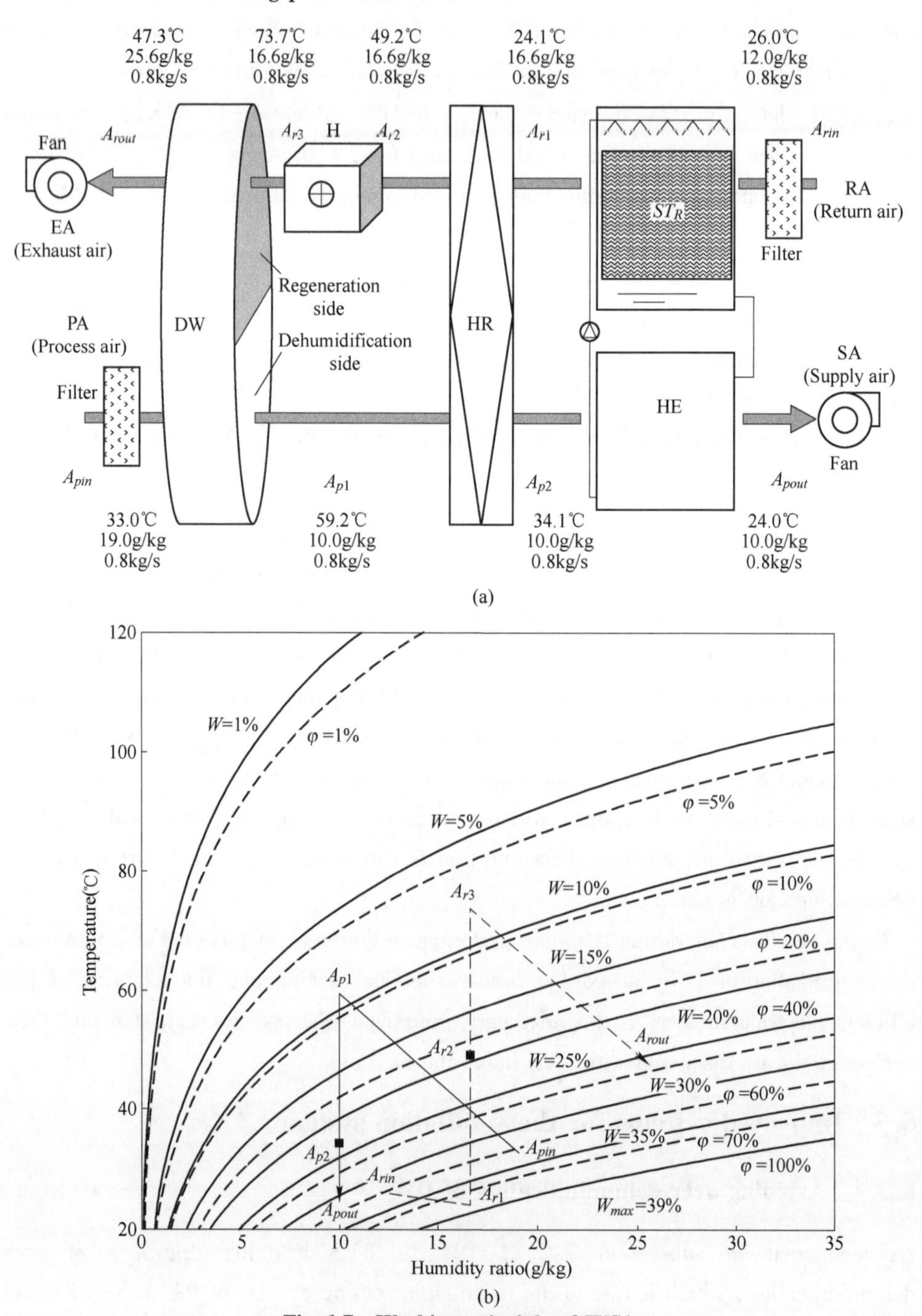

Fig. 6-7 Working principle of IVS1

(a) Schematic of IVS1; (b) Air handling process on psychrometric chart

The water used to cool PA in HE is produced by ST_R. For IVS1, NTU of HE (NTU_{HE}) is designed to be 2.5, and NTU of ST_R (NTU_{ST}) is calculated to be 1.30 for the required supply temperature (24℃) under Beijing summer condition. Parameters of other components are listed in Table 6-2. The air states after each component under Beijing summer condition are shown in Fig. 6-7.

The performances of IVS1 under various process air inlet states are shown in Fig. 6-8. Supply air temperature is lower than 24℃ for all cases. Compared with BVS, regeneration air temperature is significantly reduced. COP and exergy efficiency are increased.

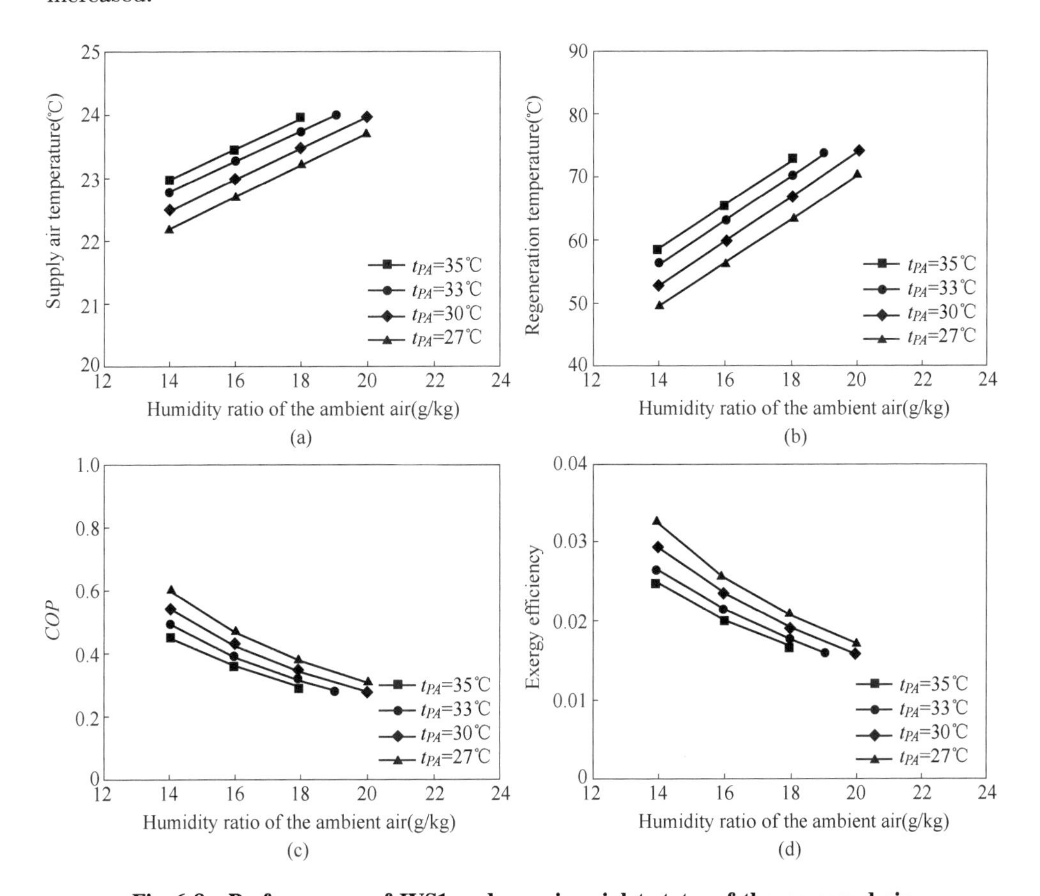

Fig. 6-8 Performances of IVS1 under various inlet states of the processed air

(a) Supply air temperature; (b) Regeneration temperature; (c) COP; (d) Exergy efficiency

Detailed comparisons with BVS can be referred to Table 6-3 and Fig. 6-9 under Beijing summer condition and ARI summer condition.

Table 6-3 shows that ODR of IVS1 equals to 1, meaning that over dehumidification of

DW is eliminated. Consequently, the heat and mass transfer capacities in DW and HR are reduced as compared with those of BVS. Besides, ζ_t and ζ_ω of DW are significantly reduced without over dehumidification. This agrees with the previous result that smaller heat and mass transfer capacity is beneficial for improving the uniformity of temperature and humidity ratio differences fields. Therefore, ΔE_{DW} and ΔE_{HR} are greatly reduced as compared with those of BVS. As a result, η_e and COP are increased and t_{reg} is reduced to 73.7℃ and 59.6℃ under Beijing summer condition and ARI summer condition, respectively, as shown in Fig. 6-9. Fig. 6-9 shows that exergy destruction of the electrical heater (ΔE_H) is still the largest part, which is 17 and 26 times larger than the heat and mass transfer exergy destruction (ΔE) under Beijing summer condition and ARI summer condition for IVS1, respectively. When t_{reg} is reduced, low-grade and highly efficient heating sources can be adopted. In the next subsection, the adoption of a heat pump system in desiccant cooling systems is discussed.

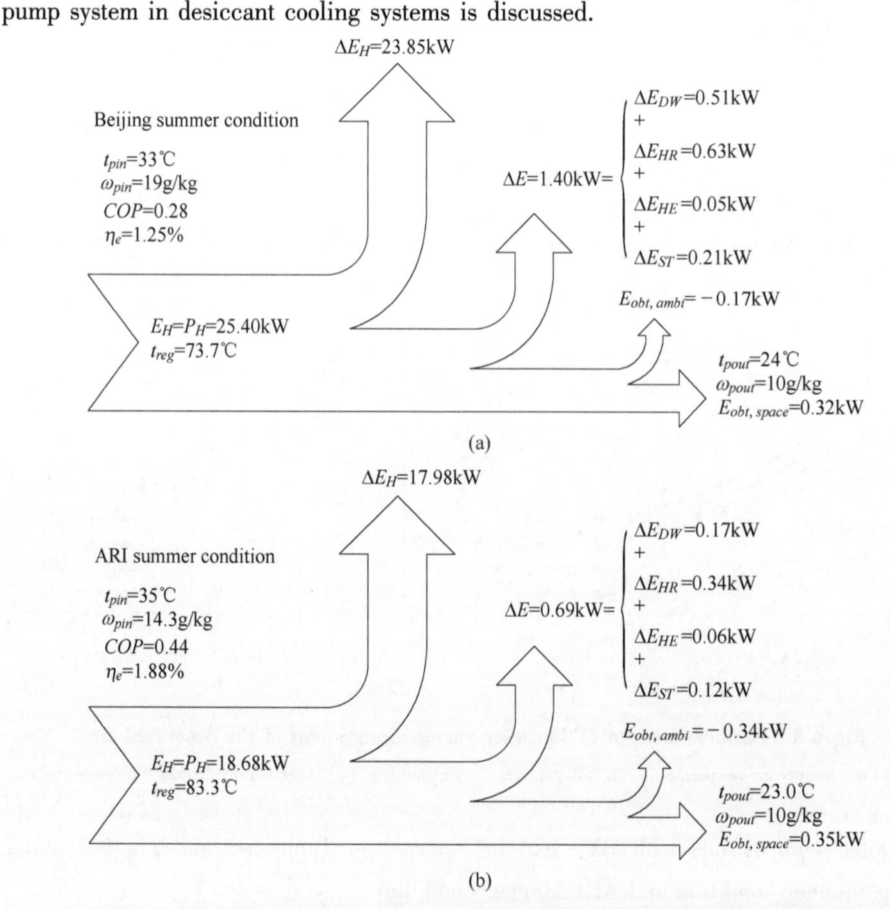

Fig. 6-9 Exergy flow chart of IVS1 under Beijing and ARI summer conditions

(a) Beijing summer condition; (b) ARI summer condition

6.3.2 Adopting the heat pump as the heating source

In this subsection, performances of the second improved ventilation system (IVS2), which uses a heat pump (HP) to replace the electrical heater, are discussed. The schematic of IVS2 can be referred to in Fig. 6-10 (a). The evaporator of HP is used to

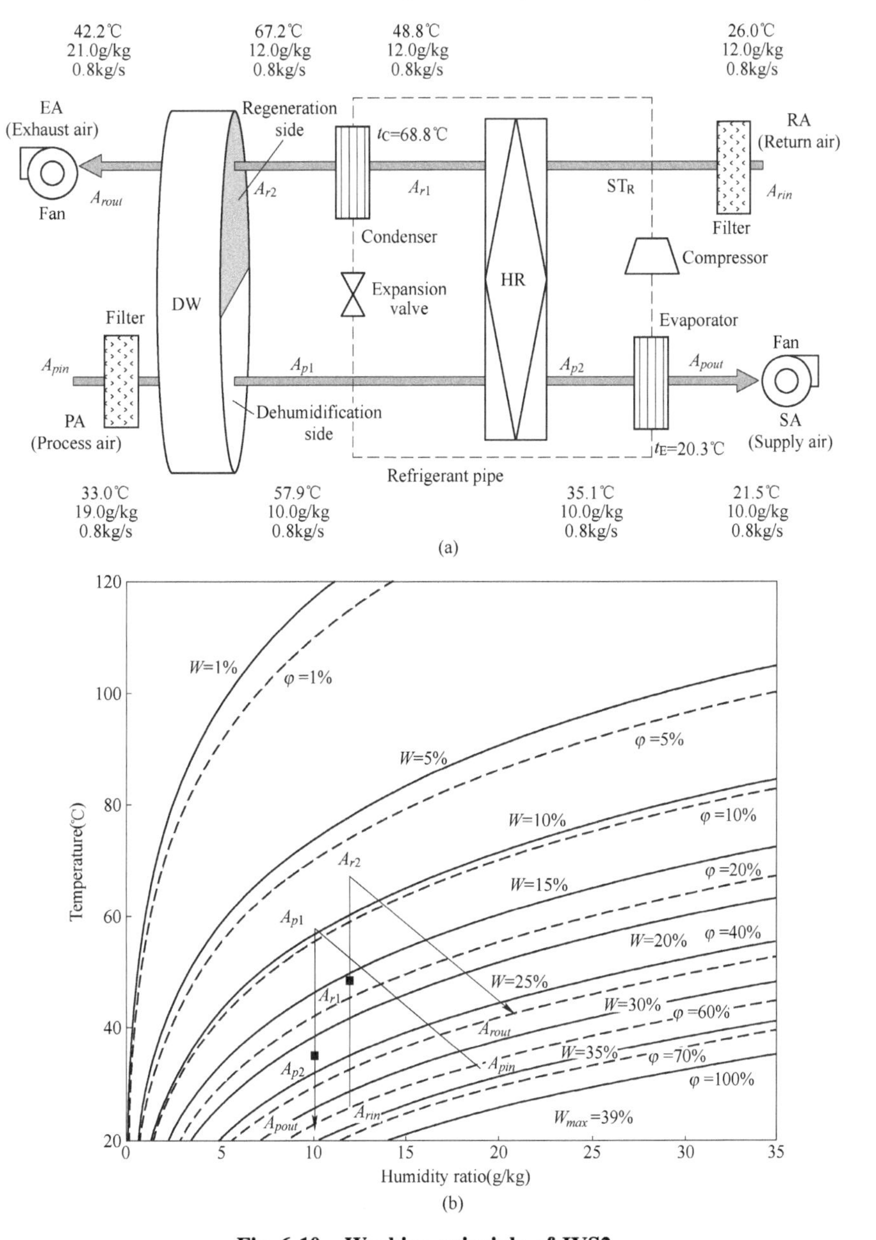

Fig. 6-10 Working principle of IVS2

(a) Schematic of IVS2; (b) Air handling process on psychrometric chart cool the process air and the condenser of HP is used to heat the regeneration air. Therefore, heat of the process air is recovered to heat the regeneration air under the assistant of the compressor, which reduces power input compared with that of the electrical heater in BVS and IVS1. Besides, this system is of more compact structure.

For IVS2, *COP* and exergy efficiency (η_e) can be written as Eqs. (6-14) ~ (6-15):

$$COP = \frac{\dot{m}_p(i_{space} - i_{pout})}{P_{HP}} \tag{6-14}$$

$$\eta_e = \frac{\dot{m}_p(e_{pout} - e_{space})}{E_{HP}} \tag{6-15}$$

From the perspective of exergy destructions, exergy provided by the heat pump (E_{HP}, which equals to P_{HP}) can be written as Eq. (6-16):

$$E_{HP} = P_{HP} = \Delta E_{HP} + \Delta E + E_{obt,\ ambi} + E_{obt,\ space} \tag{6-16}$$

Similar with Eq. (6-9), ΔE in Eq. (6-16) is the total exergy destruction of DW and HR; ΔE_{HP} is the exergy destruction of the heat pump, expressed as Eq. (6-17):

$$\begin{aligned} \Delta E_{HP} &= P_{HP} - \dot{m}_p(e_{pout} - e_{p2}) - \dot{m}_r(e_{r2} - e_{r1}) \\ &= \Delta E_E + \Delta E_C + T_0(Q_C/T_C - Q_E/T_E) \end{aligned} \tag{6-17}$$

where $\dot{m}_p(e_{pout} - e_{p2})$ and $\dot{m}_r(e_{r2} - e_{r1})$ are the exergy obtained by PA and RA, respectively, from the evaporator and the condenser, respectively; ΔE_C and ΔE_E are exergy destructions of the condenser and the evaporator, respectively, which are caused by heat transfer processes between the air and the refrigerant. ΔE_C and ΔE_E can be expressed in the form of Eq. (6-12) or as Eq. (6-18):

$$\Delta E_{E(C)} = \dot{m}_{p(r)}\, c_{pa} T_0 \left[\ln \frac{T_{pout(r2)}}{T_{p2(r1)}} - \frac{T_{pout(r2)} - T_{p2(r1)}}{T_{E(C)}} \right] \tag{6-18}$$

$T_0\left(\frac{Q_C}{T_C} - \frac{Q_E}{T_E}\right)$ is the exergy destruction caused by the thermodynamic efficiency (ε) of the heat pump being less than 1. ε is defined as the real *COP* of the heat pump system divided by the Carnot *COP*, shown in Eq. (6-19):

$$\varepsilon = \frac{COP_{HP}}{COP_{Carnot}} = \frac{\dfrac{Q_C}{P_{HP}}}{\dfrac{T_C}{T_C - T_E}} \tag{6-19}$$

The performances of IVS2 are simulated based on the parameters listed in Table 6-3. *NTU* of the evaporator (NTU_E) and the condenser (NTU_C) are designed to be 2.5. ε is fixed at 0.5. Under Beijing summer condition, the air states after each component and the air handling processes drawn on the psychrometric chart are shown in Fig. 6-10 (b).

Performances under various inlet states of the process air are shown in Fig. 6-11. Detailed results from exergy analysis and comparisons with BVS and IVS1, under Beijing summer condition and ARI summer condition, are shown in Fig. 6-12 and Table 6-3.

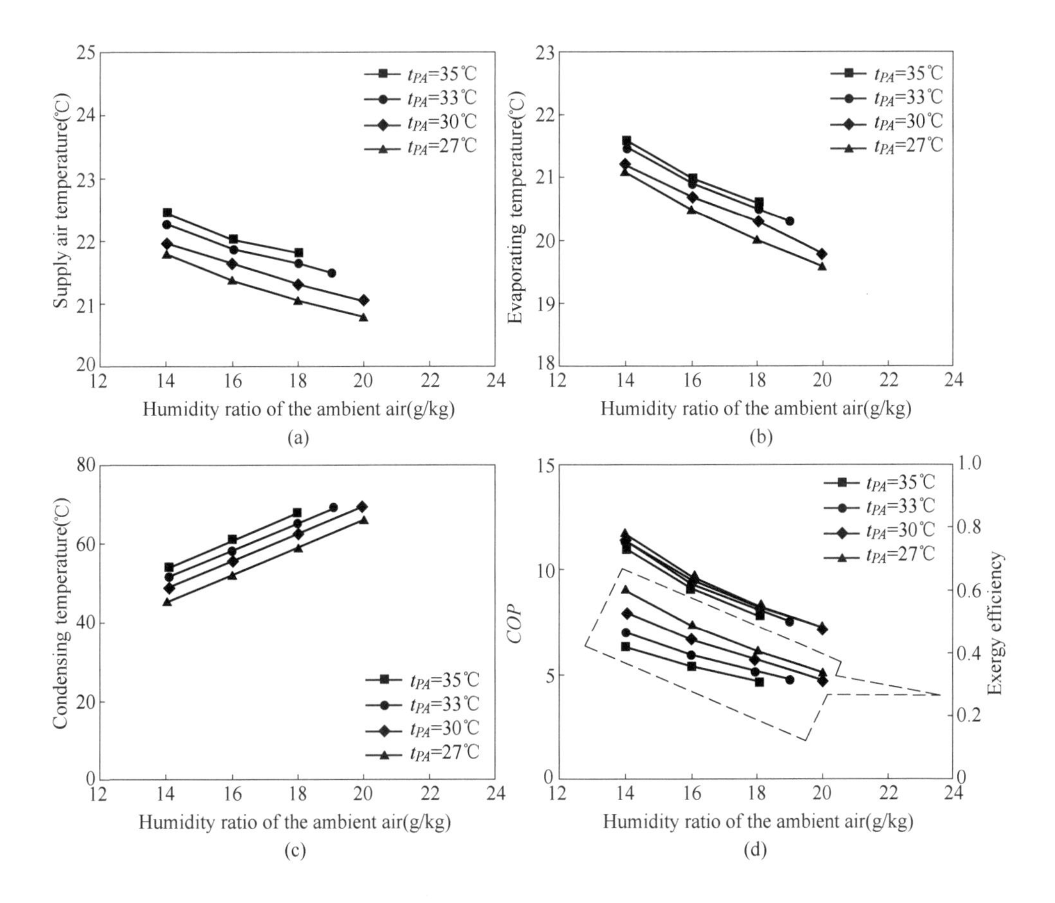

Fig. 6-11 Performances of IVS2 under various inlet states of the processed air

(a) Supply air temperature; (b) Evaporating temperature;

(c) Condensing temperature; (d) *COP* and exergy efficiency

It can be seen in Fig. 6-12 that ΔE_{DW} and ΔE_{HR} of IVS2 are decreased as compared with BVS and IVS1, which contribute to the increase of *COP* and η_e. However, the main improvement is the reduction of ΔE_{HP}, compared with ΔE_H for BVS and IVS1. This significantly decreases the power input of the compressor (P_{HP}) as compared with that of the electrical heater (P_H).

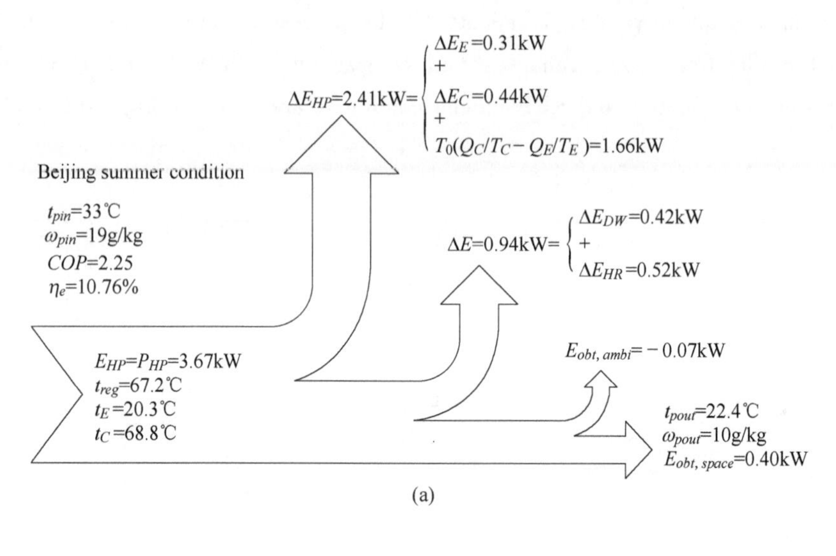

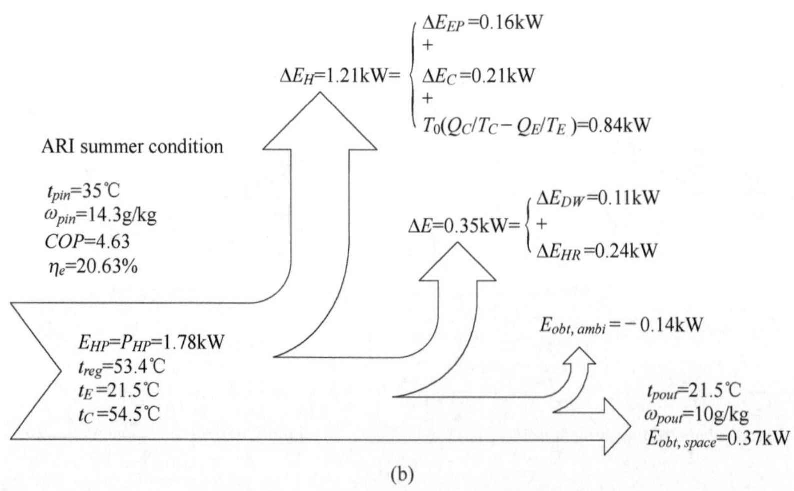

Fig. 6-12 Exergy flow chart of IVS2 under Beijing and ARI summer conditions

(a) Beijing summer condition; (b) ARI summer condition

6.3.3 Performance comparison of the three systems

Performance comparisons among the three systems are shown in Figs. 6-13~6-14, which are based on the same humidity ratio of the supply air (10g/kg) under Beijing summer condition and ARI summer condition, respectively.

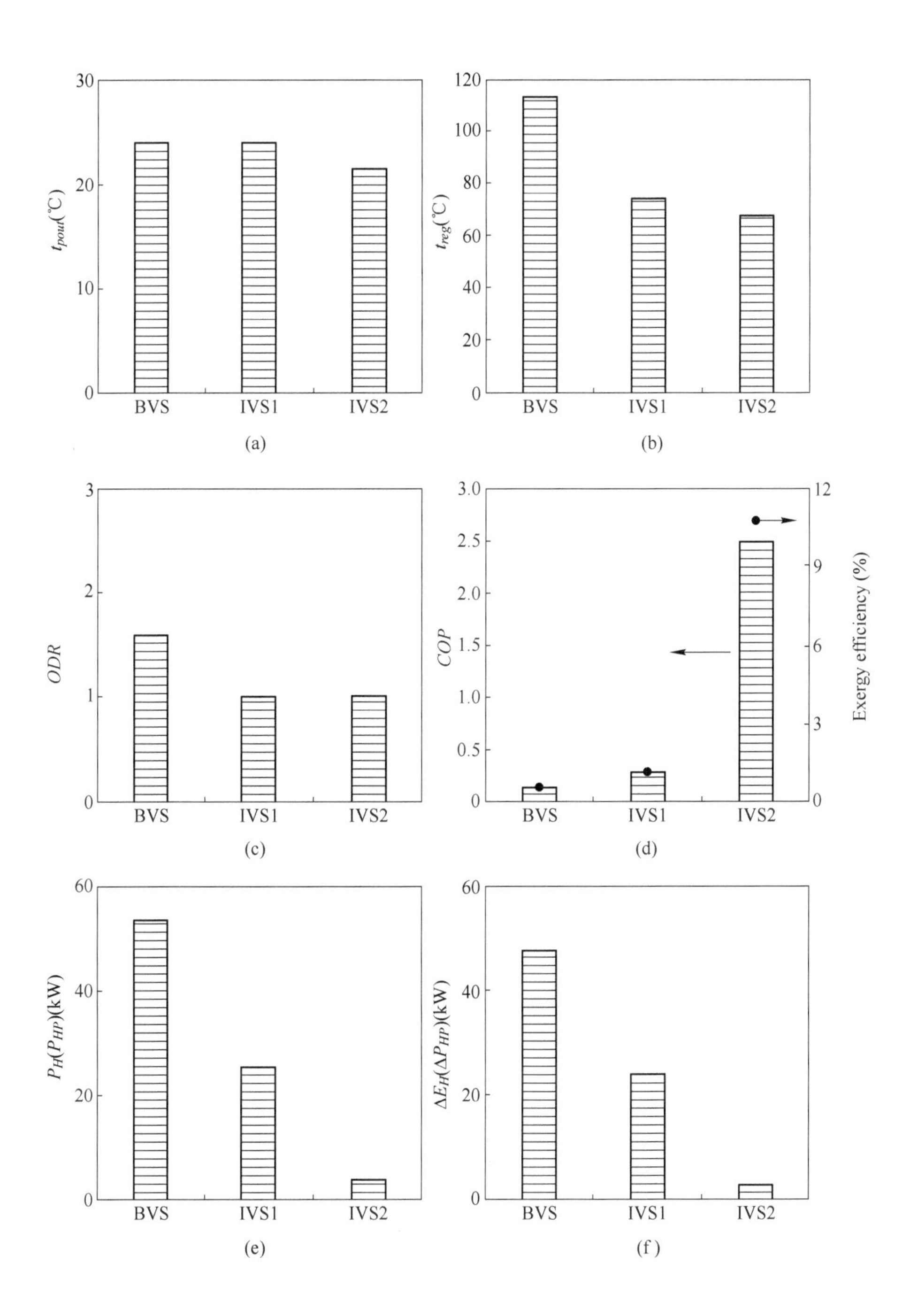

30
20
10
0
t_{pout}(℃)
BVS
IVS1
IVS2
(a)
120
100
80
60
40
20
0
t_{reg}(℃)
BVS
IVS1
IVS2
(b)
3
2
1
0
ODR
BVS
IVS1
IVS2
(c)
3.0
2.5
2.0
1.5
1.0
0.5
0
COP
12
9
6
3
0
Exergy efficiency (%)
BVS
IVS1
IVS2
(d)
60
40
20
0
$P_H(P_{HP})$(kW)
BVS
IVS1
IVS2
(e)
60
40
20
0
$\Delta E_H(\Delta P_{HP})$(kW)
BVS
IVS1
IVS2
(f)

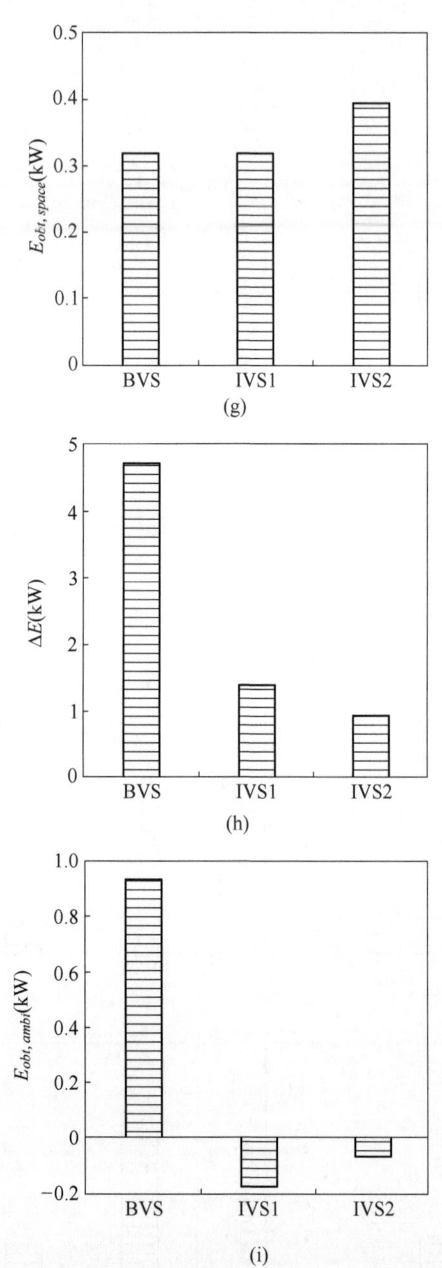

Fig. 6-13 Comparison among BVS, IVS1 and IVS2 under Beijing summer condition

(a) Supply air temperature; (b) Regeneration temperature; (c) Over dehumidification ratio;
(d) *COP* and exergy efficiency; (e) Power consumptions of the electrical heater or the heat pump system;
(f) Exergy destructions of the electrical heater or the heat pump system;
(g) Exergy obtained by the air conditioning space;
(h) Heat and mass transfer exergy destructions; (i) Exergy obtained by the ambient air

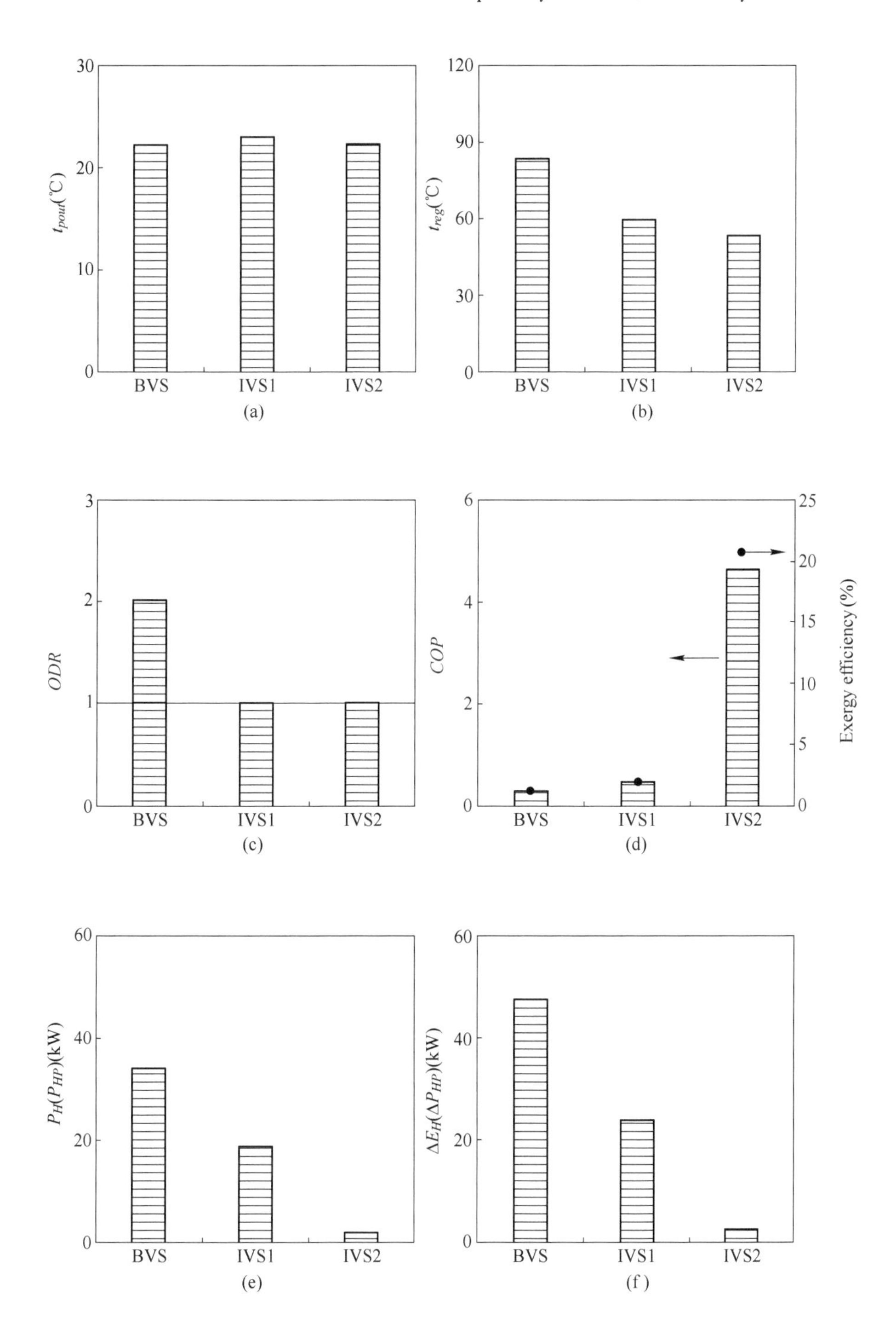

30
20
10
0
t_{pout}(℃)
BVS
IVS1
IVS2
(a)
120
90
60
30
0
t_{reg}(℃)
BVS
IVS1
IVS2
(b)
3
2
1
0
ODR
BVS
IVS1
IVS2
(c)
6
4
2
0
COP
25
20
15
10
5
0
Exergy efficiency (%)
BVS
IVS1
IVS2
(d)
60
40
20
0
$P_H(P_{HP})$(kW)
BVS
IVS1
IVS2
(e)
60
40
20
0
$\Delta E_H(\Delta P_{HP})$(kW)
BVS
IVS1
IVS2
(f)

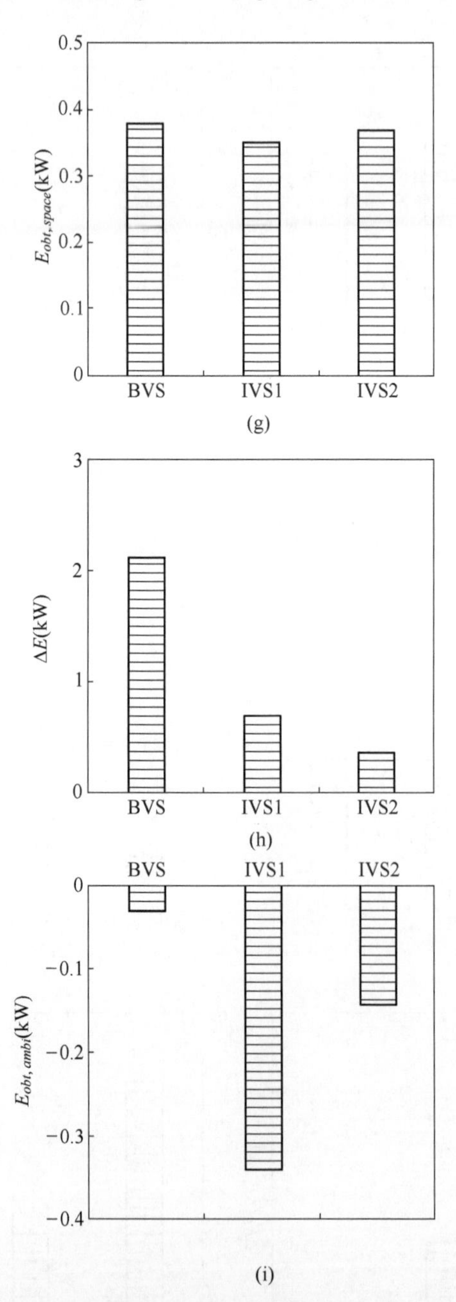

Fig. 6-14 Comparison among BVS, IVS1 and IVS2 under ARI summer condition

(a) Supply air temperature; (b) Regeneration temperature;
(c) Over dehumidification ratio; (d) *COP* and exergy efficiency;
(e) Power consumptions of the electrical heater or the heat pump system;
(f) Exergy destructions of the electrical heater or the heat pump system;
(g) Exergy obtained by the air conditioning space; (h) Heat and mass transfer exergy destructions;
(i) Exergy obtained by the ambient air

It can be seen that the supply air temperature (t_{pout}) and exergy obtained by the air conditioning space ($E_{obt,space}$) are almost the same for different systems. However, the required regeneration temperature (t_{reg}) and COP vary greatly. With over dehumidification eliminated, t_{reg} of IVS1 and IVS2 are 39.3℃ and 45.8℃ respectively lower than that of BVS under Beijing summer condition, and are 23.7℃ and 29.9℃ respectively lower than that of BVS under ARI summer condition. COP of IVS1 is 0.28 and 0.44 while that for BVS is 0.13 and 0.26, under Beijing summer condition and ARI summer condition, respectively. When the electrical heater is replaced by the heat pump in IVS2, COP is increased to 2.49 and 4.63 under Beijing summer condition and ARI summer condition, respectively. The improving rate of COP and η_e for IVS1 and IVS2, defined as the performance improvement compared with BVS divided by the performance of BVS, are listed in Table 6-3.

It is found based on exergy analysis that the lower ΔE and ΔE_H, the lower t_{reg} and higher COP and η_e will be. ΔE is influenced by ODR and the unmatched coefficients. For IVS1, the direct evaporative cooler (ST_P in BVS) is replaced by a sensible heat exchanger (HE), and ODR reduces from 1.6 and 2.0 for BVS to 1, as seen in Figs. 6-13~6-14. At the same time, decreased unmatched coefficients for DW in IVS1 are observed, too. Therefore, ΔE is decreased, leading to lower $E_H(P_H)$, lower t_{reg} and higher COP and η_e for IVS1. For IVS2, the electrical heater and cooling devices, i.e. HE and ST_R, in IVS1 are replaced by a heat pump system. ΔE_{HE} and ΔE_{ST} no longer exist for IVS2. Besides, as seen in Figs. 6-13 ~ 6-14, exergy destruction of HP (ΔE_{HP}) in IVS2 is significantly reduced as compared with ΔE_H in IVS1, leading to a dramatic reduction of E_{HP} (P_{HP}) and an apparent enhancement of COP and η_e.

6.4 Conclusions

This paper analyzed the performances of ventilation systems with desiccant wheel cooling from exergy destruction point of view. The inherent performance influencing factors for ventilation systems are identified. Based on this, two improved systems with enhanced performances are proposed. The main conclusions are as follows:

(1) Exergy analysis of BVS shows that performances of desiccant wheel cooling systems are mainly influenced by exergy destructions of the heater (ΔE_H) and the heat and mass transfer components (ΔE). ΔE is inherently influenced by heat and mass transfer capacities (Q and M), heat and mass transfer areas (hF and h_mF) and unmatched coefficients. When geometrical dimensions of components are fixed, ΔE of desiccant wheel cooling systems is influenced by ODR and unmatched coefficients.

(2) BVS has drawbacks of over dehumidification which is caused by the direct evaporate cooling. Over dehumidification leads to larger Q and M for DW, larger Q for HR, and high unmatched coefficients for DW. This is the reason for large ΔE and high t_{reg} of BVS. Therefore, BVS has relatively low COP and η_e.

(3) IVS1, with ST_P in BVS being replaced by HE, has no over dehumidification issue. With the reduction of Q and M for DW, Q for HR and unmatched coefficients for DW, ΔE and t_{reg} of IVS1 are decreased. The reduction of ΔE and t_{reg} lead to lower E_H (P_H) and higher COP and η_e.

(4) For IVS2, the electrical heater and cooling devices, i. e. HE and ST_R are replaced by a heat pump system, which recovers heat from the process air to heat the regeneration air. The reductions of ΔE and t_{reg} are not obvious as compared with IVS1. However, the reduction of exergy destruction at HP is significant as compared with IVS1, resulting in a great reduction of E_{HP}(P_{HP}) and a large enhancement of COP and η_e.

(5) For efficient system design, it is essential to avoid over dehumidification by adopting sensible heat exchangers rather than the direct evaporative cooler, and adopt high efficient heat sources such as heat pump systems.

Nomenclature

A	air
BVS	basic ventilation system
COP	coefficient of performance
e	specific exergy, kJ/kg
E	exergy rate, kW
ΔE	exergy destruction rate, kW
EA	exhaust air
IVS	improved ventilation system
$\dot{m}$	mass flow rate, kg/s
M	mass transfer capacity, kg/s
NTU	number of heat transfer units
ODR	overdehumidification ratio, dimensionless
P_a	standard atmospheric pressure, Pa
P	power consumption, kW

PA	process air
Q	heat transfer capacity, kW
R_a	gas constant for air, kJ/(mol · K)
RA	return air
SA	supply air
t	Celsius temperature, ℃
T	Kelvin temperature, K
u	velocity, m^2/s
W	water content, dimensionless

Greek symbols

ω	humidity ratio, g/kg
η_e	exergy efficiency
ζ	unmatched coefficient
ρ	density, kg/m^3
φ	relative humidity ratio of the humid air
τ	time, s
ε	thermodynamic efficiency of the heat pump system

Subscripts

a	air
C	condenser
E	evaporator
s	solid
w	water
p	process air
r	regeneration air
in	inlet
out	outlet
t	temperature
ω	humidity ratio
reg	regeneration

HE	heat exchanger
DW	desiccant wheel
HR	heat recovery unit
HP	heat pump
ST	spray tower
H	heater
obt	obtain
max	maximum
ambi	ambient air
space	air conditioning space
0	dead state

References

[1] S Jain, P L Dhar, S C Kaushik. Evaluation of solid-desiccant-based evaporative cooling cycles for typical hot and humid climates [J]. Int. J. Refrig, 1995, 18 (5): 287-296.

[2] D La, Y Li, Y J Dai, T S Ge, R Z Wang. Development of a novel rotary desiccant cooling cycle with isothermal dehumidification and regenerative evaporative cooling using thermodynamic analysis method [J]. Energy, 2012, 44 (1): 778-791.

[3] D La, Y Li, Y J Dai, T S Ge, R Z Wang. Effect of irreversible processes on the thermodynamic performance of open-cycle desiccant cooling cycles [J]. Energy Conver. Manage, 2013, 67: 44-56.

[4] L Zhang, X H Liu, Y Jiang. Exergy analysis of parameter unmatched characteristic in coupled heat and mass transfer between humid air and water [J]. Int. J. Heat and Mass Transfer, 2015, 84: 327-338.

[5] R Tu, X H Liu, Y Jiang. Performance analysis of a two-stage desiccant cooling system [J]. Appl. Energy, 2014, 113 (1): 1562-1574.

[6] J P Holman. Heat transfer [M]. 10th ed. New York: McGraw-Hill Companies, 2009.

[7] A Bejan. Advanced engineering thermodynamics [M]. 3rd ed. Hoboken NJ: John Wiley & Sons, 2006.

[8] R Tu,X H Liu, Y Jiang. Lowering the regeneration temperature of a rotary wheel dehumidification system using exergy analysis [J]. Energy Conver. and Managet, 2015, 89 (1): 162-174.

[9] Z Y Guo, S Q Zhou, Z X Li, L G Chen. Theoretical analysis and experimental confirmation of the uniformity principle of temperature difference field in heat exchanger [J]. Int. J. Heat and Mass Transfer, 2002, 45 (10): 2117-2119.

[10] T Zhang, X H Liu, L Zhang, Y Jiang. Match properties of heat transfer and coupled heat and mass transfer processes in air-conditioning system [J]. Energy Conver. Manage, 2012, 59:

103-113.

[11] T Zhang, X H Liu, L Zhang, Y Jiang. Performance comparison of liquid desiccant air handling processes from the perspective of match properties [J]. Energy Conver. Manage, 2013, 75: 51-60.

[12] R Tu, X H Liu, Y Jiang, F Ma. Influence of the number of stages on the heat source temperature of desiccant wheel dehumidification systems using exergy analysis [J]. Energy, 2015, 85: 379-391.

Chapter 7 Performance Analyses of an Advanced System with Single-Stage Desiccant Wheel

Normally, the humidity ratio of indoor air is lower than that of the ambient air. When indoor air is used as the regeneration air (A_{reg}), and ambient air is used as the processed air (A_{deh}), a significant heat recovery potential exists between A_{reg} and A_{deh}. This can be realized using enthalpy wheels. This chapter presents a novel outdoor air handling unit based on traditional outdoor air dehumidification systems, the dehumidification wheel of which is evenly divided into a dehumidification wheel and an enthalpy wheel, without the additional investment of the desiccant wheels. The energy-saving potentials of the novel system were studied under different ambient conditions and desiccant wheel thickness. Through exergy analysis, the main exergy destruction is determined, providing directions for performance improvement. The performance of this system is compared with that of traditional outdoor air dehumidification systems from t_{reg}, energy efficiency (η_{th}), exergy efficiency (η_{ex}), and system COP(COP_{sys}). Efficient system design parameters and operation instructions for the proposed system are summarized for practical applications.

7.1 Methodology

7.1.1 Working principles of the proposed dehumidification systems

Schematics of a typical traditional outdoor air dehumidification system (traditional system: TS) and the advanced outdoor air dehumidification system (advanced system: AS) are shown in Fig. 7-1. Both systems are used to dehumidify and cool the outdoor air, which is called dehumidification air (A_{deh}). Indoor air is used to regenerate the dehumidification wheel, which is called the regeneration air (A_{reg}). Fig. 7-2 shows typical air handling processes of both systems on the psychrometric chart under the Beijing (China) summer working condition (BSC: $t_{deh,\ in}$ = 33℃, $\omega_{deh,\ in}$ = 19g/kg)[1], which represents a humid climate, and the Washington (USA) summer working condition (WSC: $t_{deh,\ in}$ = 33℃, $\omega_{deh,\ in}$ = 14g/kg)[1], which represents a mild climate. Temperature and humidity of the indoor air ($A_{reg,in}$) are $t_{reg,\ in}$ = 26℃ and $\omega_{reg,\ in}$ = 12g/kg. Air handling processes were calculated based on the parameters listed in Table 7-1.

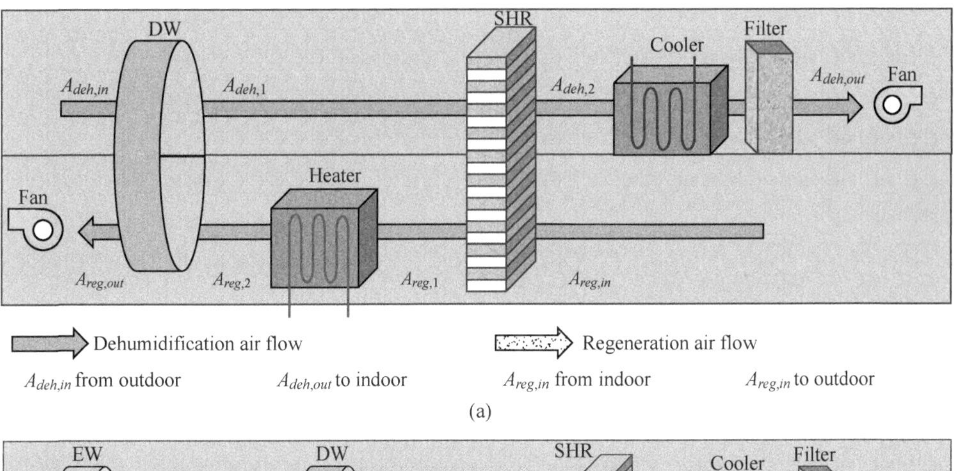

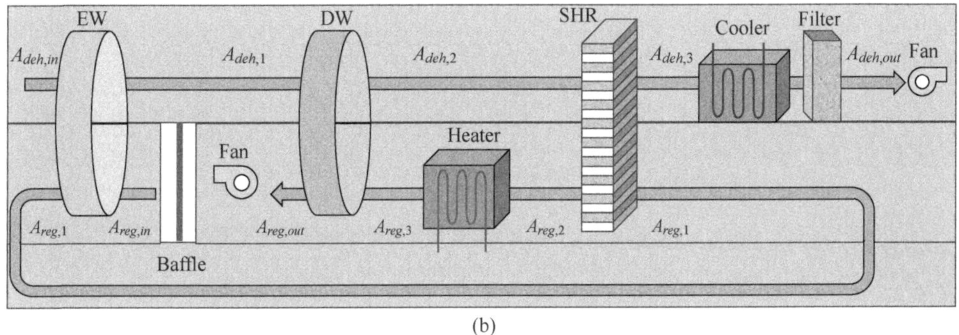

(b)

Fig. 7-1 Schematic of the two outdoor air handling units
(a) Schematic of a typical traditional dehumidification system (TS);
(b) Schematic of the advanced system (AS)

The TS, as shown in Fig. 7-1 (a), comprises a dehumidification wheel (DW), sensible heat recovery unit (SHR), and cooling/heating units. A_{deh} enters the DW to be dehumidified to the required supply air humidity ratio ($A_{deh,in}-A_{deh,1}$). Meanwhile, its temperature increased during dehumidification. Subsequently, it passes the SHR, where it is cooled by A_{reg} ($A_{deh,1}-A_{deh,2}$). Finally, it is cooled by the cooler to the required supply air temperature ($A_{deh,2}-A_{deh,out}$) and introduced into the occupant room. In the case of A_{reg}, heat is first exchanged with the A_{deh} in the SHR, where its temperature is increased ($A_{reg,in}-A_{reg,1}$). Subsequently, A_{reg} is further heated to the required regeneration temperature (t_{reg}) by the heater ($A_{reg,1}-A_{reg,2}$). Finally, A_{reg} enters the DW to dry the DW ($A_{reg,2}-A_{reg,out}$). As shown in Fig. 7-1 (b), the DW in the TS is divided into a DW and an enthalpy wheel (EW). A_{deh} and A_{reg} firstly exchange humid and sensible heat in the

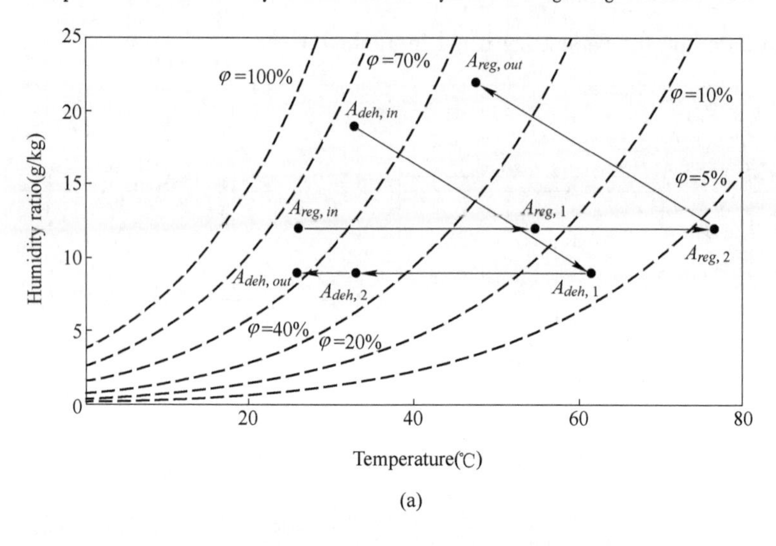
Humidity ratio(g/kg)
Temperature(℃)
φ=100%
φ=70%
$A_{reg, out}$
φ=10%
$A_{deh, in}$
φ=5%
$A_{reg, in}$
$A_{reg, 1}$
$A_{reg, 2}$
$A_{deh, out}$
$A_{deh, 2}$
$A_{deh, 1}$
φ=40%
φ=20%
0
5
10
15
20
25
20
40
60
80

(a)

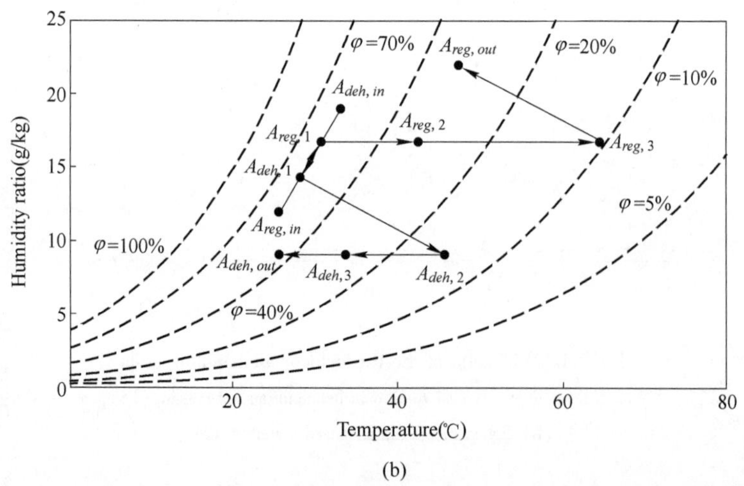
Humidity ratio(g/kg)
Temperature(℃)
φ=70%
$A_{reg, out}$
φ=20%
φ=10%
$A_{deh, in}$
$A_{reg, 1}$
$A_{reg, 2}$
$A_{reg, 3}$
$A_{deh, 1}$
φ=5%
$A_{reg, in}$
φ=100%
$A_{deh, out}$
$A_{deh, 3}$
$A_{deh, 2}$
φ=40%
0
5
10
15
20
25
20
40
60
80

(b)

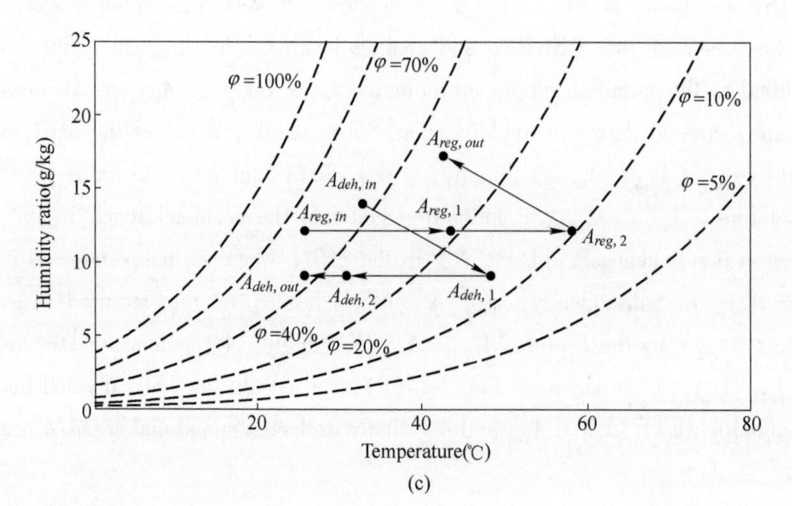
Humidity ratio(g/kg)
Temperature(℃)
φ=100%
φ=70%
φ=10%
$A_{reg, out}$
φ=5%
$A_{deh, in}$
$A_{reg, in}$
$A_{reg, 1}$
$A_{reg, 2}$
$A_{deh, out}$
$A_{deh, 2}$
$A_{deh, 1}$
φ=40%
φ=20%
0
5
10
15
20
25
20
40
60
80

(c)

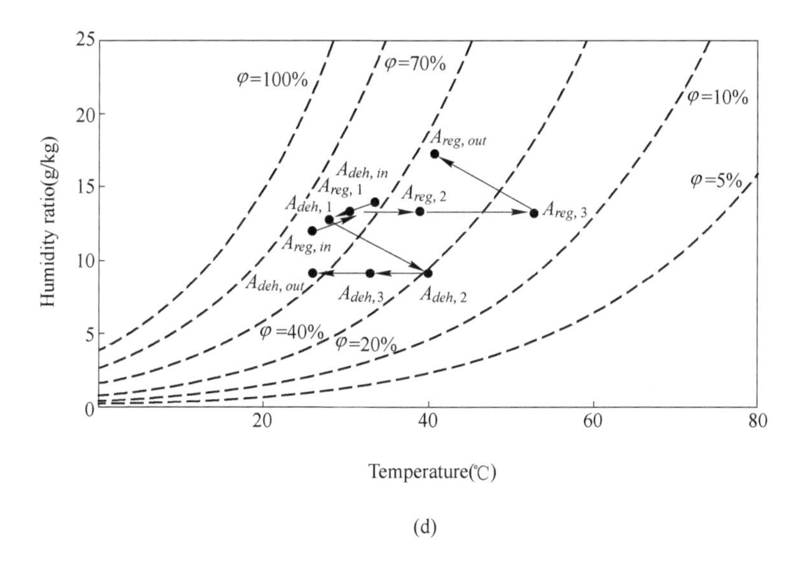

(d)

Fig. 7-2 Air handing processes of A_{deh} and A_{reg} for AS and TS under BSC and WSC

(a) Air handling process of TS under BSC; (b) Air handling process of AS under BSC;

(c) Air handling process of TS under WSC; (d) Air handling process of AS under WSC

EW, where A_{deh} is pre-cooled and pre-dehumidified (A_{deh}: $A_{deh,in}-A_{deh,1}$) and A_{reg} is heated and humidified (A_{reg}: $A_{reg,in}-A_{reg,1}$). After leaving the EW, the following air handling processes were the same as those of the TS.

Fig. 7-2 shows that, when the humidity ratio of supply air ($A_{deh,out}$) is 9g/kg, under BSC, t_{reg} values of TS and AS are 76.5℃ and 64.5℃, respectively, and the difference is 12℃. Under WSC, t_{reg} of TS and AS are 58.2℃ and 52.7℃, respectively, and the difference is 5.5℃. t_{reg} of AS is lower than that of TS under both working conditions, which is beneficial for energy saving in heating and cooling systems. However, it can also be found that differences in t_{reg} between the two systems are smaller in dry ambient conditions.

Table 7-1 Parameters of the two systems

Dehumidification wheel (DW):
Air channel: sinusoidal shape, 2mm high and 2mm wide, Nu = 2.463
Mass flow rates of dehumidification air (m_{deh}) and regeneration air (m_{reg}): m_{deh} = 0.89kg/s, m_{reg} = 0.89kg/s, $Fr = m_{deh} / m_{reg}$ = 1 : 1
The wheel's radius is 0.5m
The wheel's thickness (L_{DW}) is 0.2m and 0.1m for the TS and for the AS, respectively
The wheel's rotation speed (RS) is 25r/h and 35r/h for the TS and the AS, respectively

Continued Table 7-1

Enthalpy wheel (EW):
Air channel: sinusoidal shape, 2mm high and 2mm wide, Nu = 2.463 m_{deh} = 0.89kg/s, m_{reg} = 0.89kg/s, Fr = 1 : 1 The wheel's radius is 0.5m The wheel's thickness (L_{EW}) is 0.1m The wheel's rotation speed (RS) is 1000r/h
Sensible heat recovery (SHR) and Heat pump system (HP):
Heat exchange efficiency (η) of SHR = 80% Utilization factor: ε = 0.5, the NTU of condenser and evaporator = 2 Heat exchange efficiency (η) of condenser and evaporator = 86.5%
Working conditions:
Supply air ($A_{deh,out}$): 9g/kg; Return air ($A_{reg,in}$): 26℃, 12g/kg Ambient air: temperature (t_{ambi}) is from 26℃ to 33℃, and humidity ratio (ω_{ambi}) is from 15g/kg to 23g/kg Mass flow rates of dehumidification air (m_{deh}): m_{deh} = 0.89kg/s Mass flow rates of regeneration air (m_{reg}): m_{reg} = 0.89kg/s

7.1.2 Performance evaluation indexes

The performance of the above dehumidification systems is related to the types of heating and cooling systems. With the reduction in t_{reg}, heat pumps driven by the vapor compression cycle can be effectively adopted. It has been reported that with the adoption of HP and SHR, the performance of the single-stage desiccant wheel dehumidification system is close to that of dehumidification systems with multi-stage desiccant wheels[2,3]. In this study, a heat pump (HP) was used to provide both heating and cooling for both systems. The evaporator of HP can be used as the cooler in Fig. 7-1, and the condenser of HP can be used as the heater in Fig. 7-1.

In this study, performances of both systems were compared based on simulation and theoretical analysis. The main components of both systems were DW, EW, SHR and HP. For SHR and HP, the heat transfer efficiency-NTU method[4] can be used to accurately calculate the outlet parameters of air. In the case of DW and EW, the one-dimensional-double-diffusion coupled heat and mass transfer model [2,5] is adopted. This model has been widely used in previous studies[1-9]. Details of the mathematical model and experimental validations of enthalpy wheels and dehumidification wheels can be found in studies by R. Tu[1,5,7]. The main parameters of DW and EW are shown in

Table 7-1. The rotation speeds of the DW are selected based on the wheel thickness, which is approximately the optimal value[1]. Using the above mathematical model, the outlet parameters of the air after each component can be calculated.

The performances of both systems were evaluated using t_{reg}, energy efficiency(η_{th}), coefficient of performance of the heat pump (COP_{HP}), coefficient of performance of the dehumidification system (COP_{sys}), and exergy efficiency of the dehumidification system (η_{ex}). These performance indices can be calculated using Eqs. (7-1) ~ (7-4).

$$\eta_{th} = \frac{Q_{deh,\ t}}{Q_{cond}} = \frac{m_{deh} \times (h_{deh,\ in} - h_{deh,\ out})}{Q_{cond}} \tag{7-1}$$

$$COP_{HP} = \frac{Q_{cond}}{P_{HP}} = \frac{T_{cond}}{T_{cond} - T_{evap}} \varepsilon \tag{7-2}$$

$$COP_{sys} = \frac{Q_{deh,\ t}}{P_{HP}} = \frac{Q_{deh,\ t}}{Q_{cond}} COP_{HP} \tag{7-3}$$

$$\eta_{ex} = \frac{\Delta E_{deh,\ t}}{P_{HP}} = \frac{E_{deh,\ out} - E_{deh,\ in}}{P_{HP}} = \frac{m_{deh} \times (e_{deh,\ out} - e_{deh,\ in})}{P_{HP}} \tag{7-4}$$

where $Q_{deh,\ t}$ is the total heat change of A_{deh}, kW; Q_{cond} is the heat provided by the condenser, kW, which equals heat gain of A_{hum} at the heater; m_{deh} is the mass flow rate of A_{deh}, kg/s; $h_{deh,\ in}$ and $h_{deh,\ out}$ are the enthalpy of $A_{deh,in}$ and $A_{deh,out}$, respectively, kJ/kg; T_{cond} and T_{evap} are the condensation and evaporation temperature, respectively, K; ε is the utilization factor of the heat pump system; P_{HP} is the power consumption of the heat pump's compressor, kW; $\Delta E_{deh,\ t}$ is exergy flow rate variation between the $A_{deh,in}$ and $A_{deh,out}$, kW; $E_{deh,\ out}$ is exergy of $A_{deh,out}$, kW; $E_{deh,\ in}$ is exergy of $A_{deh,in}$, kW.

e is the specific exergy of air, which is calculated using Eq. (7-5)[3,4]:

$$e_a = c_{pa} T_0 \left(\frac{T_a}{T_0} - 1 - \ln \frac{T_a}{T_0} \right) + R_a T_0 \left[(1 + 1.608 \times 10^{-3} \omega_a) \ln \frac{1 + 1.608 \times 10^{-3} \omega_0}{1 + 1.608 \times 10^{-3} \omega_a} + 1.608 \times 10^{-3} \omega_a \ln \frac{\omega_a}{\omega_0} \right] \tag{7-5}$$

where subscripts a and 0 represent the calculated and reference states, respectively. The reference state is selected as the saturation point under ambient temperature[2,10], which means $T_0 = T_{ambi}$, and ω_0 is the saturation humidity ratio at T_{ambi}. The units of T and ω are K and g/kg, respectively. c_{pa} is the specific heat capacity of air, kJ/(kg · K). The assumptions of Eq. (7-5) are as follows: (1) air pressure equals the atmospheric pressure, and (2) the specific volume of water vapor in the air is ignored.

7.2 Comparisons of both systems based on energy consumptions

7.2.1 Ambient working conditions

In this section, the performances of both systems are compared with those of t_{reg}, COP_{HP} and COP_{sys} under various ambient working conditions. In China, Beijing and Guangzhou are two typical hot and humid cities in the summers. The ambient conditions of Guangzhou and Beijing during the cooling period were selected for the study. The hourly ambient temperature and humidity ratio from May 1st to October 30th are shown in Fig. 7-3. The comfortable indoor thermal conditions range from 24℃ to 26℃ and 9g/kg to 12g/kg, as shown in the black box in Fig. 7-3. Outdoor air needs to be cooled and dehumidified if its temperature and humidity ratio are higher than those of comfortable indoor thermal conditions. Consequently, 11 working conditions were selected, as shown by the hollow red dots in Fig. 7-3. For these conditions, the temperature ranged from 26℃ to 33℃, and the humidity ratio ranged from 15g/kg to 23g/kg.

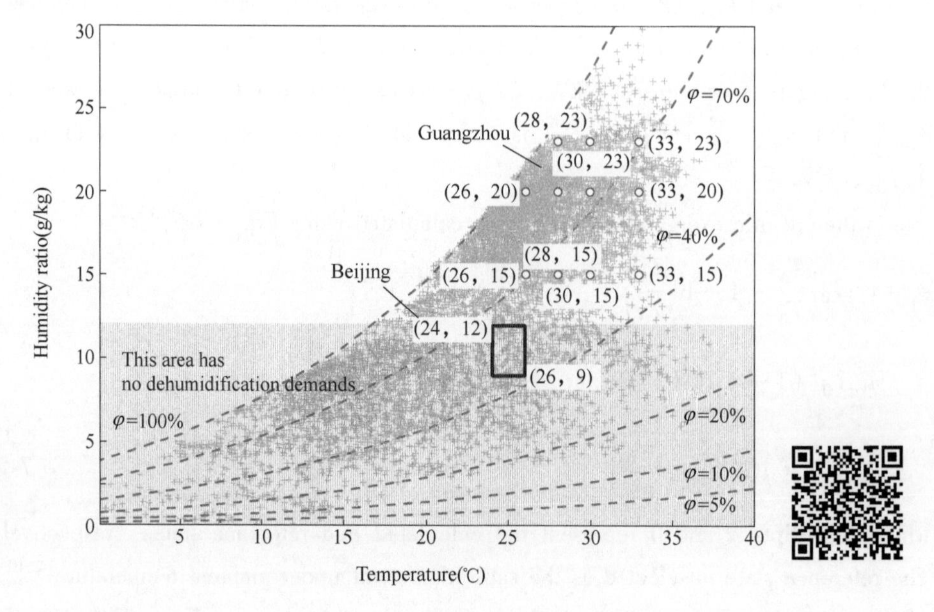

Fig. 7-3 The hourly ambient conditions of Guangzhou and Beijing during cooling period, and discussed ambient working conditions

7.2.2 Results and discussions

The performances of the two systems under the 11 working conditions are shown in Fig. 7-4. $t_{deh,\ out}$ is different under these working conditions and the both systems, as shown in Fig. 7-4 (a). The values of $t_{deh,\ out}$ were lower than 24℃. The fresh air carried no extra cooling load. $Q_{deh,\ t}$ and $\Delta E_{deh,\ t}$ are related to both $t_{deh,\ out}$ and working conditions. $Q_{deh,\ t}$ and $\Delta E_{deh,\ t}$ are shown in Fig. 7-4 (b) ~ (c). In the case of AS, the vapor in A_{deh} is transferred to A_{reg} through the EW. The reduction in A_{deh}'s humidity is beneficial for the reduction in t_{reg}, whereas the increase in A_{reg}'s humidity is not. As shown in Fig. 7-4 (d), the t_{reg} values of AS are lower than those of TS. Under cool and mild working conditions, such as 26℃ and 15g/kg, the difference between AS and TS is small (2℃). Under hot and humid working conditions, such as 33℃ and 23g/kg, the difference between AS and TS is large (19.3℃). The heat provided by the heater was related to t_{reg}. Q_{cond} values of TS and AS are shown in Fig. 7-4 (e). Both Q_{cond} and t_{reg} increased with higher ambient humidity ratio. When the ambient temperature was higher than 28℃, Q_{cond} of the AS was lower than that of the TS.

η_{th} values of the TS and AS are shown in Fig. 7-4 (f). η_{th} is related to both Q_{cond} and $Q_{deh,\ t}$, the results of which are shown in Fig. 7-4 (e) and Fig. 7-4 (b), respectively. When the ambient humidity ratio is higher than 15g/kg, η_{th} of the AS is always higher than that of the TS. The higher the ambient temperature and humidity ratio, the larger the differences. However, when the ambient humidity ratio is low, such as 15g/kg, η_{th} of the TS is similar or even higher than that of the AS, which results from the lower Q_{cond} and $Q_{deh,\ t}$ of the TS, as shown in Fig. 7-4(b) and Fig. 7-4(e). Therefore, as compared with TS, AS shows better performance under hot ambient conditions.

COP_{HP} and the power consumption of the heat pump system (P_{HP}) of both systems are shown in Fig. 7-4 (g) and Fig. 7-4 (h), respectively. COP_{HP} of the AS is higher than that of the TS under the discussed working conditions (except 26℃、15g/kg). In addition, the difference of COP_{HP} between the both system is higher when the ambient conditions are heater and drier. Based on Q_{cond} and COP_{HP}, P_{HP} can be calculated, and the results are plotted in Fig. 7-4 (h). With P_{HP}, $Q_{deh,\ t}$ and $\Delta E_{deh,\ t}$ being known, COP_{sys} and η_{ex} can be calculated, as shown in Figs. 7-4 (i) ~ (j), respectively. Under the WSC, COP_{sys} of the TS and AS are 7.2 and 7.9, respectively. η_{ex} of the TS and the AS are 0.40 and 0.49 respectively. Under the BSC, COP_{sys} of the TS and the AS are 5.1 and 6.0, respectively. Moreover, η_{ex} of the TS and the AS are

0.19 and 0.23 respectively. COP_{sys} and η_{ex} of the AS are higher than those of the TS under the discussed working conditions, except at 15g/kg and temperatures lower than 27℃. Both COP_{sys} and η_{ex} of the two systems increased with an increase in the ambient temperature and a decrease in the ambient humidity ratio. The differences in both COP_{sys} and η_{ex} between the two systems were enlarged under higher ambient temperatures.

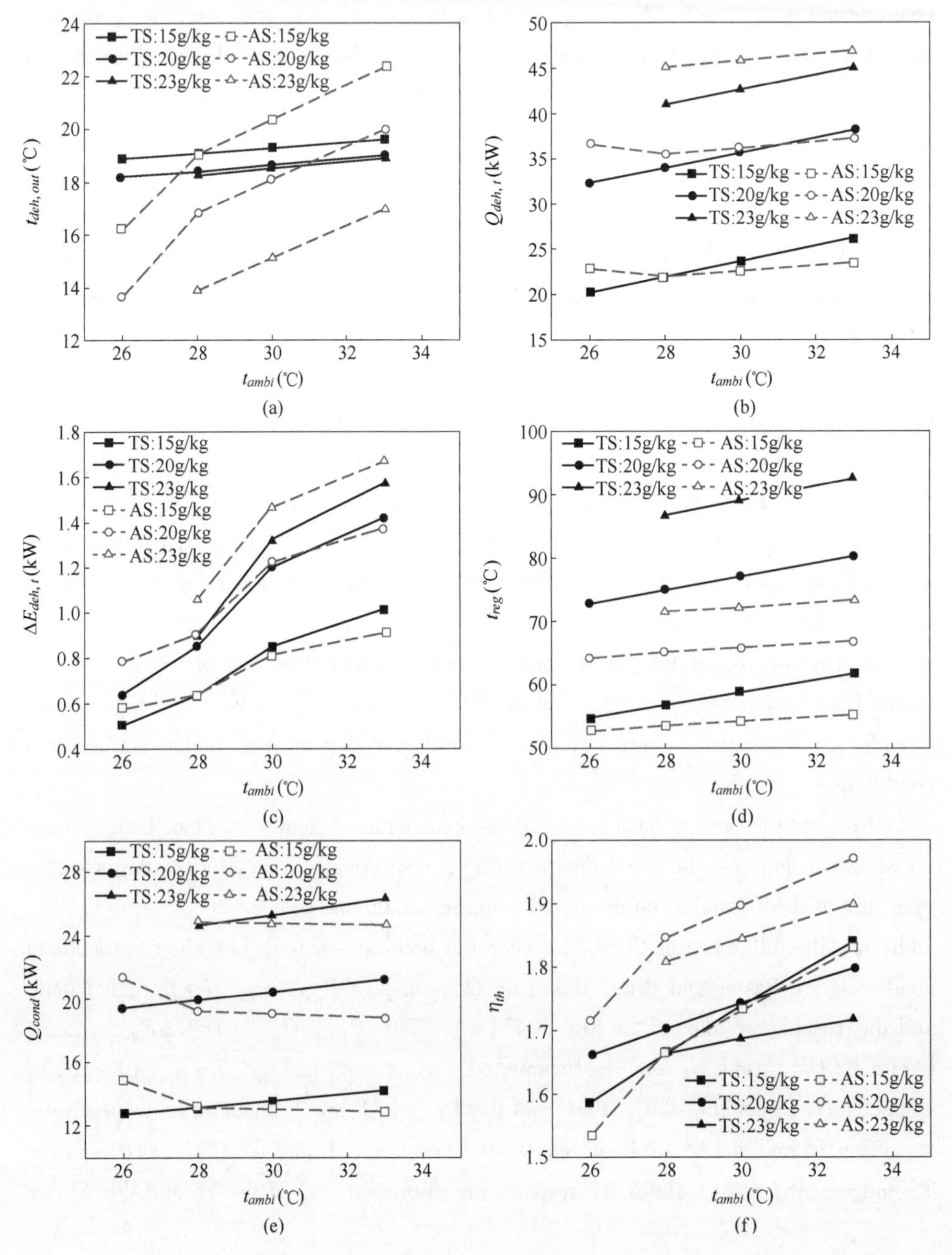

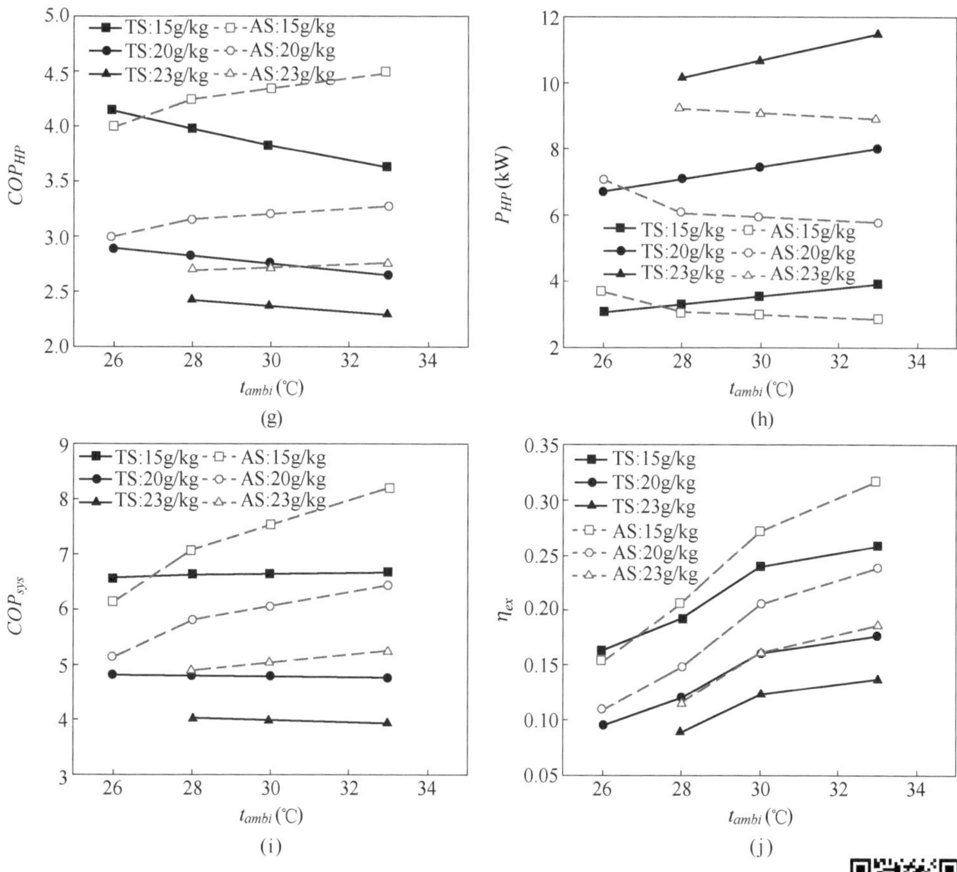

Fig. 7-4 Comparisons between TS and AS under various working conditions

(a) $t_{deh,\ out}$; (b) $Q_{deh,\ t}$; (c) $\Delta E_{deh,\ t}$; (d) t_{reg}; (e) Q_{cond};
(f) η_{th}; (g) COP_{HP}; (h) P_{HP}; (i) COP_{sys}; (j) η_{ex}

Based on the above analyses, it can be concluded that the AS shows better performance than the TS, especially under hot and humid working conditions. When the ambient temperature is lower than 26℃ and the humidity ratio is lower than 15g/kg, which accounts for 5.9% and 22.0% of the hours with dehumidification demand (ambient humidity is higher than 12g/kg) for Guangzhou and Beijing, respectively, COP_{sys} of the AS are lower than those of TS. Otherwise, COP_{sys} of the AS is 6.4% ~ 26.1% higher than that of the TS.

7.3 Comparisons of both systems based on exergy destructions

Here, exergy destruction of the two systems is investigated to explain why AS outperforms TS. In addition, the main exergy destruction is determined, which provides guidance for performance improvement.

7.3.1 Exergy analysis

Based on the system configurations in Fig. 7-1, schematics of exergy flow for both systems can be drawn as shown in Fig. 7-5, including exergy flows into and out of the control surface and the destruction of exergy in the system. Three exergy flows exist in the system, namely exergy provided by the heat pump system (P_{HP}), exergy of $A_{reg,in}$ and $A_{deh,in}$ ($E_{reg,\ in}$ and $E_{deh,\ in}$). There are two exergy flows out of the system, namely $E_{reg,\ out}$ and $E_{deh,\ out}$. They can be calculated using the equations shown in Fig. 7-5 (b).

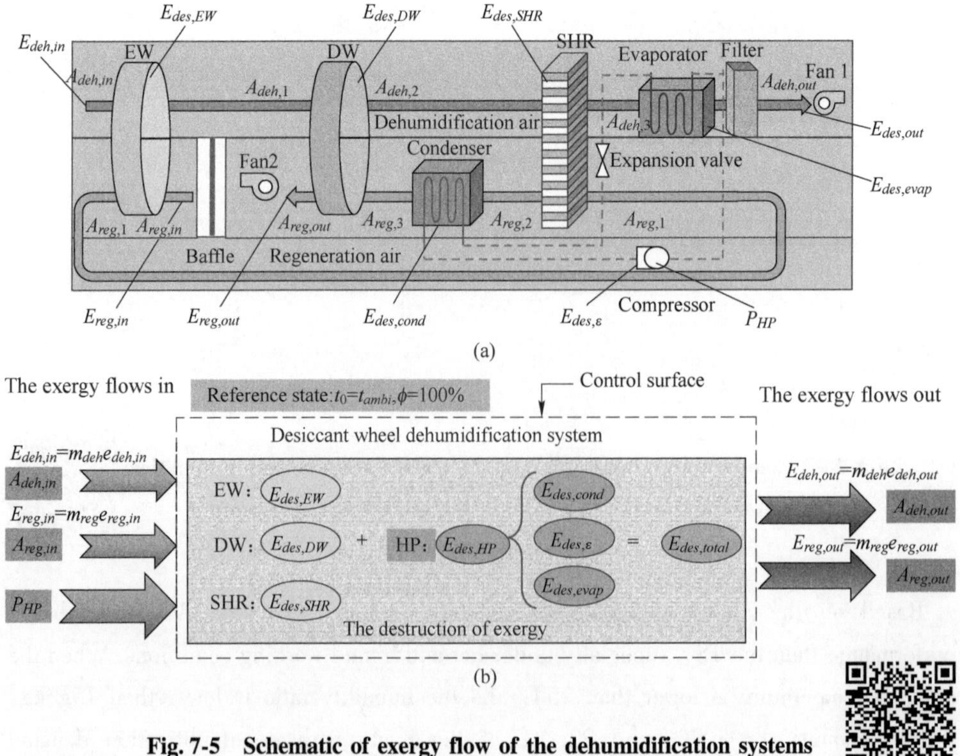

Fig. 7-5 Schematic of exergy flow of the dehumidification systems

(a) Exergy input and exergy loss of each part; (b) Schematic of exergy flow

The exergy destruction in the system comprises four parts. The exergy destruction of the DW ($E_{des,\ DW}$), EW ($E_{des,\ EW}$) and SHR ($E_{des,\ SHR}$) are caused by irreversible heat and mass transfer processes. The exergy destruction of the heat pump ($E_{des,\ HP}$) is primarily caused by irreversible heat transfer processes at the condenser ($E_{des,\ cond}$) and evaporator ($E_{des,\ evap}$), as well as the utilization factor being less than 1($E_{des,\ \varepsilon}$).

The exergy balance equation of the dehumidification system is given by Eq. (7-6).

$$\begin{aligned}&(P_{HP}+E_{reg,\ in}+E_{deh,\ in})-(E_{reg,\ out}+E_{deh,\ out})\\&=E_{des,\ DW}+E_{des,\ SHR}+E_{des,\ EW}+E_{des,\ HP}\end{aligned} \tag{7-6}$$

$E_{des,\ DW}$, $E_{des,\ EW}$ and $E_{des,\ SHR}$ can be easily calculated based on the air inlet and outlet exergies, as shown in Eq. (7-7):

$$E_{des,\ i}=\sum_{i=1}^{n}m_i e_{in,\ i}-\sum_{i=1}^{n}m_i e_{out,\ i} \tag{7-7}$$

where n is the number of air streams that exchange heat and mass in the component; e_{in} is the specific exergy of the inlet air; and e_{out} is the specific exergy of the outlet air. e can be calculated using Eq. (7-5).

For $E_{des,\ HP}$, it can be calculated with Eq. (7-8):

$$E_{des,\ HP}=E_{des,\ \varepsilon}+E_{des,\ cond}+E_{des,\ evap} \tag{7-8}$$

Because the utilization factor is less than 1($\varepsilon<1$), $E_{des,\ \varepsilon}$ is calculated using Eq. (7-9).

$$E_{des,\ \varepsilon}=T_{ambi}\left(\frac{Q_{cond}}{T_{cond}}-\frac{Q_{evap}}{T_{evap}}\right) \tag{7-9}$$

When the heat transfer efficiencies (η) of the evaporators and condensers are fixed, T_{eavp} and T_{cond} can be calculated using Eqs. (7-10) ~ (7-11).

$$T_{evap}=T_{in,\ evap}-\frac{T_{in,\ evap}-T_{out,\ evap}}{\eta} \tag{7-10}$$

$$T_{cond}=T_{in,\ cond}+\frac{T_{out,\ cond}-T_{in,\ cond}}{\eta} \tag{7-11}$$

where $T_{in,\ evap,\ 1}$ and $T_{out,\ evap,\ 1}$ are the inlet and outlet temperatures of air at evaporator 1, respectively. For the TS in Fig. 7-1 (a), $T_{in,\ evap}$ and $T_{out,\ evap}$ are $T_{deh,\ 2}$ and $T_{deh,\ out}$, respectively. For the AS in Fig. 7-1 (b), $T_{in,\ evap}$ and $T_{out,\ evap}$ are $T_{deh,\ 3}$ and $T_{deh,\ out}$, respectively. $T_{in,\ cond}$ and $T_{out,\ cond}$ are the inlet and outlet temperatures of the air at the condenser, respectively. For the TS in Fig. 7-1 (a), $T_{in,\ cond}$ and $T_{out,\ cond}$ are $T_{reg,\ 1}$ and $T_{reg,\ 2}$, respectively. For the AS in Fig. 7-1 (b), $T_{in,\ cond}$ and $T_{out,\ cond}$ are $T_{reg,\ 2}$ and $T_{reg,\ 3}$, respectively.

$E_{des,\ cond}$ and $E_{des,\ evap}$ can be calculated using Eqs. (7-12) ~ (7-13):

$$E_{des,\ cond}=T_{ambi}m_{reg}c_{pa}\left(\ln\frac{T_{out,\ cond}}{T_{in,\ cond}}-\frac{T_{out,\ cond}-T_{in,\ cond}}{T_{cond}}\right) \tag{7-12}$$

$$E_{des,\ evap}=-T_{ambi}m_{deh}c_{pa}\left(\ln\frac{T_{in,\ evap}}{T_{out,\ evap}}-\frac{T_{in,\ evap}-T_{out,\ evap}}{T_{evap}}\right) \tag{7-13}$$

Based on the above equations, the exergy flow of the two systems under a typical hot and humid ambient condition (33℃, 23g/kg) can be drawn, as shown in Fig. 7-6.

The sum of $E_{des,\ DW}$ and $E_{des,\ DW}$ in the AS is 64.0 % of $E_{des,\ DW}$ in the TS. $E_{des,\ SHR}$ in the AS is 17.4 % of $E_{des,\ SHR}$ in the TS. In addition, for both systems, over 60% of the total exergy supply ($E_{deh,\ in}$, $E_{reg,\ in}$ and P_{HP}) is used to compensate for $E_{des,\ HP}$. In addition, $E_{des,\ \varepsilon}$ takes 72.3% and 77.5% of $E_{des,\ HP}$ for the AS and TS, respectively. It can be concluded that $E_{des,\ HP}$ has the highest exergy destruction, which mainly results from the irreversible reverse Carnot cycle. In addition, the total exergy destruction of $E_{des,\ DW}$, $E_{des,\ EW}$ and $E_{des,\ SHR}$ in the AS is much lower than that in the TS, leading to improved performance of the AS.

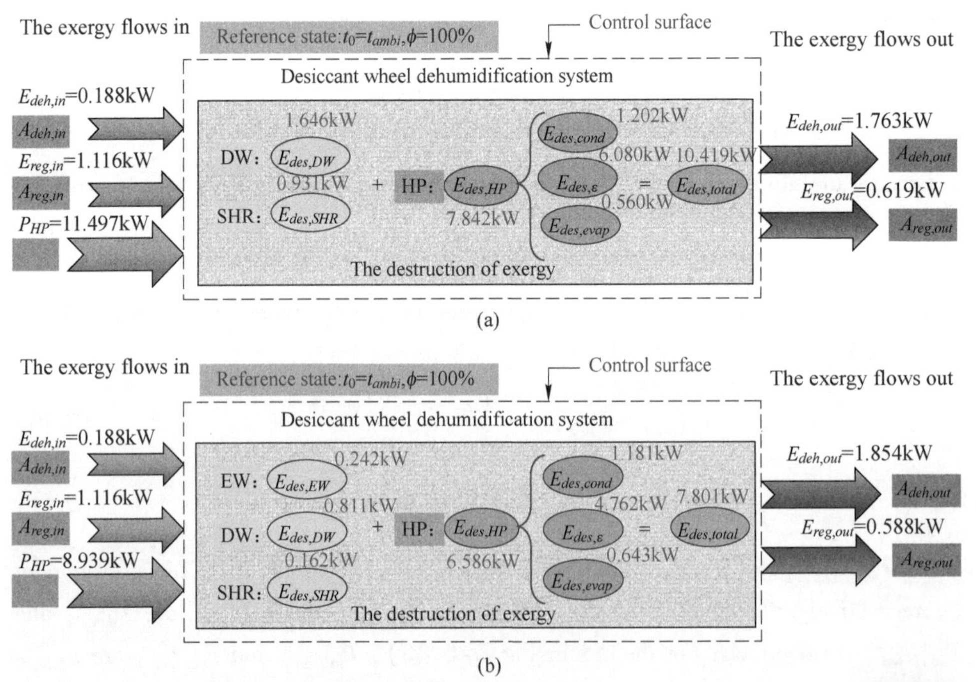

Fig. 7-6 Exergy flow of the two system: ambient condition is 33℃ and 23g/kg

(a) Exergy flow of TS under $t_{deh,\ in}$ = 33℃, $\omega_{deh,\ in}$ = 23g/kg;

(b) Exergy flow of AS under $t_{deh,\ in}$ = 33℃, $\omega_{deh,\ in}$ = 23g/kg

Furthermore, the AS and TS are compared under other working conditions, and the results of the exergy destruction are shown in Table 7-2. As shown in Table 7-2, splitting the DW in the TS into one DW and one EW can greatly reduce the total exergy destruction of $E_{des,\ DW}$, $E_{des,\ EW}$ and $E_{des,\ SHR}$. Under the discussed working conditions, the total exergy destruction of $E_{des,\ DW}$ and $E_{des,\ SHR}$ in the TS ranges from 0.39 to 2.58. However, the total exergy destruction of $E_{des,\ DW}$, $E_{des,\ EW}$ and $E_{des,\ SHR}$ in the AS ranges only from 0.25kW to 1.22kW. The ratios of the total exergy destruction of the wheels and the sensible heat exchanger in the AS and TS $\left(R=\frac{[E_{des,DW}+E_{des,EW}+E_{des,SHR}]_{AS}}{[E_{des,DW}+E_{des,SHR}]_{TS}}\right)$ range from 0.47 to 0.79.

Further, $E_{des,\ HP}$ ranges from 2.09kW to 7.84kW for the TS, whereas it ranges from 2.09kW to 6.72kW for the AS. Except for the working conditions of 26℃, the ratio of $E_{des,\ HP}$ in the AS and TS $\left(R_{HP}=\frac{[E_{des,\ HP}]_{AS}}{[E_{des,\ HP}]_{TS}}\right)$ ranged from 0.76 to 0.97. Similarly, $E_{des,\ HP}$ is primarily caused by $E_{des,\ \varepsilon}$, which is 70%~78% of $E_{des,\ HP}$ for the AS, and the TS, respectively.

Table 7-2 Exergy destructions of the AS and TS under the discussed working conditions

Condition	$E_{des,\ DW}+E_{des,\ EW}+E_{des,\ SHR}$		R	$E_{des,\ HP}$		R_{HP}	$E_{des,\ \varepsilon}/E_{des,\ HP}$	
	TS	AS		TS	AS		TS	AS
26℃, 15g/kg	0.52kW	0.41kW	0.79	2.16kW	2.71kW	1.26	0.73	0.71
26℃, 20g/kg	1.44kW	0.68kW	0.59	4.60kW	5.19kW	1.12	0.76	0.72
28℃, 15g/kg	0.55kW	0.43kW	0.79	2.32kW	2.23kW	0.96	0.74	0.72
28℃, 20g/kg	1.50kW	0.92kW	0.61	4.85kW	4.35kW	0.90	0.76	0.73
28℃, 23g/kg	2.33kW	1.19kW	0.51	6.90kW	6.72kW	0.97	0.77	0.73
30℃, 15g/kg	0.58kW	0.43kW	0.74	2.49kW	2.17kW	0.87	0.74	0.72
30℃, 20g/kg	1.56kW	0.92kW	0.59	5.11kW	4.28kW	0.84	0.76	0.73
30℃, 23g/kg	2.42kW	1.20kW	0.49	7.52kW	6.66kW	0.92	0.76	0.73
33℃, 15g/kg	0.64kW	0.44kW	0.69	2.75kW	2.10kW	0.76	0.75	0.71
33℃, 20g/kg	1.66kW	0.92kW	0.56	5.51kW	4.18kW	0.76	0.77	0.73
33℃, 23g/kg	2.58kW	1.22kW	0.47	7.84kW	6.59kW	0.84	0.78	0.72
33℃, 19g/kg (BSC)	1.41kW	0.78kW	0.55	4.88kW	4.46kW	0.91	0.76	0.71
33℃, 14g/kg (WSC)	0.39kW	0.25kW	0.64	2.30kW	2.09kW	0.91	0.75	0.70

7.3.2 Discussions

In this subsection, the reasons behind the lower exergy destruction, including $E_{des,\ \varepsilon}$,

$E_{des,\ DW}$, $E_{des,\ EW}$ and $E_{des,\ SHR}$ of the AS are discussed. As seen in Eq. (7-9), $E_{des,\ \varepsilon}$ is related to T_{evap}, T_{cond}, Q_{cond} and Q_{evap}. Details of the corresponding parameters, which were calculated under typical ambient conditions (33℃, 23g/kg), are listed in Table 7-3.

Table 7-3 $E_{des,\varepsilon}$ of the heat pump systems for the TS and the AS

Parameters	TS	AS
T_{ambi} (K)	306.15 (33℃)	306.15 (33℃)
T_{cond} (K)	370.32 (97.17℃)	350.74 (77.59℃)
T_{evap} (K)	289.47 (16.32℃)	287.36 (14.21℃)
Q_{cond} (kW)	26.33	24.73
Q_{evap} (kW)	14.8	15.8
Q_{cond}/T_{cond} (kW/K)	0.0711	0.0512
Q_{evap}/T_{evap} (kW/K)	0.0536	0.0550
$E_{des,\ \varepsilon}$ (kW)	6.08	4.76

Compared to the TS, T_{cond} of the AS is low owing to the reduction in t_{reg}. Because of the reduction in t_{reg}, Q_{cond} of the AS is lower than that of the TS. In addition, T_{evap} values of the two systems were similar, and Q_{evap} values of the two systems were similar. Therefore, it has been calculated that the difference between Q_{cond}/T_{cond} and Q_{evap}/T_{evap} of the AS is smaller than that of the TS. Therefore, as compared to the TS, $E_{des,\ \varepsilon}$ of the AS is smaller.

In the above analyses, exergy destruction of the DW, EW and SHR were calculated using the specific exergy of air. This method is relatively simple. However, it cannot identify the essential factors that cause exergy destruction. The DW, EW and SHR are heat and mass transfer components, the exergy destruction of which is caused by temperature difference fields and humidity ratio difference fields[6,9]. In the case of SHR, only heat transfer occurs, whereas in the case of the DW and EW, both heat and mass transfer occur concurrently. The exergy destruction during heat and mass transfer can be calculated using Eqs. (7-14) ~ (7-15), respectively[11], and the derivation process was analyzed in Chapter 4:

$$E_{des,\ HT} \approx \frac{T_0}{\overline{T_1} \times \overline{T_2}} \frac{Q^2}{hF} \frac{\overline{\Delta T^2}}{\overline{\Delta T}^2} = \frac{T_0}{\overline{T_1} \times \overline{T_2}} \frac{Q^2}{hF} \xi_T,\ \xi_T = \frac{(\int_0^F \Delta T^2 dF)/F}{[(\int_0^F \Delta T dF)/F]^2} \tag{7-14}$$

$$E_{des,\ MT} \approx \frac{T_0 S_l i_v}{\overline{T_{1,\ dew}} \times \overline{T_{2,\ dew}}} \frac{M^2}{h_m F} \xi_\omega,\ \xi_\omega = \frac{\left(\int_0^F \Delta\omega^2 dF\right)/F}{\left[\left(\int_0^F \Delta\omega dF\right)/F\right]^2} \tag{7-15}$$

where $\overline{T}$ and $\overline{T_{dew}}$ are the average temperature and average dew point temperature, respectively, of the same heat transfer media on the entire heat transfer area, K; $\overline{\Delta T}$ and $\overline{\Delta T_{dew}}$ are the respectively average values of temperature difference (ΔT) and dew point temperature difference (ΔT_{dew}) between two heat transfer media on the entire heat transfer area, K; Q and M are the heat and mass transfer rates, respectively, kW and kg/s; h and h_m are the heat and mass transfer coefficients, respectively, kW/(m^2/℃) and kg/(m^2/s); F is the heat and mass transfer area, m^2; i_v is vaporization heat, kJ/kg; S_l can be regarded as a constant [6]. ξ_T and ξ_ω are unmatched coefficients, indicating the uniformity of ΔT and $\Delta\omega$ fields, respectively. Normally, ξ_T and ξ_ω are higher than 1. When the ΔT field or the $\Delta\omega$ field is uniform, ξ_T or $\xi_\omega = 1$.

This is indicated by Eqs. (7-14) ~ (7-15), the exergy destruction of heat (mass) transfer processes is primarily influenced by $h(h_m)$, F and $\xi_T(\xi_\omega)$. With the same Q or M, to reduce exergy destruction, except from enlarging h or h_m and increasing F, it is also important to make the ΔT or $\Delta\omega$ field uniform. In the case of sensible heat exchangers, such as the SHRs in both systems, if the heat capacities of the two heat transfer fluids are equal ($m_1 c_{p1} = m_2 c_{p2}$), then $\xi_T = 1$[2]. For DW and EW, it has been proven that if the mass flow rates of A_{deh} and A_{reg} are equal and the wheel is evenly divided between both systems, the smallest ξ_T and ξ_ω can be realized[6,10].

In this study, A_{deh} and A_{reg} had the same mass flow rate. Therefore, the lowest ξ_T and ξ_ω values have already been realized. Under the same working conditions (33℃, 23g/kg), the important parameters in Eqs. (7-14) ~ (7-15) are compared between AS and TS, as listed in Table 7-4. ξ_T and ξ_ω of the DW in the AS were lower than those of the DW in the TS. ξ_T and ξ_ω of the EW in the AS were smaller than those of the DW in the TS. In addition, the dehumidification rates of DW and EW in the AS are similar. However, the heat exchange rate of the EW is much lower than that of the DW for the AS. Therefore, the total exergy destruction of the DW and EW in the AS was much lower than that of the DW in the TS. Moreover, the heat transfer rate of the SHR in the AS was much lower than that of the SHR in the TS. Therefore, the exergy destruction of the SHRs in the AS was lower than that in the TS.

Table 7-4 Exergy destruction analyses of EW, DW and SHR

Parameters	EW		DW		SHR	
	TS	AS	TS	AS	TS	AS
Q(kW)	—	4.2	36.2	19.2	33.7	13.7
M(g/s)	—	6.7	12.6	6.0	—	—
ξ_h	—	1.8	2.1	1.8	1.00	1.00
ξ_m	—	1.8	2.1	1.9	—	—
E_{des} (kW)	—	0.24	1.65	0.81	0.93	0.16

Based on the above analyses, when half of the DW is used as an EW, such as the AS, the total exergy destruction during dehumidification can be reduced. Taking both exergy destruction of $E_{des,\ DW} + E_{des,\ SHR} + E_{des,\ EW}$ and $E_{des,\ HP}$, it is advisable to use the AS when ambient conditions are hot to improve dehumidification performance. When the ambient condition is cooler than 27℃ and drier than 15g/kg, it is advised to use the TS instead of the AS. Further, in the case of the heat pump system, the largest exergy destruction is $E_{des,\ \varepsilon}$. In the following section, sensitivity analysis is performed to improve the performance of the AS.

7.4 Sensitivity analysis of the advanced system

Here, sensitivity studies of the AS are performed to reduce the exergy destruction of DW, EW and the heat pump system. For DW and EW, the effects of the total length of the two wheels (L_{DW+EW}) on exergy destruction are discussed. In the case of the heat pump system, since the largest proportion is $E_{des,\ \varepsilon}$, the effects of ε will be discussed. Furthermore, when the effects of L_{DW+EW} on the power consumption of fans and the compressor are considered, the optimal L_{DW+EW} is recommended for practical applications.

7.4.1 Effects of ε

Utilization factor is the second-law efficiency of thermodynamics[4], which reflects the degree to which the equipment deviates from the ideal state. So, ε refers to the ratio of the heat pump system efficiency of the equipment under a certain working condition to the efficiency of the ideal Reverse Carnot cycle under this working condition.

Under the typical ambient condition (33℃ and 23g/kg), the effect of ε, which varies from 0.5 to 1, on exergy destruction and system performance is shown in Fig. 7-7.

When ε is 1, the vapor compression cycle is an ideal Reverse Carnot cycle. With the increase in ε from 0.5 to 1, P_{HP} decreases gradually from 8.9kW to 4.8kW. Meanwhile, the proportion of $E_{des,\ \varepsilon}$ in $E_{des,\ HP}$ and in the total exergy destruction ($E_{des,\ HP} + E_{des,\ DW} + E_{des,\ SHR} + E_{des,\ EW}$) decreased from 0.72 to 0 and from 0.61 to 0, respectively. COP_{HP}, COP_{sys}, and η_{ex} also increase with an increase in ε. Under the ideal condition ($\varepsilon = 1$), COP_{HP}, COP_{sys}, and η_{ex} are 5.1, 10.6 and 0.40 respectively.

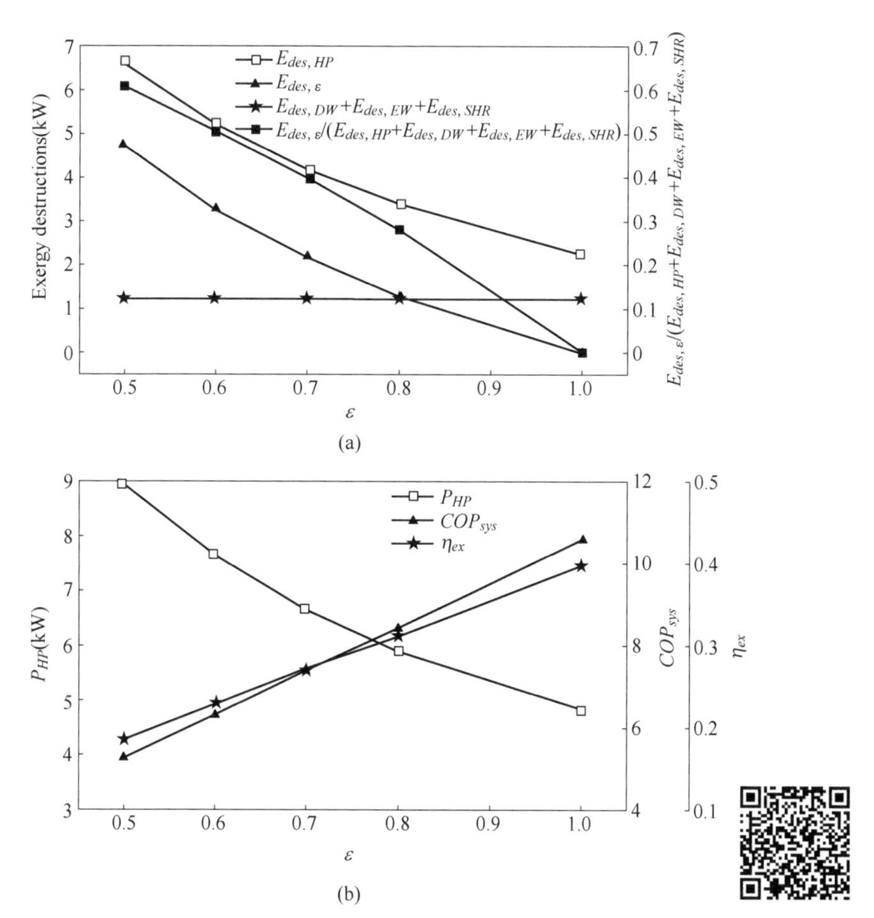

Fig. 7-7 Effects of ε on performances of AS: L_{DW+EW} = 200mm; ambient condition is 33℃ and 23g/kg

(a) Effects of ε on exergy destructions; (b) Effects of ε on P_{HP}, COP_{sys} and η_{ex}

7.4.2 Effects of L_{DW+EW}

Fig. 7-8 shows the influence of L_{DW+EW} on the performance of the AS under typical ambient conditions (33℃ and 23g/kg). Fig. 7-8 (a) shows that, as the L_{DW+EW} increases

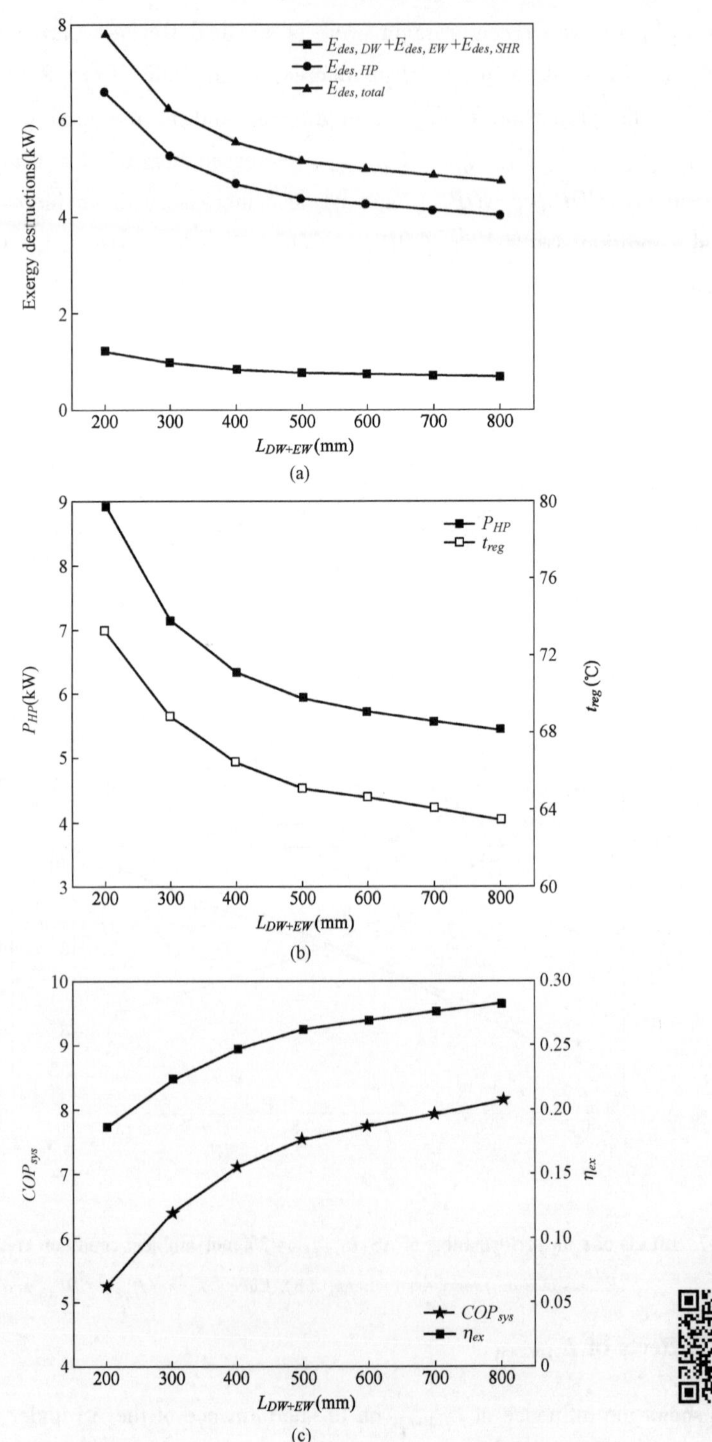

Fig. 7-8 Effects of L_{DW+EW} on performances of AS: ε=0.5

(a) Effects of L_{DW+EW} on exergy destruction; (b) Effects of L_{DW+EW} on P_{HP} and t_{reg}; (c) Effects of L_{DW+EW} on COP_{sys} and η_{ex}

from 200mm to 800mm, the exergy destruction of both the heat pump system and the other heat and mass transfer components decreases gradually. The reduction in exergy destruction leads to a lower t_{reg} and smaller P_{HP}, as seen in Fig. 7-8 (b). Consequently, COP_{HP}, COP_{sys} and η_{ex} increase with an increase in L_{DW+EW}, as shown in Fig. 7-8 (c).

In addition, Fig. 7-8 shows that the rate of improvement in the system's performance decreases with an increase in L_{DW+EW} When L_{DW+EW} is greater than 500mm, the reduction in P_{HP} is not obvious. It is calculated that when the L_{DW+EW} increases from 200mm to 500mm, P_{HP} reduces from 8.9kW to 5.9kW. When L_{DW+EW} increases from 500mm to 800mm, P_{HP} reduces from 5.9kW to 5.4kW. However, the increase in L_{DW+EW} simultaneously leads to an increase in the pressure drop, which increases the power consumption of the fans (P_{fan}). Therefore, both P_{fan} and P_{HP} must be considered to analyze the effects of L_{DW+EW} on the system's performance. The energy efficiency ratio (EER) is used as the performance indicator, which can be calculated using Eq. (7-16).

$$EER = \frac{Q_{deh}}{P_{HP} + P_{fan}} \tag{7-16}$$

P_{fan} can be calculated based on the pressure drop of each component (ΔP_i) along the direction of air flow, as shown in Eq. (7-17).

$$P_{fan} = \frac{G \sum_i \Delta P_i}{3600 \times \eta_{fan}} \tag{7-17}$$

where G is the air volumetric flow rate, m^3/h; and η_{fan} represents the efficiency of the fan, which is 60% in the study.

The pressure-drop components of each fan, pressure drop, and rates of airflow are represented in Table 7-5.

Table 7-5 Estimated pressure drops of the heat and mass transfer units

Fan	Description	Pressure drop (Pa)
Fan_1	Desiccant wheel (EW and DW)	Related to thickness
	Evaporator	125
	SHR	150
	Filter	100
	air duct of A_{deh}	100
	The static pressure	50

Continued Table 7-5

Fan	Description	Pressure drop (Pa)
Fan_2	Wheel (EW and DW)	Related to thickness
	Condenser	125
	SHR	150
	air duct of A_{reg}	150
	pressure drop caused by changing flow direction of A_{reg}	100

For desiccant wheels, the pressure drop is related to the thickness. However, for other components, the pressure drops are fixed, the values of which can be referred to inprevious studies[9,12,13]. The pressure drop of the desiccant wheels is estimated based on the test results from the literature[14]. Fig. 7-9 shows the variations in the pressure drop and fan power consumption with L_{DW+EW}.

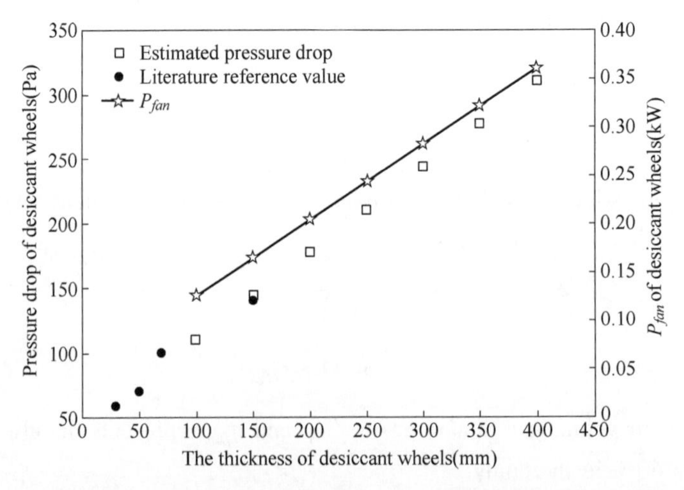

Fig. 7-9 Pressure drops and fan power consumption under different thicknesses of wheels

The total energy consumption of the AS can be calculated by varying the L_{DW+EW} from 200mm to 800mm. The effects of L_{DW+EW} on the total energy consumption ($P_{HP}+P_{fan}$) and the *EER* are shown in Fig. 7-10. Both temperature and moisture content affect the performance of the system. The higher the temperature is, the higher the *EER* is. When the moisture content is constant (15g/kg), the *EER* is 4. 20～4. 55 and 4. 95～5. 40 at 26℃ and 33℃, respectively. The higher the moisture content is, the more energy consumption is required. But there was little change at *EER*. The *EER* values of 33℃, 15g/kg and 33℃, 23g/kg are 4. 95 ～ 5. 40 and 4. 40 ～ 5. 60, respectively. Four working conditions are selected to calculate P_{HP} + P_{fan} and *EER* . With an increase in

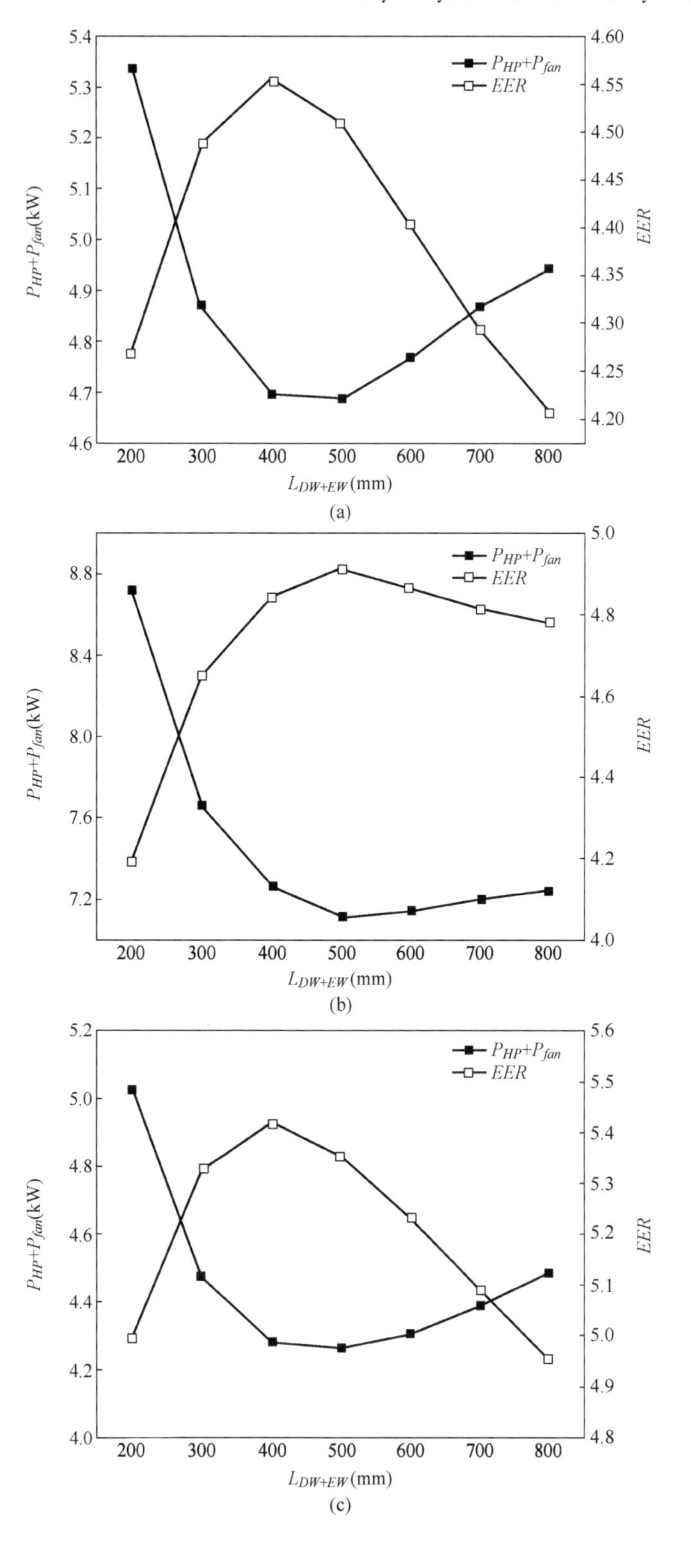

5.4
5.3
5.2
5.1
5.0
4.9
4.8
4.7
4.6
$P_{HP}+P_{fan}$(kW)
4.60
4.55
4.50
4.45
4.40
4.35
4.30
4.25
4.20
EER
$P_{HP}+P_{fan}$
EER
200
300
400
500
600
700
800
L_{DW+EW}(mm)
(a)
8.8
8.4
8.0
7.6
7.2
$P_{HP}+P_{fan}$(kW)
5.0
4.8
4.6
4.4
4.2
4.0
EER
$P_{HP}+P_{fan}$
EER
200
300
400
500
600
700
800
L_{DW+EW}(mm)
(b)
5.2
5.0
4.8
4.6
4.4
4.2
4.0
$P_{HP}+P_{fan}$(kW)
5.6
5.5
5.4
5.3
5.2
5.1
5.0
4.9
4.8
EER
$P_{HP}+P_{fan}$
EER
200
300
400
500
600
700
800
L_{DW+EW}(mm)
(c)

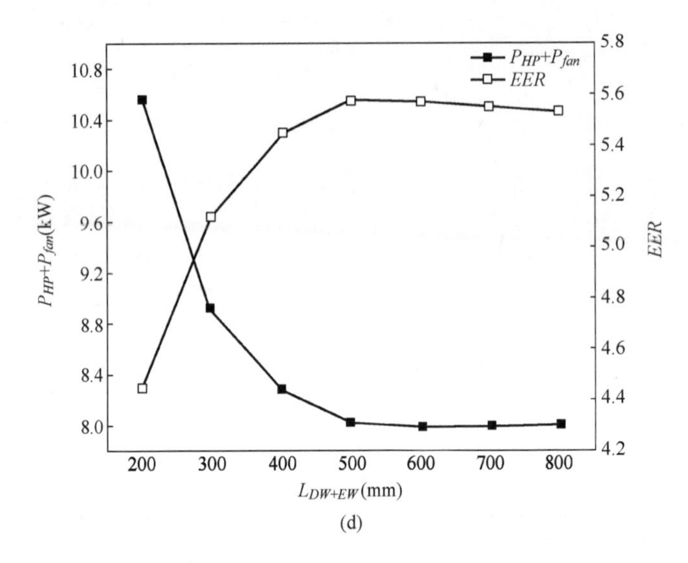

(d)

Fig. 7-10 Effects of L_{DW+EW} on $P_{HP}+P_{fan}$ and *EER* under different working conditions

(a) 26℃, 15g/kg; (b) 26℃, 20g/kg; (c) 33℃, 15g/kg; (d) 33℃, 23g/kg

L_{DW+EW}, $P_{HP}+P_{fan}$ first decreases and then increases. Meanwhile, *EER* increases and then decreases with an increase in L_{DW+EW}. Hence, an optimal L_{DW+EW} occurs with the highest *EER*, which varies from 400mm to 500mm for the four working conditions.

7.5 Conclusions

An AS is proposed, which separates the DW in the TS into a DW and an EW. Performances of the two systems are compared under various ambient conditions, and sensitivity studies are performed based on the AS to improve its performance. Performance evaluation indexes are t_{reg}, η_{th}, COP_{sys} and η_{ex}. Main conclusions are as follows.

(1) Under the discussed working conditions (t_{ambi} = 26 ~ 33℃, ω_{ambi} = 15 ~ 23g/kg), t_{reg}s of the AS are lower than those of the TS, with the maximum difference being 19.3℃ at 33℃ and 23g/kg. COP_{sys} of the AS ranged from 4.88 to 8.21, which is 1.07 to 1.35 times that of the TS (expect 26℃, 15g/kg). And η_{ex} of the AS ranged from 0.11 to 0.32, which is 1.07 to 1.36 times of the TS (expect 26℃, 15g/kg).

(2) Exergy analyses show that the exergy destruction of AS is much smaller, which mainly results from the reduction in the total exergy destruction of DW, EW, and SHR as compared with that of DW and SHR in the TS. The total exergy destruction of DW, EW, and SHR in the AS is 47% ~ 79% of those in the TS.

(3) The sensitivity study showed that when ε and the total thickness of EW and DW (L_{DW+EW}) in the AS are increased, exergy destruction can be effectively reduced. When the energy consumption of fans is considered, the optimal L_{DW+EW} is advised to be 400~500mm for a wide range of working conditions.

Nomenclature

A_r	facial area ratio of the dehumidification air to the humidification air
AS	advanced system
COP	coefficient of performance
c_p	specific heat of the air, kJ/(kg · K)
E	exergy value in some state, kW
E_{des}	exergy destruction, kW
e	specific exergy, kJ/kg
F_r	flow ratio of the dehumidification air to the humidification air
h	specific enthalpy, kJ/kg
L	the length of wheels, mm
m	mass flow rate, kg/s
NTU	number of heat transfer units
P	power consumption of compressor, kW
Q	heat transfer capacity, kW
R	the ratios of the total exergy destruction of the wheels and the sensible heat exchanger in the AS and TS
R_{HP}	the ratios of the total exergy destruction of the heat pump in the AS and TS
RS	rotate speed, r/h
t	temperature, ℃
T	temperature in Kelvin, K
TS	traditional desiccant wheel dehumidification system
ΔE	exergy flow rate variation, kW

Greek symbols

ω	humidity ratio, g/kg

η	heat transfer effectiveness
η_{th}	energy efficiency
η_{ex}	exergy efficiency, dimensionless
φ	relative humidity ratio of the humid air
ε	utilization factor

Subscripts

ambi	ambient air
cond	condenser
deh	dehumidification air of desiccant wheel
ex	exergy
evap	evaporator
HP	Heat pump
in	inlet
out	outlet
reg	regeneration air of desiccant wheel
sys	system

References

[1] R Tu, Y Hwang, T Cao, M D Hou, H S Xiao. Investigation of adsorption isotherms and rotational speeds for low temperature regeneration of desiccant wheel systems [J]. Int. J. Refrig, 2018, 86: 495-509.

[2] R Tu, X H Liu, Y H Hwang, F Ma. Performance analysis of ventilation systems with desiccant wheel cooling based on exergy destruction [J]. Energy Convers. Manage, 2016, 123: 265-279.

[3] R Tu, X H Liu, Y Jiang, F Ma. Influence of the number of stages on the heat source temperature of desiccant wheel dehumidification systems using exergy analysis [J]. Energy, 2015, 85: 379-391.

[4] A Bejan. Advanced engineering thermodynamics [M]. 3rd ed. Hoboken (NJ): John Wiley & Sons, 2006.

[5] R Tu, X H Liu, Y Jiang. Performance comparison between enthalpy recovery wheels and dehumidification wheels [J]. Int. J. Refrig, 2013, 36: 2308-2322.

[6] R Tu, X H Liu, Y Jiang. Lowering the regeneration temperature of a rotary wheel dehumidification system using exergy analysis [J]. Energy Conver. Manage, 2015, 89: 162-174.

[7] R Tu, X H Liu, Y Jiang. Performance analysis of a two-stage desiccant cooling system [J].

Appl. Energy, 2014, 13: 1562-1574.

[8] R Tu, Y Hwang. Performance analyses of a new system for water harvesting from moist air that combines multi-stage desiccant wheels and vapor compression cycles [J]. Energy Conver. Manage, 2019, 198: 111811.

[9] R Tu, X H Liu, Y Jiang. Irreversible processes and performance improvement of desiccant wheel dehumidification and cooling systems using exergy [J]. Appl. Energy, 2015, 145: 331-344.

[10] J Lin, D Bui, R Z Wang, K Chua. On the exergy analysis of the counter-flow dew point evaporative cooler [J]. Energy, 2018, 165: 958-971.

[11] X H Liu, T Zhang, Y W Zheng, R Tu. Performance investigation and exergy analysis of two-stage desiccant wheel systems [J]. Renew. Energy, 2016, 86: 877-888.

[12] R Tu, Y Hwang, F Ma. Performance analysis of a new heat pump driven multi-stage fresh air handler using solid desiccant plates [J]. Appl. Therm. Eng, 2017, 117: 553-567.

[13] L Z Zhang. Total Heat Recovery: Heat and Moisture Recovery from Ventilation Air [M]. first ed. New York: Nova Science Publisher, 2008.

[14] T Cao, H Lee, Y H Hwang, R Radermacher, H Chun. Experimental investigations on thin polymer desiccant wheel performance [J]. Int. J. Refrig, 2014, 44: 1-11.

Chapter 8 Performance Analyses of Dehumidification Systems with Multi-Stage Desiccant Wheels

In this chapter, multi-stage desiccant wheel dehumidification and cooling systems are designed, and the influence of the number of stages on t_{hs} is examined from the perspective of exergy. First, the thermodynamic method is used to determine the key factors that influence t_{hs}. Next, a simulation method is used to validate the results. Finally, the optimal number of stages for both water- and refrigerant- based heating/cooling systems is recommended.

8.1 Exergy analysis of multi-stage wheel system

8.1.1 Description of water and refrigerant driven systems

Water and refrigerant represent two types of heating/cooling media and have finite and infinite specific heat capacities, respectively. Fig. 8-1 and Fig. 8-2 illustrate the corresponding multi-stage desiccant wheel dehumidification and cooling systems, taking two-stage systems as examples.

In Fig. 8-1, water is used as the heating and cooling medium. For the two-stage system, there are two desiccant wheels (DW_1 and DW_2), two counter-flow air-water coolers, and two counter-flow air-water heaters. The first stage is composed of DW_1, $Cooler_1$ and $Heater_1$, and the second stage is composed of DW_2, $Cooler_2$ and $Heater_2$. The processed air flows in sequence through the first stage and the second stage. After being dehumidified by DW_1 (A_{pin} to A_{p1}) and DW_2 (A_{p2} to A_{p3}), the air is cooled down by the chilled water in $Cooler_1$ (A_{p1} to A_{p2}) and $Cooler_2$ (A_{p3} to A_{pout}), before finally being introduced into the occupied room (A_{pout}). The regeneration air flows through the second stage and the first stage in sequence. After being heated by the hot water to the required t_{reg} in $Heater_1$ (A_{rin} to A_{p1}) and $Heater_2$ (A_{r2} to A_{p3}), the regeneration air is used to regenerate DW_2 (A_{r1} to A_{p2}) and DW_1 (A_{r3} to A_{pout}), before finally being exhausted (A_{pout}). The chilled water (W_{cin}) and the hot water (W_{hin}) flow in parallel into all the coolers and heaters, respectively. Chilled water

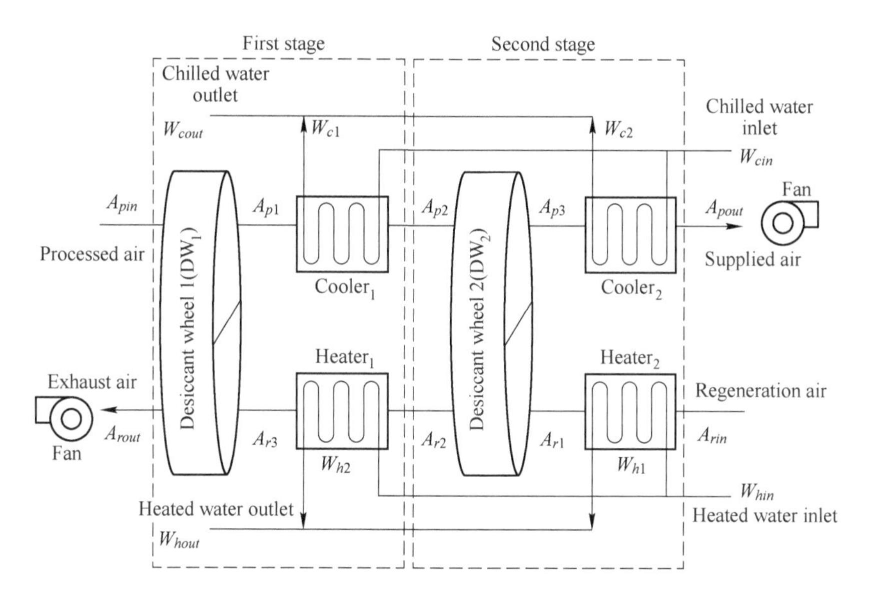

Fig. 8-1 Schematic of a desiccant dehumidification and cooling system using water as the cooling and heating medium

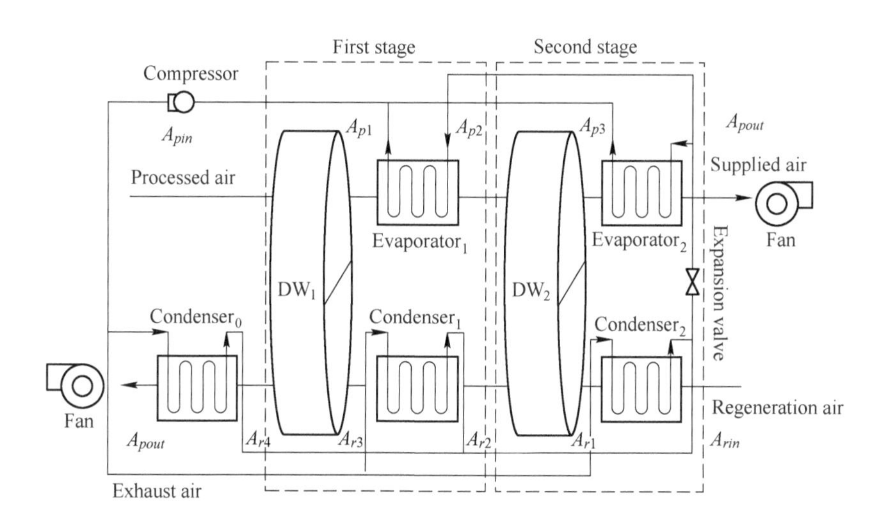

Fig. 8-2 Schematic of a heat pump-driven desiccant dehumidification and cooling system using refrigerant as the cooling and heating medium

from all the coolers (W_{c1} and W_{c2}) mixes together and proceeds to the chilled water outlet (W_{cout}); similarly, hot water from all the heaters (W_{h1} and W_{h2}) mixes together and proceeds to the heated water outlet (W_{hout}).

In Fig. 8-2, refrigerant is used as the cooling and heating medium. The heat pump system, with one compressor and one expansion valve, is integrated with the desiccant wheels. For the two-stage system, there are two evaporators used as coolers and three condensers used as heaters. The first stage is composed of DW_1, $Evaporator_1$ and $Condenser_1$, and the second stage is composed of DW_2, $Evaporator_2$ and $Condenser_2$. The air handling processes are the same as those of Fig. 8-1, except that the regeneration air leaving DW_1 is heated by $Condenser_0$ (A_{r4} to A_{pout}) to dissipate the extra heat of the heat pump system. All evaporators and condensers are linked in parallel. Thus, all the evaporators are of the same evaporating temperature, and all the condensers are of the same condensing temperature.

8.1.2 Performance indicators of the systems

For the system shown in Fig. 8-1, the thermal energy from water is used to heat or cool the air. The reduction of t_{hs} (temperature of the hot water entering the heaters) is beneficial for the adoption of solar energy and low-temperature waste heat, etc. For the system shown in Fig. 8-2, the reduction of t_{hs} (condensing temperature) is beneficial for the reduction of the power of the compressor (P).

COP can be expressed as COP_W for the system in Fig. 8-1 and as COP_R for the system in Fig. 8-2, shown as Eq. (8-1) and Eq. (8-2), respectively:

$$COP_W = \frac{m_p(i_{pin} - i_{pout})}{Q_c + Q_h} = \frac{m_p(i_{pin} - i_{pout})}{m_c c_{pw}(t_{cout} - t_{cin}) + m_h c_{pw}(t_{hin} - t_{hout})} \tag{8-1}$$

$$COP_R = \frac{m_p(i_{pin} - i_{pout})}{P} = \frac{m_p(i_{pin} - i_{pout})}{Q_c} \frac{T_{evap}}{T_{cond} - T_{evap}} \varepsilon \tag{8-2}$$

where Q_c is the cooling capacity provided by the chilled water or refrigerant in Fig. 8-1 or Fig. 8-2, respectively; Q_h is the heat capacity provided by the hot water in Fig. 8-1; T_{evap} and T_{cond} are evaporating and condensing temperature in Kelvin, respectively; and ε is the thermodynamic perfectness of the compressor, which is lower than 1.

Exergy efficiency (η_e) can be expressed as $\eta_{e,W}$ for the system in Fig. 8-1 and as $\eta_{e,R}$ for the system in Fig. 8-2, shown as Eq. (8-3) and Eq. (8-4), respectively:

$$\eta_{e,W} = \frac{m_p(e_{pout} - e_{pin})}{E_c + E_h} = \frac{m_p(e_{pout} - e_{pin})}{m_c(e_{cin} - e_{cout}) + m_h(e_{hin} - e_{hout})} \tag{8-3}$$

$$\eta_{e,R} = \frac{m_p(e_{pout} - e_{pin})}{P} \tag{8-4}$$

where E_c and E_h represent the exergy provided by the chilled water and the hot water, respectively, in Fig. 8-1; e_p is the exergy of the processed air, expressed as e_a in Eq. (8-5)[1]; and e_c and e_h represent the exergy of the chilled water and the hot water, respectively, expressed as e_w in Eq. (8-6).

$$e_a = c_{pa}T_0\left(\frac{T_a}{T_0} - 1 - \ln\frac{T_a}{T_0}\right) +$$
$$R_aT_0\left[(1 + 1.608 \times 10^{-3}\omega_a)\ln\frac{1 + 1.608 \times 10^{-3}\omega_0}{1 + 1.608 \times 10^{-3}\omega_a} + 1.608 \times 10^{-3}\omega_a\ln\frac{\omega_a}{\omega_0}\right] \tag{8-5}$$

$$e_w = c_{pw}T_0\left(\frac{T_w}{T_0} - 1 - \ln\frac{T_w}{T_0}\right) \tag{8-6}$$

where T is the temperature in Kelvin; and ω has the unit of g/kg.

According to Eqs. (8-1) ~ (8-4), when the inlet and outlet states of the processed air (A_{pin} and A_{pout}) are fixed, the lower Q_c+Q_h and P are, the higher COP_W and COP_R will be; similarly, the lower E_c+E_h and P are, the higher $\eta_{e,W}$ and $\eta_{e,R}$ will be.

Exergy takes into account both the heat or cooling capacity and temperature level of the heat or cooling media. Therefore, the factors that influence system performance and t_{hs} will be examined from the perspective of exergy.

8.1.3 Factors that influence exergy efficiency and t_{hs}

When the air is handled in the desiccant wheels and heat exchangers (i.e., coolers and heaters in Fig. 8-1; evaporators and condensers in Fig. 8-2), there exists heat and mass transfer exergy destruction because of the temperature and humidity ratio differences. E_c+E_h and P can be written as Eq. (8-7) and Eq. (8-8), respectively:

$$E_c + E_h = m_p(e_{pout} - e_{pin}) + m_r(e_{rout} - e_{rin}) + \Delta E_{DW} + \Delta E_{HE} + \Delta E_{mix} \tag{8-7}$$

$$P = m_p(e_{pout} - e_{pin}) + m_r(e_{rout} - e_{rin}) + \Delta E_{DW} + \Delta E_{HE} + \Delta E_{\varepsilon} \tag{8-8}$$

where ΔE_{DW} and ΔE_{HE} are the exergy destruction of all the desiccant wheels and all the heat exchangers, respectively; ΔE_{mix} is the exergy destruction resulting from chilled water or hot water of different outlet temperatures mixing together; and ΔE_{ε} is the exergy destruction of the compressor, which is equal to $T_0(Q_{cond}/T_{cond} - Q_{evap}/T_{evap})$, resulting from ε being lower than 1.

Eq. (8-7) and Eq. (8-8) show that under the fixed inlet and outlet states of the processed air, when the exergy destruction decreases, E_c+E_h and P can decrease and the system's exergy efficiency can increase.

For the system in Fig. 8-1, t_{hs} is normally higher than the reference temperature (t_0, chosen here to be the ambient dry-bulb temperature), and the temperature change of the hot water is almost constant under fixed working conditions. The relationship between t_w and e_w is illustrated in Fig. 8-3. When the mass flow rate of the hot water (m_h) is fixed, according to Eq. (8-3), the decrease of E_h leads to the reduction of the hot water inlet exergy (e_{hin}). And according to Fig. 8-3, when e_{hin} is reduced, t_{hs} is also reduced. In conclusion, under the fixed inlet and outlet states of the processed air, the hot water inlet temperature (t_{hs}) can be reduced when E_h is lower, which can be realized with lower exergy destruction and a higher E_c. For the system in Fig. 8-2, when

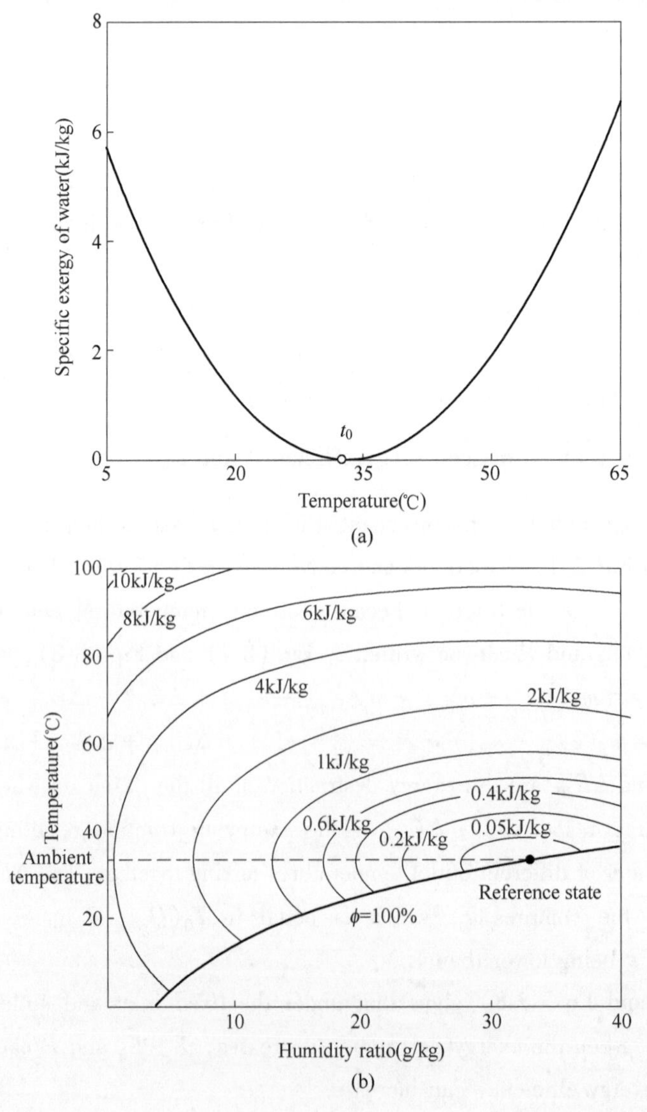

Fig. 8-3 Specific exergy of (a) water and (b) humid air in a psychrometric chart (the ambient saturated condition is chosen as the reference state)

exergy destruction decreases, P can also decrease. According to Eq. (8-2), the condensing temperature (T_{cond}), which is t_{hs} in Fig. 8-2 is reduced, and the evaporating temperature (T_{evap}), which is t_{cs} in Fig. 8-2 increases.

In summary, the heat source temperature (t_{hs}) is determined by the required E_h or P. In order to reduce t_{hs} and increase exergy efficiency, the systems' exergy destruction should be reduced. In the next section, the factors that influence the heat and mass transfer exergy destruction will be discussed.

8.2 Key performance influencing factors

There are three types of exergy destruction for the systems shown in Fig. 8-1 and Fig. 8-2: ΔE_{mix}, ΔE_{ε} and heat and mass transfer exergy destruction (ΔE_{DW}, ΔE_{HE}). ΔE_{mix} represents only a small part of the total exergy destruction, and ΔE_{ε} can be improved by using a compressor with a higher ε. Thus, the factors that influence heat and mass transfer exergy destruction are mainly discussed in this part.

Exergy destruction of heat transfer (ΔE_{htr}) and that of mass transfer (ΔE_{mtr}) are caused by the temperature differences (T_2-T_1) and humidity ratio differences ($\omega_2-\omega_1$) between the two heat and mass transfer media. The differential forms of heat and mass transfer exergy destruction ($d\Delta E_{htr}$ and $d\Delta E_{mtr}$, respectively) can be expressed as Eq. (8-9) and Eq. (8-10), respectively [2,3]:

$$d\Delta E_{htr} = T_0\left(\frac{1}{T_1} - \frac{1}{T_2}\right)\delta Q,\quad d\delta Q = h(T_2 - T_1)dF \tag{8-9}$$

$$d\Delta E_{mtr} = T_0 r\left(\frac{1}{T_{1,\,dew}} - \frac{1}{T_{2,\,dew}}\right)\delta M,\quad d\delta M = h_m(\omega_2 - \omega_1)\times 10^{-3}dF \tag{8-10}$$

where $d\Delta E_{htr}$ and $d\Delta E_{mtr}$ have the unit of kW; δQ (unit: kW) and δM (unit: kg/s) are the heat and mass transfer capacities, respectively, of area dF; r is the vaporization heat (unit: kJ/kg); T is the dry-bulb temperature; and T_{dew} is the dew point temperature.

Based on Eq. (8-9) and Eq. (8-10), heat and mass transfer exergy destruction (ΔE_{htr} and ΔE_{mtr}, respectively) of total area F can be approximately written as Eq. (8-11)[2]:

$$\Delta E_{htr} \approx \psi\frac{Q^2}{hF}\zeta_t,\quad \Delta Ex_{mtr} \approx \sigma\frac{M^2}{h_m F}\zeta_\omega \tag{8-11}$$

where ψ (unit: 1/K) and σ (unit: kJ/kg) can be assumed as constants [2]; Q and M represent the total heat and mass transfer capacities, respectively; and ζ_t and ζ_ω are the unmatched coefficients, which evaluate the uniformity of the temperature differences

($\Delta t = T_2 - T_1$) and humidity ratio differences ($\Delta\omega = \omega_2 - \omega_1$), respectively, between the two heat and mass transfer media.

ζ_t and ζ_ω can be commonly expressed as Eq. (8-12):

$$\zeta_t = \frac{\left(\int_0^F \Delta t^2 \mathrm{d}F\right)/F}{\left[\left(\int_0^F \Delta t \mathrm{d}F\right)/F\right]^2}, \quad \zeta_\omega = \frac{\left(\int_0^F \Delta\omega^2 \mathrm{d}F\right)/F}{\left[\left(\int_0^F \Delta\omega \mathrm{d}F\right)/F\right]^2} \tag{8-12}$$

The values of ζ_t and ζ_ω are greater than or equal to 1. The less uniform the Δt and $\Delta\omega$ fields are, higher the values of ζ_t and ζ_ω will be. ζ_t (or ζ_ω) will equal 1 when the field of Δt (or $\Delta\omega$) is uniform.

It can be seen from Eq. (8-11) that ΔE_{htr} (or ΔE_{mtr}) is influenced not only by Q (or M) and hF (or $h_m F$), but also by ζ_t (or ζ_ω). Thus, when Q and M are fixed, in order to reduce ΔE_{htr} and ΔE_{mtr}, apart from providing sufficient hF and $h_m F$, the uniformity of the Δt and $\Delta\omega$ fields have to be improved.

For the desiccant wheels, the heat transfer and mass transfer processes exist simultaneously. Thus, $\Delta E_{DW} = \Delta E_{DW,\ htr} + \Delta E_{DW,\ mtr}$. It has been proven that when A_r (facial area ratio of the processed air to the regeneration air) and F_r (air flow rate ratio of the processed air to the regeneration air) of the wheel are both equal to 1, the Δt and $\Delta\omega$ fields in the desiccant wheel are relatively uniform, and ζ_t and ζ_ω are close to 1[2]. When Q, M, hF and $h_m F$ are fixed, the lowest ΔE_{DW} (total exergy destruction of all the desiccant wheels) can be obtained when $A_r = F_r = 1$, and the influence of the number of stages on ζ_t and ζ_ω of each desiccant wheel, as well as that on ΔE_{DW}, is insignificant [2]. Thus, in the following analysis, A_r and F_r are both equal to 1 for all desiccant wheels.

For the heat exchangers, only heat transfer exists. For the coolers and heaters in Fig. 8-1, heat is transferred between two fluids with finite specific heat capacity. Thus, the temperature of the two fluids changes along the flow direction, as shown in Fig. 8-4 (a). For the evaporators and condensers in Fig. 8-2, one fluid is of finite specific heat capacity and the other is of infinite specific heat capacity. Thus, the temperature of one fluid changes and that of the other fluid stays the same along the flow direction, as shown in Fig. 8-4 (b). Based on Eq. (8-9), considering the counter-flow pattern, the heat transfer exergy destruction and the unmatched coefficient of the heat exchangers in Figs. 8-1 and 8-2 can be expressed as Eq. (8-13) and Eq. (8-14), respectively:

$$\Delta E_{HE} = \begin{cases} T_0\left(m_1 c_{p1} \ln \dfrac{T_{1out}}{T_{1in}} + m_2 c_{p2} \ln \dfrac{T_{2out}}{T_{2in}}\right), & \text{finite } c_{p1} \text{ and } c_{p2} \\ T_0 m_1 c_{p1}\left(\ln \dfrac{T_{1out}}{T_{1in}} - \dfrac{T_{1out} - T_{1in}}{T_2}\right), & \text{finite } c_{p1},\ c_{p2} \to +\infty \end{cases} \tag{8-13}$$

$$\zeta_{t,\ HE} = \begin{cases} \begin{cases} 1, & C_r = 1 \\ \dfrac{1/\eta_h - (1 + C_r)/2}{C_r - 1} \ln \dfrac{1/\eta_h - 1}{1/\eta_h - C_r}, & C_r \neq 1 \end{cases}, & \text{finite } c_{p1} \text{ and } c_{p2} \\ \left(\dfrac{1}{\eta_h} - \dfrac{1}{2}\right) \ln \dfrac{1}{1 - \eta_h}, & \textit{finite } c_{p1},\ c_{p2} \to +\infty \end{cases} \tag{8-14}$$

where subscripts 1 and 2 represent the air and the other fluid, respectively; η_h is the heat transfer efficiency of the air[4]; and $C_r = (m_1 c_{p1})/(m_2 c_{p2})$.

It can be seen in Eq. (8-14) that when c_{p1} *and* c_{p2} are finite, $\zeta_{t,\ HE}$ is influenced by C_r and η_h; when c_{p1} is finite and c_{p2} is infinite, $\zeta_{t,\ HE}$ is only influenced by η_h.

In conclusion, when Q (or M) is fixed, hF (or $h_m F$) and ζ_t (or ζ_ω) are two essential factors that influence heat (or mass) transfer exergy destruction. The number of stages has little influence on the exergy destruction of the desiccant wheel when A_r and F_r are equal to 1. Therefore, in the following two sections, ΔE_{HE} and $\zeta_{t,\ HE}$ are adopted to analyze the influence of the number of stages on t_{hs} when water and refrigerant are used as cooling/heating media. Eq. (8-13) and Eq. (8-14) are used to calculate ΔE_{HE} and $\zeta_{t,\ HE}$, respectively.

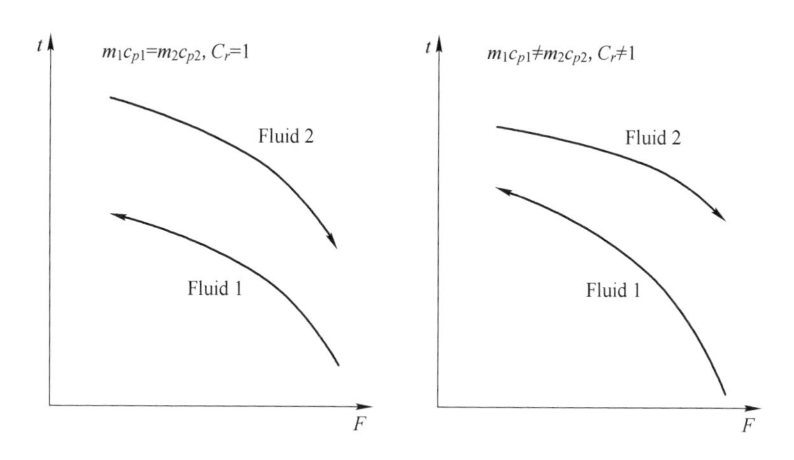

(a)

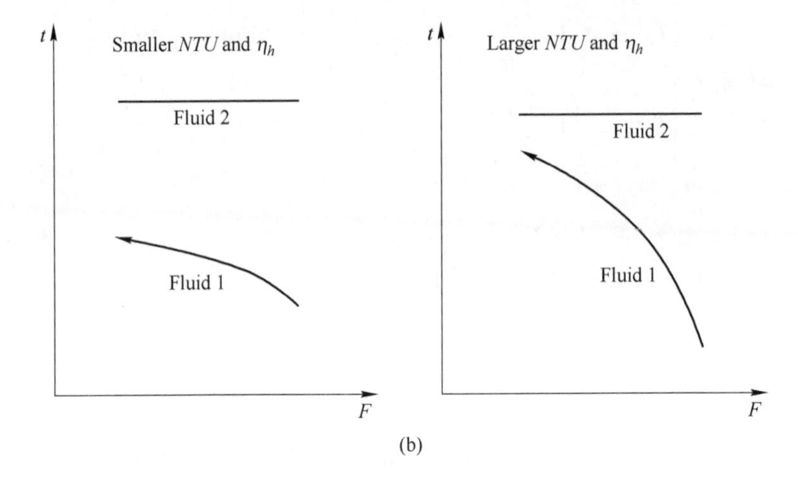

(b)

Fig. 8-4 Diagrams of the temperature difference distribution of the heat exchangers

(a) Two fluids with finite heat capacity; (b) Fluid 1 with finite heat capacity and fluid 2 with infinite heat capacity

8.3 Design criterion of high efficiency multi-stage wheel considering the form of cold and heat sources

8.3.1 Influence of the number of stages on t_{hs} of water-driven systems

8.3.1.1 Theoretical analysis

Fig. 8-1 is an example of a two-stage desiccant wheel dehumidification and cooling system. For each stage, there is one cooler and one heater. For a system with N stages, assume that the cooling and heating capacities (Q_c and Q_h, respectively), total heat transfer areas of the coolers and heaters (F_c and F_h, respectively), and total water flow rates of the chilled water and hot water (m_c and m_h, respectively) are all evenly distributed among all the coolers and heaters. Taking the cooling side as an example, the cooling capacity of each cooler is equal to Q_c/N; the area of each cooler is equal to F_c/N; and the water flow rate is equal to m_c/N. According to Eq. (8-14), the unmatched coefficients of each cooler ($\zeta_{t,\ HE}^{N}$) in an N-stage system are the same. Based on Eq. (8-11), for an N-stage system, the total exergy destruction of all the coolers ($\Delta E_{HE,\ c}^{N}$) can be expressed as Eq. (8-15):

$$\Delta E_{HE,\ c}^{N} \approx \sum_{i=1}^{N} \psi_{(i)} \frac{(Q_c/N)^2}{hF_c/N} \zeta_{t,\ HE}^{N} \approx \psi \frac{Q_c^2}{hF_c} \zeta_{t,\ HE}^{N} \tag{8-15}$$

It can be seen from Eq. (8-15) that when total Q_c and F_c are fixed, $\Delta E_{HE,\ c}^{N}$ is influenced by $\zeta_{t,\ HE}^{N}$. According to Eq. (8-14), when F_c is fixed, $\zeta_{t,\ HE}^{N}$ is influenced by the η_h and C_r of each cooler, which vary according to the number of stages.

When F_c and the mass flow rate of the air are fixed, the total air-side NTU of all the coolers (NTU_c), expressed as $(hF)/(m_a c_{pa})$, is also fixed. The influence of the number of stages on $\zeta_{t,\ HE}^{N}$ can be calculated for different values of m_c; the results are shown in Fig. 8-5. C_r of each cooler of an N-stage system is equal to $(m_a c_{pa})/(m_c c_{p_w}/N)$ during the calculation of η_h [4]. It can be seen from Fig. 8-4 (a) and Fig. 8-5 that for different m_c values, when C_r of each cooler is equal to 1, the temperature difference is uniform and $\zeta_{t,\ HE}^{N}=1$ can be obtained. According to Eq. (8-15), the lowest exergy destruction can thus be obtained. The same conclusions apply to the heaters.

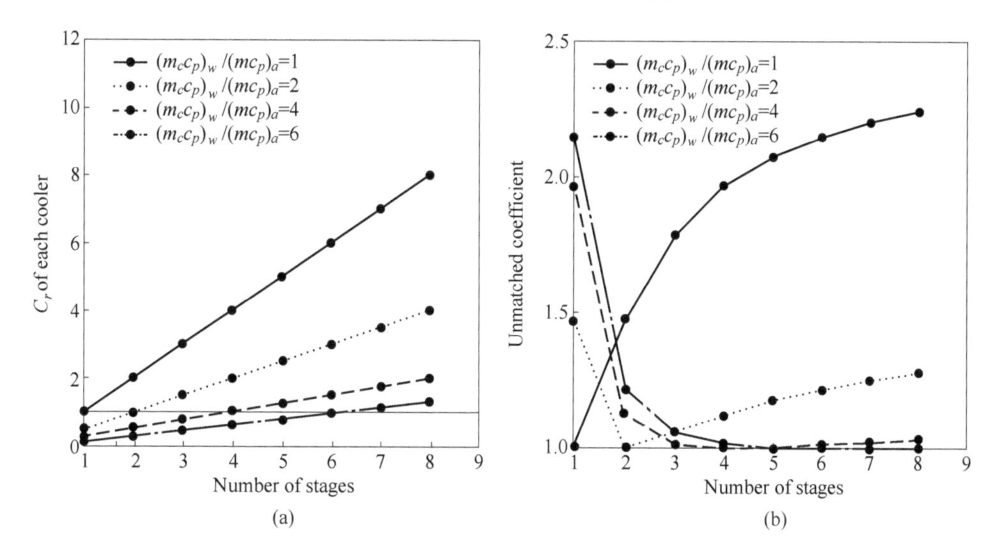

Fig. 8-5 Influence of the number of stages when NTU_c is fixed (NTU_c = 5 in this case)

(a) C_r of each cooler; (b) The unmatched coefficient of each cooler

With the lowest ΔE_{HE}, according to Eq. (8-3) and Eq. (8-7), the highest $\eta_{e,\ W}$ and the lowest E_c+E_h can be obtained, and the lowest t_{hs} can also possibly be obtained.

8.3.1.2 Simulation analyses

Performance of the desiccant wheel systems was calculated through numerical simulation. The desiccant wheels and heat exchangers are the main components. The mathematical model of the desiccant wheels utilized here has been established and validated in previous research [5,6]. This model considers the resistance of both gas and solid sides. For the solid side, the ordinary diffusion of the vapor, surface diffusion of the absorbed water, and heat conductivity of the desiccant material in the wheel thickness direction are all considered. The η_h -NTU method [4] is used to simulate the heat exchangers.

The influence of the number of stages on t_{hs} of the system in Fig. 8-1 was simulated under the working conditions listed in Table 8-1. The mass flow rate of the air (m_a) is fixed. The structure of the desiccant wheel for the single-stage system is described in Table 8-1; the wheel's total thickness is equal to 0.2m. The total air-side NTU of the coolers (NTU_c) and that of the heaters (NTU_h) are both equal to 5. The total mass flow rate of the chilled water (m_c) is the same as that of the hot water (m_h). For multi-stage systems, wheel thickness, NTU_c, NTU_h, m_c and m_h are evenly divided among all the stages. To achieve the required supplied air state (24℃, 10g/kg), the following cases are analyzed: $(m_c c_{p_w})/(m_a c_{pa})$ values of 1, 2 and 4, and the number of stages being 1, 2 and 4.

Table 8-1 Information related to working conditions, desiccant wheels, and heating/cooling systems

Working conditions	Desiccant wheels	Heating and cooling systems
Processed air inlet: 33℃, 19g/kg, 0.8kg/s; Regeneration air: 26℃, 12g/kg, 0.8kg/s; Supplied air humidity ratio: 10g/kg; Supplied air temperature: 24℃ for the system in Fig. 8-1. Reference state: 33℃, φ = 100%	Radius: 0.5m; Thickness: 0.2m; $A_r = F_r = 1$; Material: silica gel; Air channel structure: sinusoidal shape, 2mm high, and 2mm wide	Air-side NTU for coolers and evaporators (NTU_c): 5; Air-side NTU for heaters and condensers (NTU_h): 5; All heat exchangers are evenly divided under different numbers of stages; The cold water and hot water have the same mass flow rate; Thermodynamic perfectness of the compressor: 0.5

For cases with different $(m_c c_{p_w})/(m_a c_{pa})$ values, the air states along the air flow direction almost coincide with each other when the number of stages is identical. Fig. 8-6 shows the air handling process of the processed air and the regeneration air when $(m_c c_{p_w})/(m_a c_{pa}) = 1$ for different numbers of stages. It can be seen that as the number of stages increases, t_{reg} (the temperature of the regeneration air entering the wheel) decreases from 68.6℃ in the single-stage system to 50.5~53.5℃ in the two-stage system, and then to 40.4 ~ 48.8℃ in the four-stage system. This shows that multiple stages and inter-stage cooling are effective ways to reduce t_{reg}.

The unmatched coefficients of each desiccant wheel and heat exchanger are listed in Table 8-2; it can be seen that the unmatched coefficients of the desiccant wheels ($\zeta_{t,\ DW}$)

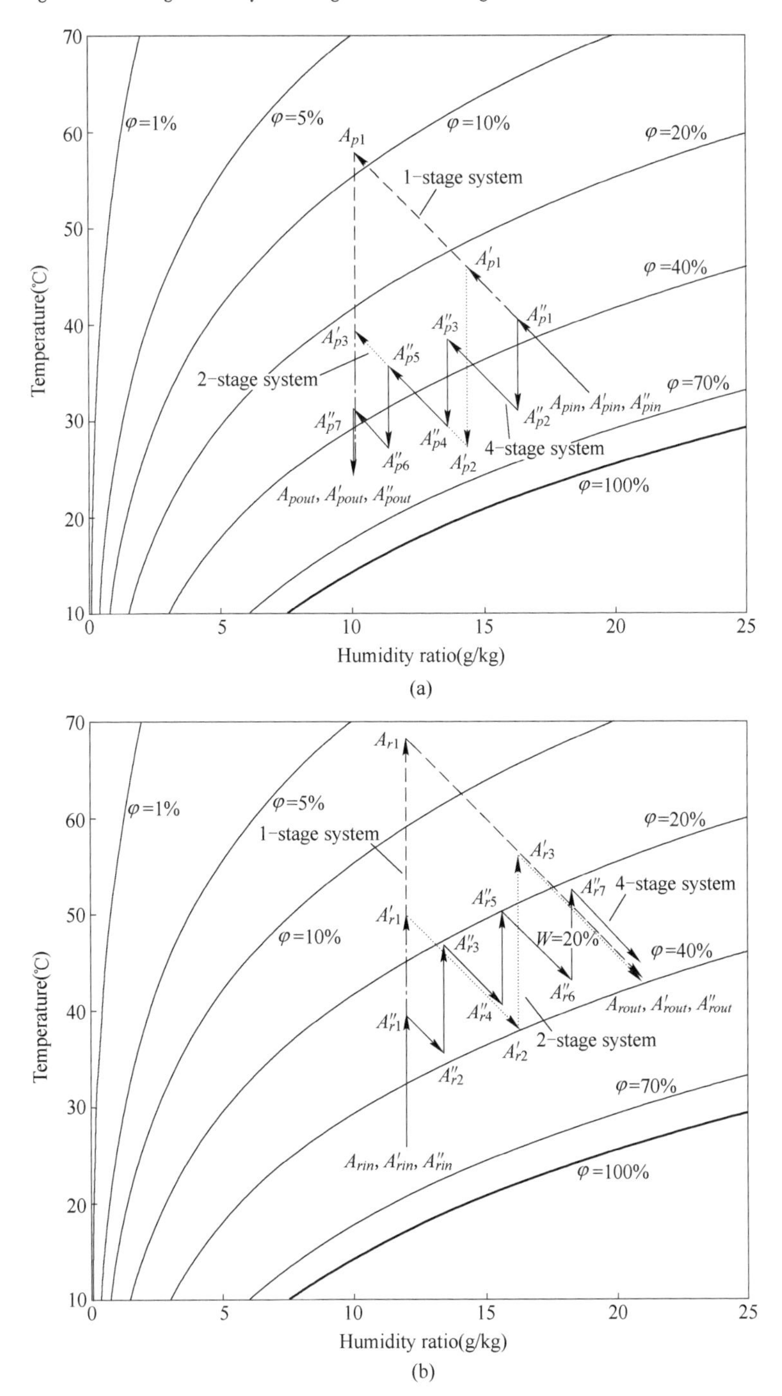

Fig. 8-6 Air handling process when $(m_c c_{pw})/(m_a c_{pa}) = 1$ for systems with different stages

(a) Processed air; (b) Regeneration air

Table 8-2 Unmatched coefficients of each heater, cooler, and desiccant wheel for different numbers of stages when $NTU_c = NTU_h = 5$

$(m_c c_{pw})/(m_a c_{pa})$	Component	1-stage system	2-stage system	4-stage system
1	Heater/cooler	1.00	1.47	1.97
	Desiccant wheel	1.12	1.10	1.16
2	Heater/cooler	1.47	1.00	1.13
	Desiccant wheel	1.12	1.10	1.15
4	Heater/cooler	1.97	1.13	1.00
	Desiccant wheel	1.12	1.10	1.15

are all close to 1. $(m_c c_{pw})/(m_a c_{pa})$ and the number of stages have little influence on $\zeta_{t,\ DW}$. However, they have tremendous influence on the unmatched coefficients of the coolers and heaters.

Figs. 8-7, 8-8 and 8-9 show the required heat source temperature (t_{hs}), cooling source temperature (t_{cs}), exergy efficiency, and exergy information of each component for different $(m_c c_{pw})/(m_a c_{pa})$ values and numbers of stages. During the calculation of exergy, the ambient saturated condition is chosen as the reference state[2], as shown in Fig. 8-3 (b).

Fig. 8-7 shows the results when $(m_c c_{pw})/(m_a c_{pa}) = 1$. The unmatched coefficients of each exchanger ($\zeta_{t,\ HE}$) are equal to 1 when the number of stages is equal to 1. It can be seen from Fig. 8-7 (a) and Fig. 8-7 (b) that the exergy obtained by the processed air, the exergy provided by the regeneration air, and the exergy destruction of the desiccant wheel (ΔE_{DW}) do not change much for the different cases; the exergy destruction resulting from mixing (ΔE_{mix}) is so small that it can be ignored. The variance of the number of stages mainly influences the exergy destruction of the heat exchangers (ΔE_{HE}). It can be seen from Table 8-2 that $\zeta_{t,\ HE}$ increases from 1 in the single-stage system to 1.47 in the two-stage system, and then to 1.97 in the four-stage system. According to Eq. (8-15), a higher $\zeta_{t,\ HE}$ leads to a higher ΔE_{HE}. As seen in Fig. 8-7 (b), the exergy destruction of the heat exchangers (ΔE_{HE}) increases from 1.41kW in the single-stage system to 2.19kW in the two-stage system, and then to 3.15kW in the four-stage system. Similarly, the total exergy provided by the cooling and heating sources ($E_c + E_h$) increases from 2.21kW to 3.05kW and then to 4.16kW in the three systems. COP_W does not change much, but the exergy efficiency decreases from 41.3% to 30.0% and then to 22.0% in the three systems.

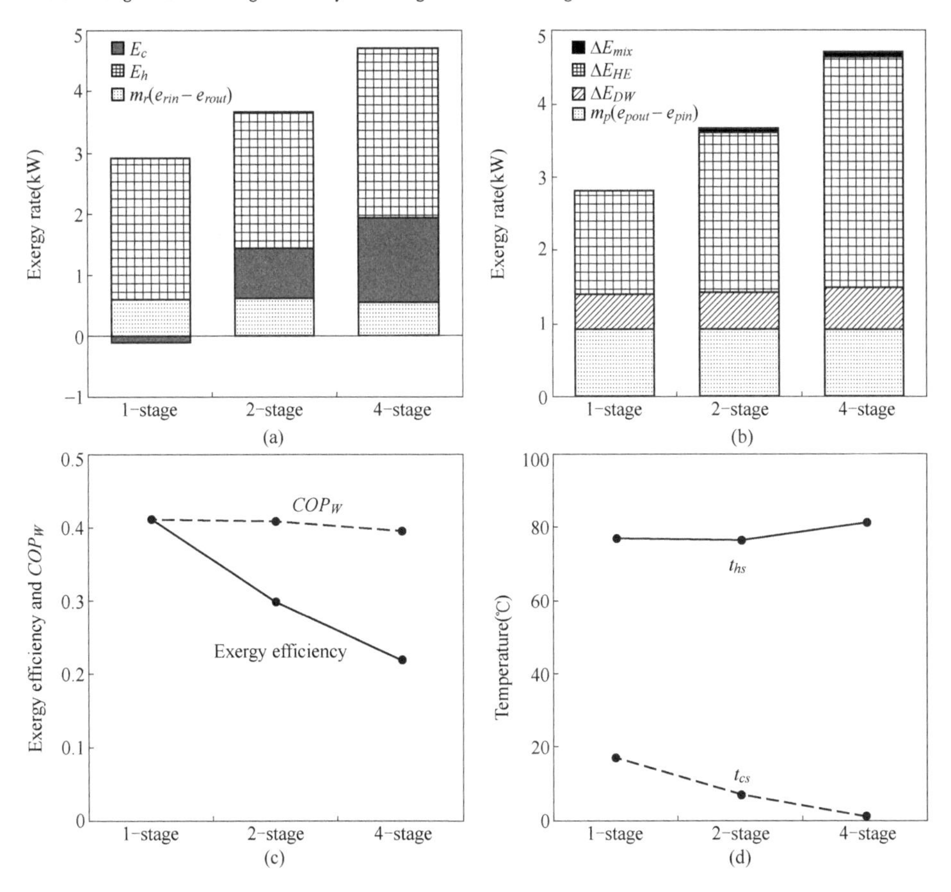

Fig. 8-7 Effect of the number of stages on performance when $(m_c c_{pw})/(m_a c_{pa})=1$

(a) Exergy provided by the cooling and heat sources and the regeneration air; (b) Exergy destruction and exergy obtained by the processed air; (c) Exergy efficiency and COP_W; (d) t_{hs} and t_{cs}

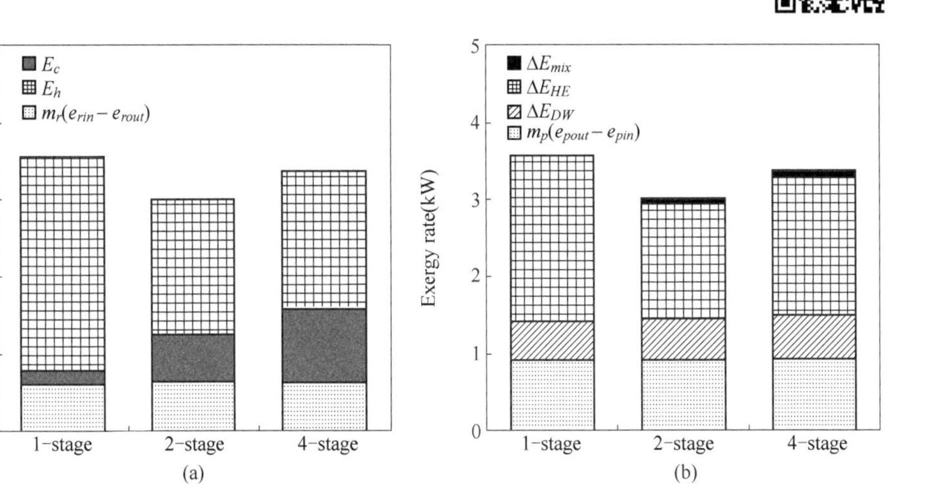

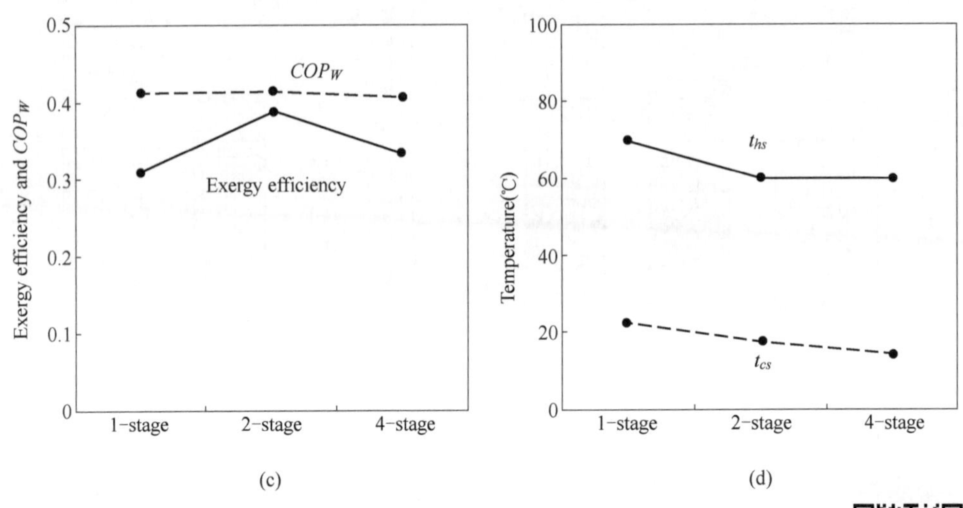

Fig. 8-8 Effect of the number of stages on performance when $(m_c c_{pw})/(m_a c_{pa})=2$

(a) Exergy provided by the cooling and heat sources and the regeneration air; (b) Exergy destruction and exergy obtained by the processed air; (c) Exergy efficiency and COP_W; (d) t_{hs} and t_{cs}

For the processed air side, the supplied air temperature is fixed. As the number of stages increases, the heat transfer efficiency of each cooler is reduced, and t_{cs} decreases from 17.2℃ in the single-stage system to 7.1℃ in the two-stage system, and then to 1.5℃ in the four-stage system, leading to the increase of E_c from −0.11kW to 0.83kW, and then to 1.38kW, respectively.

For the regeneration air side, although t_{reg} decreases as the number of stages increases, the heat source temperature is not always reduced. When the system changes from a single-stage system to a two-stage system, t_{hs} decreases slightly from 77.0℃ to 76.1℃. This is because the increase of E_c (0.94kW) is larger than that of ΔE_{HE} (0.78kW); thus, E_h is slightly reduced from 2.32kW to 2.23kW, resulting in the decrease of t_{hs}. However, t_{hs} increases back to 81.5℃ when the system becomes a four-stage system. This is because the increase of ΔE_{HE} (0.96kW) is larger than that of E_c (0.55kW) when the system changes from a two-stage system to a four-stage system, leading to the increase of E_h from 2.23kW to 2.78kW. Thus, the two-stage system can obtain the lowest t_{hs} instead of the four-stage system, which has the lowest t_{reg}. Considering that t_{hs} of the single-stage system is close to that of the two-stage system, and that t_{cs} of the single-stage system is much higher than that of the two-stage system, one stage is recommended as the optimal number of stages.

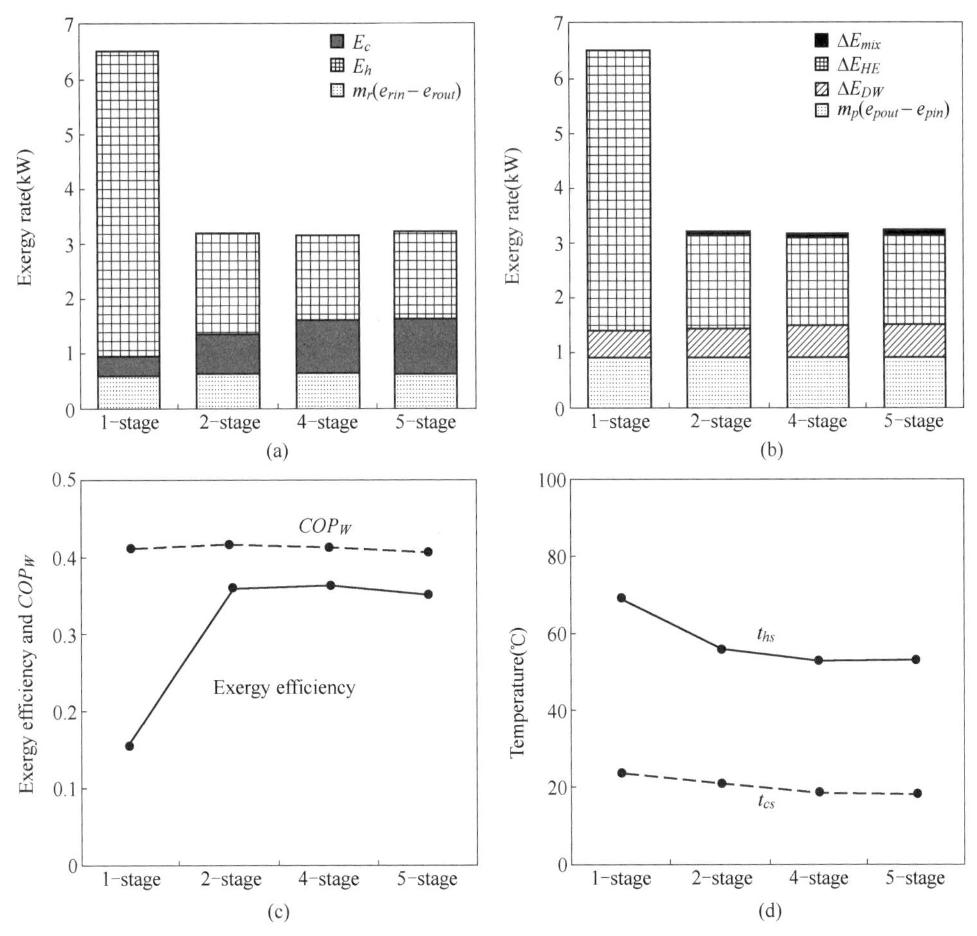

Fig. 8-9 Effect of the number of stages on performance when $(m_c c_{pw})/(m_a c_{pa}) = 4$

(a) Exergy provided by the cooling and heat sources and the regeneration air; (b) Exergy destruction and exergy obtained by the processed air; (c) Exergy efficiency and COP_W; (d) t_{hs} and t_{cs}

Fig. 8-8 shows the results when $(m_c c_{pw})/(m_a c_{pa}) = 2$. The unmatched coefficients ($\zeta_{t,\ HE}$) of each exchanger equal 1 when the number of stages is 2. As is the case when $(m_c c_{pw})/(m_a c_{pa}) = 1$, the change in the number of stages mainly influences ΔE_{HE}. It can be seen from Table 8-2 that $\zeta_{t,\ HE}$ decreases from 1.47 in the single-stage system to 1.00 in the two-stage system, and then increases to 1.13 in the four-stage system. ΔE_{HE} decreases from 2.17kW in the single-stage system to 1.51kW in the two-stage system, and then increases to 1.80kW in the four-stage system. $E_c + E_h$ decreases from 2.96kW to 2.36kW and then increases to 2.74kW in a similar progression. COP_W

does not change much, but the exergy efficiency increases from 30.9% to 38.8% and then decreases to 33.5% in the three systems.

Similar to when $(m_c c_{pw})/(m_a c_{pa}) = 1$, t_{cs} decreases from 22.5℃ to 17.9℃ and then to 17.3℃ as the number of stages increases, and E_c increases from 0.18kW to 0.60kW and then to 0.95kW. When the system changes from a single-stage system to a two-stage system, t_{hs} decreases from 70.5℃ to 60.4℃. This is because both the decrease of ΔE_{HE} (from 2.17kW to 1.51kW) and the increase of E_c lead to the decrease of E_h (from 2.78kW to 1.76kW). However, as the number of stages changes from 2 to 4, the change of t_{hs} (i.e., to 60.5℃ in the four-stage system) is insignificant. This is because for the four-stage system, the increase of $\zeta_{t, HE}$ is not significant, and the increases of both ΔE_{HE} and ΔE_{DW} are almost identical to that of E_c. Evidently, two-stage is the optimal number of stages.

Fig. 8-9 shows the results when $(m_c c_{pw})/(m_a c_{pa}) = 4$; the results for the five-stage system are given. The unmatched coefficients ($\zeta_{t, HE}$) of each exchanger are equal to 1 when the number of stages is 4. As with the other cases, the change in the number of stages mainly influences ΔE_{HE}. It can be seen from Table 8-2 that $\zeta_{t, HE}$ decreases from 1.97 in the single-stage system to 1.13 in the two-stage system, and then decreases to 1.00 in the four-stage system. The lowest ΔE_{HE} (1.61kW), the highest exergy efficiency (36.5%), and the lowest t_{hs} (53.0℃) are all obtained when the number of stages is 4. For the five-stage system, $\zeta_{t, HE} = 1.01$ and t_{hs} increases to 53.2℃. Clearly, four-stage is the optimal number of stages.

The above results prove that although the increase in the number of stages and inter-stage cooling are both effective ways to reduce t_{reg}, there exists an optimal number of stages to obtain the lowest t_{hs} that is influenced by $(m_c c_{pw})/(m_a c_{pa})$. Considering t_{hs} and t_{cs}, the optimal number of stages is chosen when C_r of each heat exchanger is equal to 1, and the lowest exergy destruction can be obtained with $\zeta_{t, HE}$ equaling to 1. When $(m_c c_{pw})/(m_a c_{pa})$ is small (e.g., equal to 1), multi-stage dehumidification is not appropriate. On the contrary, multi-stage dehumidification is preferable when $(m_c c_{pw})/(m_a c_{pa})$ is large. When $(m_c c_{pw})/(m_a c_{pa})$ is equal to 2 and 4, the lowest t_{hs} is obtained when the number of stages is equal to 2 and 4, respectively. In these optimal cases, the highest exergy efficiency is also obtained.

8.3.1.3 Discussion

When the air mass flow rate (m_a) is fixed, the proper water mass flow rate (m_w) and the number of stages should be chosen for the relatively low t_{hs} and high t_{cs}. Fig. 8-10 compares the optimal cases for different $(m_c c_{pw})/(m_a c_{pa})$ values: A presents $(m_c c_{pw})/(m_a c_{pa}) = 1$, one stage; B presents $(m_c c_{pw})/(m_a c_{pa}) = 2$, two stages; C presents $(m_c c_{pw})/(m_a c_{pa}) = 4$, four stages; and D presents $(m_c c_{pw})/(m_a c_{pa}) = 4$, two stages.

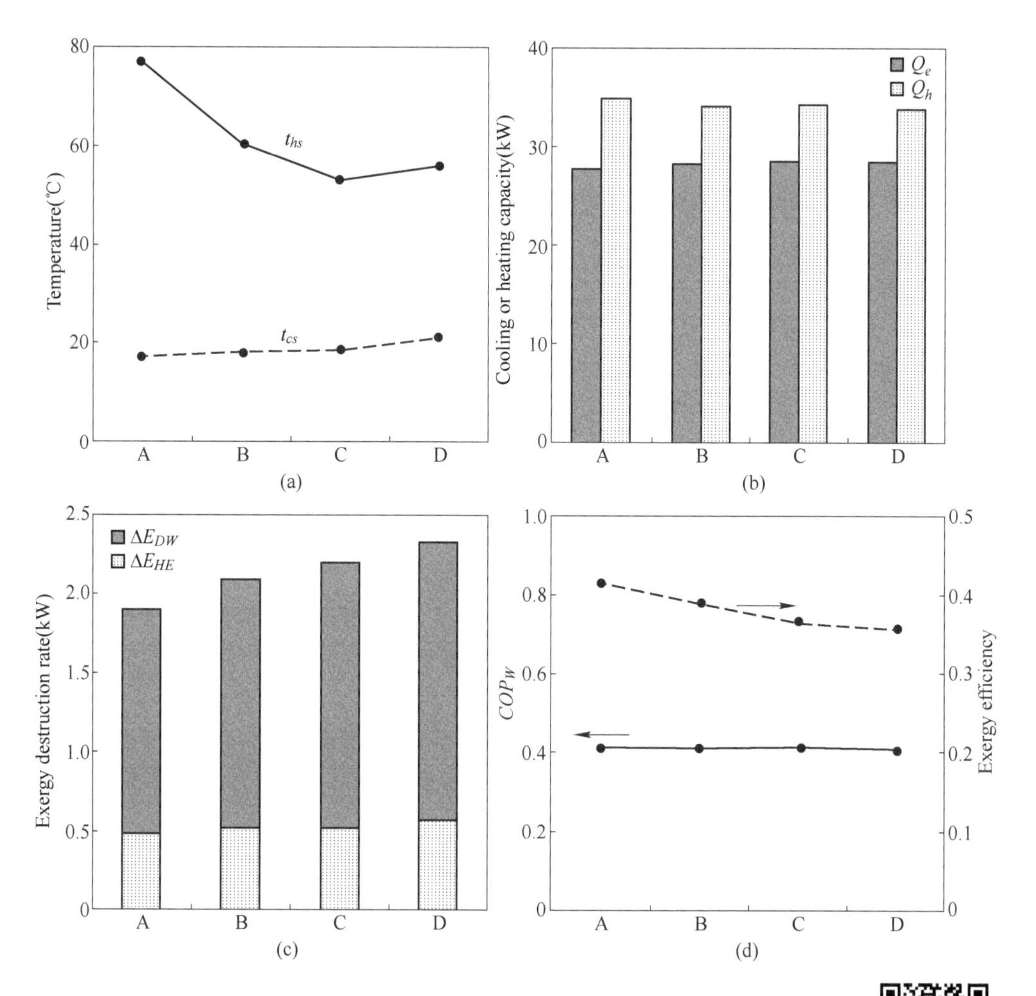

Fig. 8-10 Comparison of cases A ~ D

(a) t_{cs} and t_{hs}; (b) Cooling capacity of the cold water (Q_c) and heating capacity of the hot water (Q_h); (c) Exergy destruction of the desiccant wheels and the heat exchangers; (d) COP_W and exergy efficiency

The results show that when C_r of each stage is equal to 1 (cases A, B and C), as the number of stages increases, COP_W values for the different cases are almost identical because the changes of Q_c and Q_h are so small. Exergy efficiency is reduced because of the increase of ΔE_{HE}. However, from case A to case C, m_w increases, leading to the reduction of t_{hs} from 77.0℃ to 53.0℃ and the increase of t_{cs} from 17.2℃ to 18.6℃. Thus, case C is preferable. Case D reflects a similar performance as case C because $\zeta_{t,HE}$ of case D is close to 1, as shown in Table 8-2. Although t_{hs} of case D (55.8℃) is higher than that of case C (53.0℃), t_{cs} of case D (21.0℃) is higher than that of case C (18.6℃), and case D has a simpler structure.

In summary, to obtain a relatively low heat source temperature, a large mass flow rate of water should be guaranteed first. Accordingly, the proper number of stages should be chosen based on the rule that the unmatched coefficients of each heat exchanger should be kept close to 1. In the end, a $(m_c c_{pw})/(m_a c_{pa})$ value of around 4 and the number of stages being 2 ~ 4 are considered optimal. Under the working conditions listed in Table 8-1, t_{hs} is around 55℃ and t_{cs} is higher than 18℃, making possible the use of low-grade heat sources.

8.3.2 Influence of the number of stages on t_{hs} of refrigerant-driven systems

8.3.2.1 Theoretical analysis

Fig. 8-2 illustrates the example of a heat pump-driven two-stage desiccant wheel system. In each stage, there is one evaporator and one condenser. In addition, there is a condenser (Condenser 0) at the outlet of the regeneration air used to dissipate the extra heat from the heat pump system. Taking the cooling side of the N-stage system as an example, when Q_c and F_c are the same for different numbers of stages and are evenly divided among the N evaporators, the unmatched coefficients of each evaporator are the same. The total exergy destruction of all the evaporators ($\Delta E_{HE,c}^{N}$) can also be expressed as Eq. (8-15). $\Delta E_{HE,c}^{N}$ is influenced by the unmatched coefficient of each evaporator ($\zeta_{t,HE}^{N}$). When F_c is fixed, according to Eq. (8-14), $\zeta_{t,HE}^{N}$ is influenced by the number of stages because of the change of η_h for each evaporator.

When F_c and air mass flow rate are fixed, total air-side NTU of all the evaporators (NTU_c) is fixed. The influence of the number of stages for different NTU_c values is calculated, and the results are shown in Fig. 8-11. It can be seen that for different NTU_c values, as the number of stages increases, η_h of each evaporator decreases for the reduced heat exchange area, and the unmatched coefficient is reduced, too. This can be explained by Fig. 8-4 (b). The temperature of fluid 2 stays the same while the temperature of fluid 1 changes. When η_h increases, the temperature change of fluid 1 will be enhanced, which will lessen the uniformity of the temperature

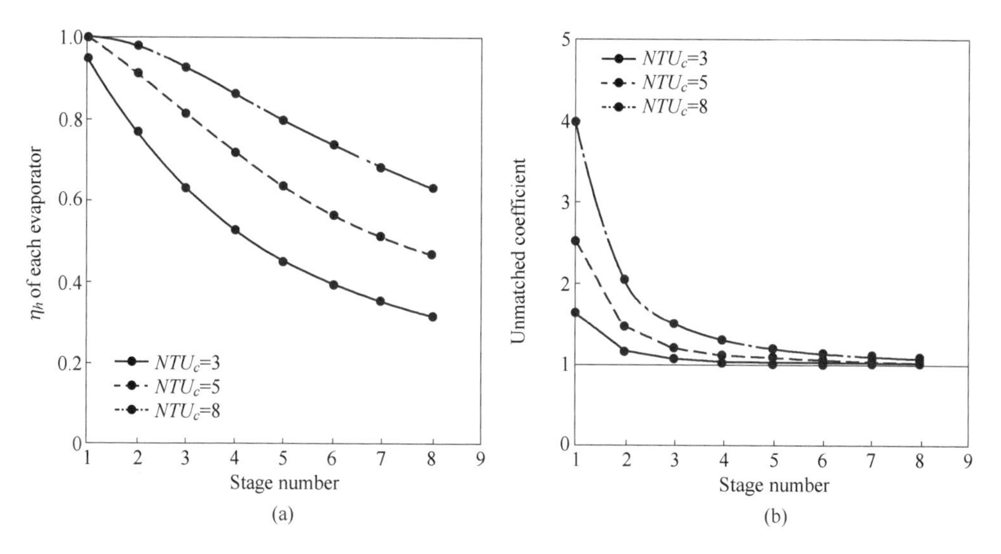

Fig. 8-11 Influence of the number of stages when NTU_c equals 3, 5 and 8 on

(a) η_h of each evaporator; (b) The unmatched coefficient of each evaporator

difference field and lead to a larger unmatched coefficient. Fig. 8-11 also shows that when NTU_c is large, the effect of increasing the number of stages will be more obvious. When the number of stages is larger than 4, the decrease of $\zeta_{t,\ HE}^{N}$ is not significant. The same conclusions apply to the condensers as well.

These results prove that increasing the number of stages is an efficient way to reduce the exergy destruction of evaporators; according to Eq. (8-8), P and t_{hs} can also be reduced.

8.3.2.2 Simulation analysis

The influence of the number of stages on t_{hs} of the system in Fig. 8-2 was simulated under the working conditions listed in Table 8-1. The mass flow rate of the air (m_a) is fixed. The structure of the desiccant wheel is described in Table 8-1; the wheel's total thickness is equal to 0.2m. The total air-side NTU of the evaporators (NTU_c) and that of the condensers (NTU_h) are both equal to 5. The supplied air humidity ratio is fixed at 10g/kg. The thermodynamic perfectness of the compressor (ε) is fixed at 0.5. 1-, 2- and 4-stage systems are analyzed.

Fig. 8-12 shows the air handling process of the processed air and the regeneration air for different numbers of stages. Because the heat has to be dissipated entirely by the regeneration air and ε is fixed, the supplied air temperature cannot be fixed. It can be

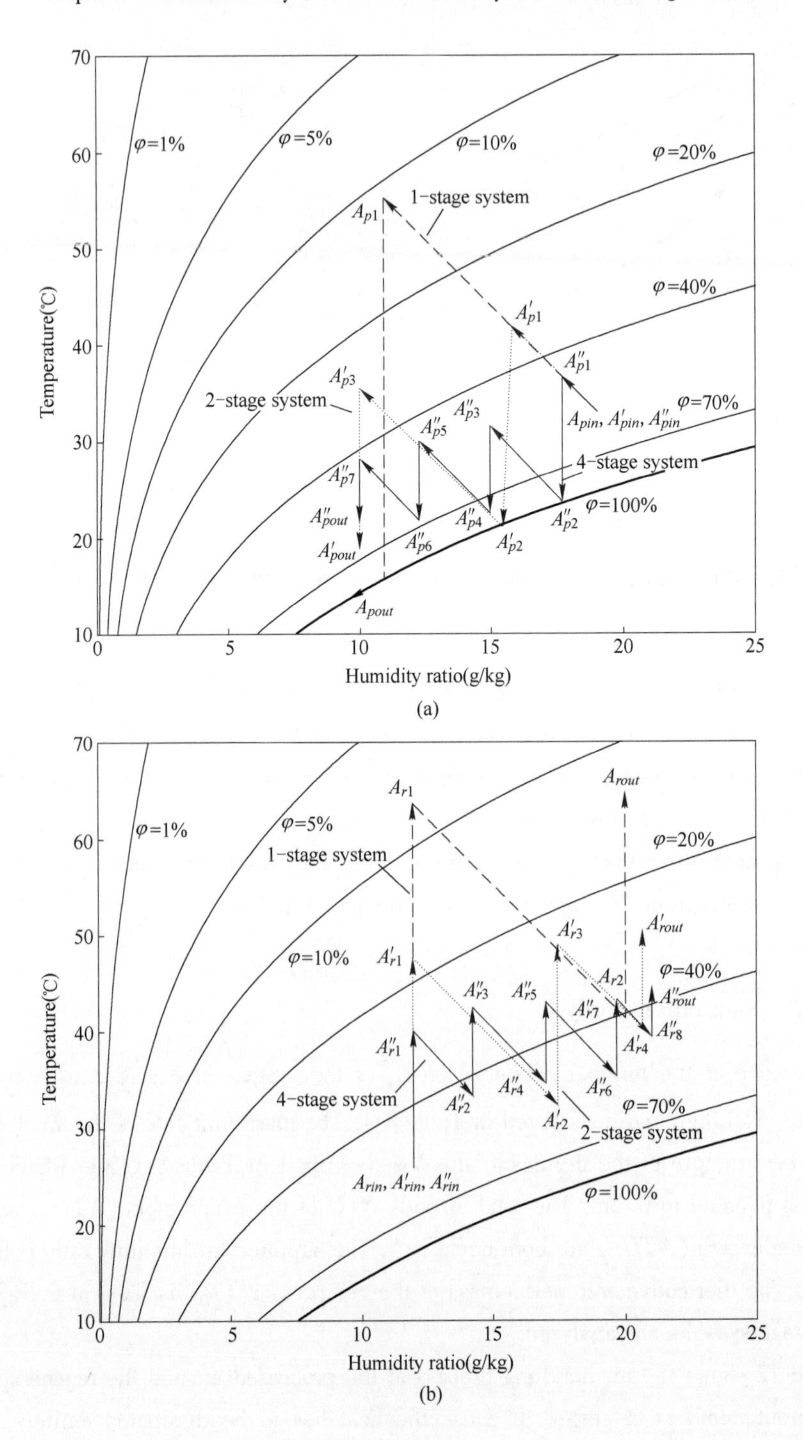

Fig. 8-12 Air handling process when $NTU_c = NTU_h = 5$ of systems with different stages

(a) Processed air; (b) Regeneration air

seen from Fig. 8-12 that for the single-stage system, condensation occurs at the evaporator, and the desiccant wheel comprises 89.8% of the total dehumidification capacity. Fig. 8-13 shows that as the number of stages increases, the supplied air temperature increases, and both the cooling capacity of the evaporators and the heating capacity of the condensers decrease.

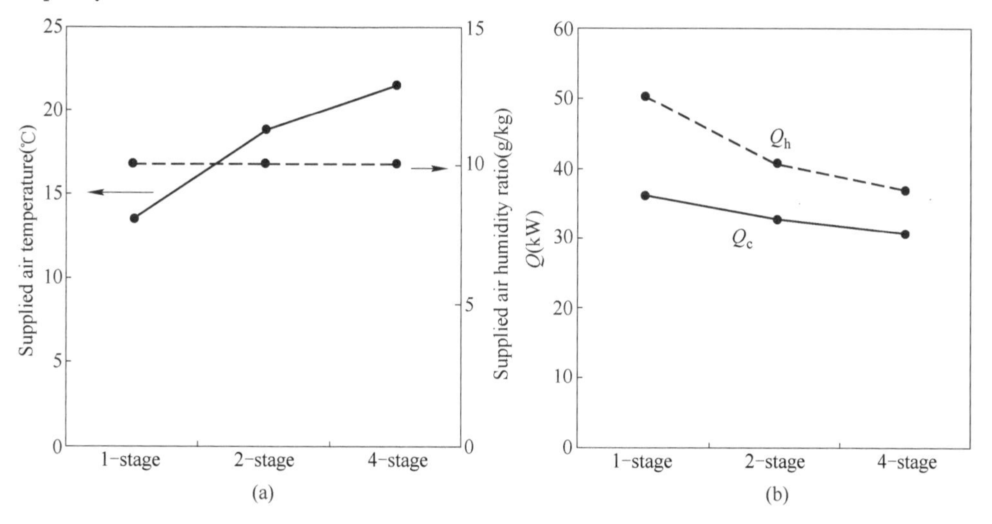

Fig. 8-13 Influence of the number of stages when $NTU_c = NTU_h = 5$

(a) Supplied air temperature and humidity ratio; (b) Cooling capacity of the evaporator (Q_c) and heating capacity of the condensers (Q_h)

Fig. 8-14 illustrates the results for exergy, evaporating temperature (t_{cs}), condensing temperature (t_{hs}), exergy efficiency and COP_R. It can be seen that for all cases, exergy is mainly provided by the compressor (P). As the number of stages increases, P decreases. Although the supplied air temperature increases, the changes of exergy and cooling capacity obtained by the processed air are not significant. $\eta_{e,\ R}$ and COP_R increase from 9.3% and 2.41, respectively, in the single-stage system to 13.5% and 3.74, respectively, in the two-stage system, and then to 16.0% and 4.50, respectively, in the four-stage system.

The reduction of P mainly results from the reduction of both the exergy destruction of the evaporators and condensers (ΔE_{HE}) and the exergy destruction of the compressor (ΔE_{ε}). Table 8-3 shows the unmatched coefficients and cooling/heating capacities of each evaporator and condenser for different numbers of stages. It can be seen that as the number of stages increases from 1 to 4, the unmatched coefficients of the evaporators and condensers decrease from 2.53 to 1.13 and then from 1.47 to 1.08.

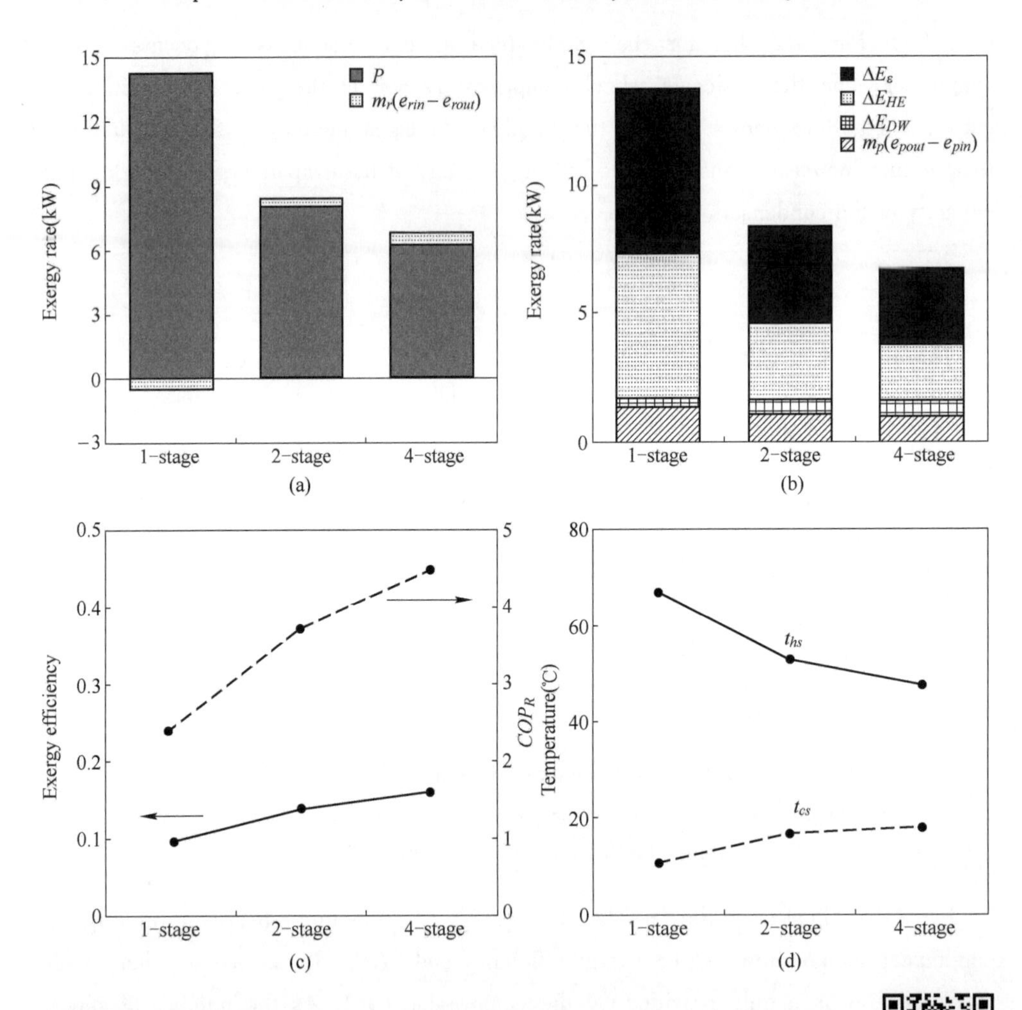

Fig. 8-14　Effect of the number of stages on performance when $NTU_c = NTU_h = 5$

(a) Exergy provided by the compressor and the regeneration air; (b) Exergy destruction and exergy obtained by the processed air; (c) Exergy efficiency and COP_R; (d) t_{hs} and t_{cs}

The exergy destruction of all the evaporators (or condensers) decreases from 2.80kW (or 2.80kW) in the single-stage system to 1.36kW (or 1.62kW) in the two-stage system, and then to 0.97kW (or 1.23kW) in the four-stage system. Thus, during heat transfer processes at the evaporators and condensers, the exergy provided by the refrigerant decreases from 2.83kW (or 5.02kW) to 1.78kW (or 2.49kW), and then to 1.52kW (or 1.72kW) in the single-stage, two-stage and four-stage systems, respectively, leading to an increase of the evaporative temperature and a reduction of the condensing temperature, as shown in Fig. 8-14 (d).

In conclusion, for the system in Fig. 8-2, as the number of stages increases, t_{reg} decreases; the unmatched coefficient and the exergy destruction decrease, too. This leads to a reduced condensing temperature (t_{hs}) and an increased evaporative temperature. So that higher $\eta_{e,R}$ and COP_R values can be obtained. When the number of stages is larger than 4, the rate of improvement slows down due to the insignificant change of the unmatched coefficient. A 4-stage system with COP_R around 4.5 is preferable.

Table 8-3 Cooling and heating capacities and unmatched coefficients of each evaporator and condenser for different numbers of stages when $NTU_c = NTU_h = 5$

Stage system	1-stage system		2-stage system			4-stage system				
Evaporator	Evaporator 1		Evaporator 1		Evaporator 2	Evaporator 1	Evaporator 2	Evaporator 3	Evaporator 4	
Q_c(kW)	36.1		18.8		13.8	10.8	7.6	6.7	5.7	
Unmatched coefficient	2.53		1.47			1.13				
Condenser	Condenser 0	Condenser 1	Condenser 0	Condenser 1	Condenser 2	Condenser 0	Condenser 1	Condenser 2	Condenser 3	Condenser 4
Q_h(kW)	19.6	30.8	8.8	14.0	17.9	4.5	6.6	6.9	7.6	11.4
Unmatched coefficient	1.47		1.22			1.08				

Nomenclature

A	air
A_r	facial area ratio of the processed air to the regeneration air
COP	coefficient of performance
C_r	the value of $(m_a c_{pa})/(m_w c_{pw})$ for each heat exchanger
c_p	specific heat capacity, kJ/kg
e	specific exergy, kJ/kg
E	exergy rate, kW
ΔE	exergy destruction rate, kW
F	area, m^2
F_r	air flow rate ratio of the processed air to the regeneration air
h	heat transfer coefficient, $kW/(m^2 \cdot K)$
h_m	mass transfer coefficient, $kg/(m^2 \cdot s)$
i	specific enthalpy, kJ/kg

m	mass flow rate, kg/s
NTU	number of heat transfer units
Q	heat or cooling rate, kW
P	power of compressor, kW
W	water
r	vaporization heat, kJ/kg
R_a	gas constant for air, kJ/(mol · K)
t	Celsius temperature, ℃
T	Kelvin temperature, K

Greek symbols

ω	humidity ratio, g/kg
η_e	exergy efficiency
η_h	heat transfer efficiency
ζ	unmatched coefficient
φ	relative humidity ratio of the humid air
ε	thermodynamic perfectness of the compressor

Subscripts

a	air
w	water
c	chilled water or cooler

References

[1] A Bejan. Advanced engineering thermodynamics [M]. 3rd ed. Hoboken, NJ: John Wiley & Sons, 2006.

[2] R Tu, X H Liu, Y Jiang. Lowering the regeneration temperature of a rotary wheel dehumidification system using exergy analysis [J]. Energy Convers. Manage, 2015, 89 (1): 162-74.

[3] L Zhang, X H Liu, Y Jiang. Ideal efficiency analysis and comparison of condensing and liquid desiccant dehumidification [J]. Energy Build, 2012, 49: 575-583.

[4] J P Holman. Heat transfer [M]. 10th ed. New York: McGraw-Hill Companies, 2009.

[5] R Tu, X H Liu, Y Jiang. Performance analysis of a two-stage desiccant cooling system [J]. Appl. Energy, 2014, 113: 1562-1574.

[6] R Tu, X H Liu, Y Jiang. Performance analysis of a new kind of heat pump-driven outdoor air processor using solid desiccant [J]. Renew. Energy, 2013, 57: 101-110.

Chapter 9 Design Criterions of Dehumidification Systems with Different Types of Heat Sources

Among the various performance indexes, $\Delta\omega$ and MRC are constant values when the process air mass flow rate and the supply air humidity ratio are fixed for each process air inlet condition. Therefore, performance indexes are t_{reg} and Q_h or COP_{latent} for desiccant wheels and COP_t for the desiccant cooling systems. The former studies showed that the optimal A_{ratio} may be different for different performance indexes. Number of stages (SN) can effectively reduce t_{reg}, whereas its effect on Q_h is not investigated. The optimal configurations of desiccant wheels, such as A_{ratio} and SN, which are obtained based on t_{reg}, Q_h or COP_{latent}, are not practically useful without considering the types of heating or cooling systems. This chapter is focused on the efficient configurations of A_{ratio} and SN for desiccant wheel systems, when the VCC, electrical heater and natural gas burner are adopted as heat sources. The scenarios of single-stage system with heat recovery were also investigated. The results are meaningful for selections of SN and A_{ratio} for different regeneration heating sources, the design of single stage with heat recovery configurations, and the preferences of heating sources to reduce standard coal consumption.

9.1 System description and performance indexes

Desiccant wheel dehumidification and cooling systems with different configurations, namely SN and A_{ratio}, and four types of heating and cooling systems are described in this part. Performance evaluation indexes are introduced afterwards.

9.1.1 System descriptions

The configuration of multi-stage desiccant wheel systems is shown in Fig. 9-1. Process air and regeneration air flow in a counter flow direction. For a n-stage configuration as shown in Fig. 9-1, each stage consists of one desiccant wheel (DW), one cooler (C) located at the downstream of process air after DW, and one heater (H) located at the upstream of regeneration air before DW. There is a pre-cooling coil (C_0) to cool down the process air before it enters the first stage of DW.

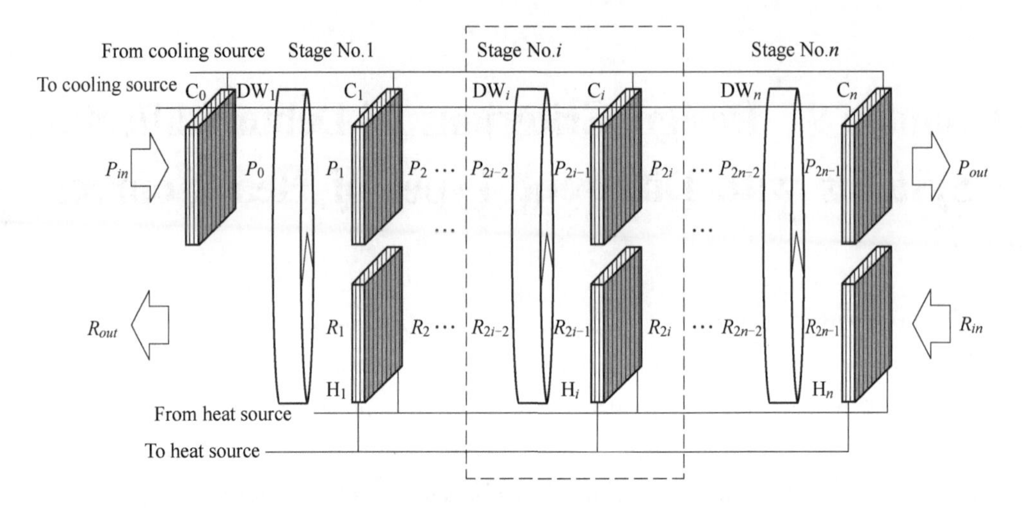

Fig. 9-1 Schematic of multi-stage desiccant wheel cooling system

For desiccant wheels, the facial area ratio (A_{ratio}) between the process air (F_P) and regeneration air (F_R) is defined as F_P/F_R. Cooling and heating fluids from cooling and heating systems flow to coolers and heaters, respectively, to cool down the process air and heat the regeneration air. The cooler is only used to remove sensible heat of processed air. Therefore, the temperature of cooling sources is relatively high, making free cooling and high evaporating temperature chillers applicable. The heater is used to heat the regeneration air to the required t_{reg}. While the efficiency of the VCC depends on operating temperature, those of natural gas burner and electricity heater are not. These three heating systems are selected as heat sources to be discussed in this chapter.

Four typical heating and cooling systems, shown in Figs. 9-2~9-5, are discussed in this chapter.

Figs. 9-2~9-4 demonstrate a n-stage system with VCC. For the system A shown in Fig. 9-2, evaporators are used as coolers at the process air side and condensers are used as heaters at the regeneration air side. There is an additional condenser (H_0) used to dissipate extra heat to outside of the system instead of the regeneration air. Fig. 9-3 shows the schematic of system B. Different from system A, cooling water produced from free cooling units is sent to the coolers. The condensers are still used as heaters in the regeneration side. The evaporator absorbs heat from ambient instead of the process air. Fig. 9-4 shows a n-stage system using hot water, which is produced by the VCC (system C-1), to heat the regeneration air. Same with system B, cooling water from

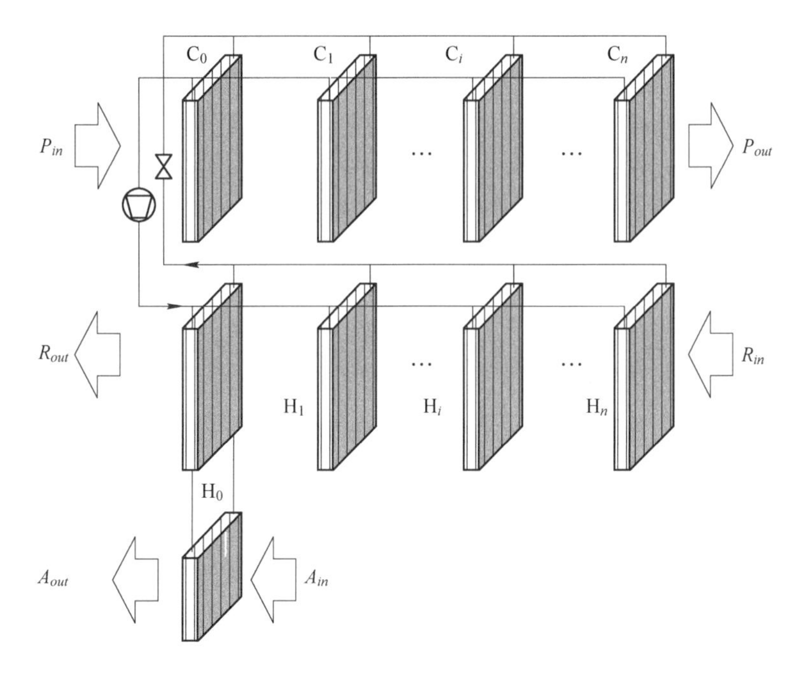

Fig. 9-2 Schematic of system A

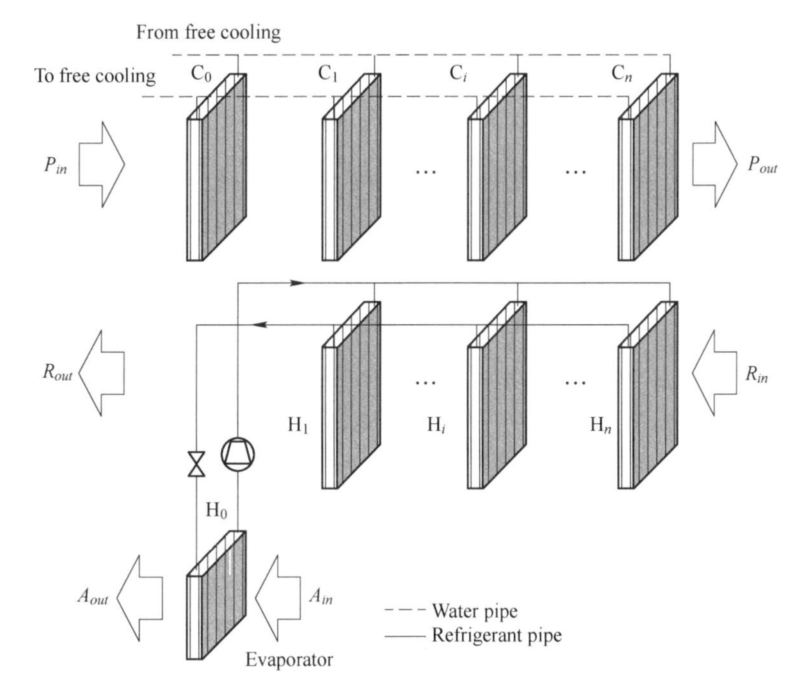

Fig. 9-3 Schematic of system B

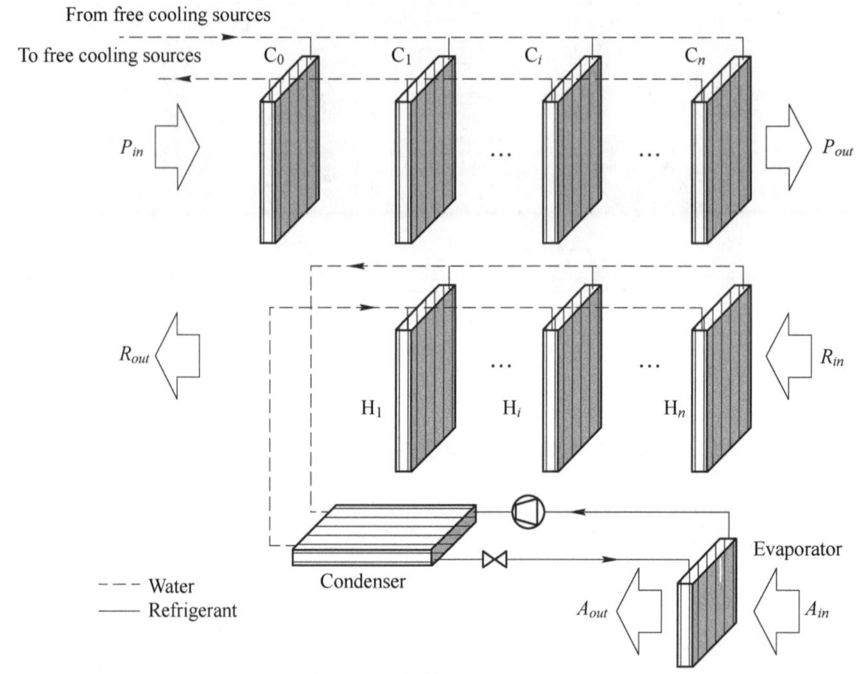

Fig. 9-4 Schematic of system C-1

free cooling is used to cool the process air. System C-2 shown in Fig. 9-5 is similar to system A but uses either natural gas burner or electrical heater to heat the hot water.

It should be noted that the scenario of using evaporator to produce cold water is not discussed. The reason is that, as compared with system A, an additional water cycle loop will reduce evaporating temperature and consequently reduce the system performances. Besides, electrical heat can be used to heat the regeneration air directly. Energy consumption of electrical heater is the same with system C-2, where electrical heater is used to heat water.

This chapter only focuses on energy consumption of heating sources, namely VCC, electrical heater and natural gas burner. Energy consumption of cooling processes and fluid machines, such as fans and water pumps, are not included.

9.1.2 Mathematic models of desiccant wheel systems

Performances of desiccant wheel systems are discussed based on simulation results of temperature and humidity ratio profiles of process air and regeneration air. A mathematic model of desiccant wheel was programmed to simulate the complex transient heat and mass transfer processes between the air and the solid desiccant in the honeycomb structure based on the following assumptions[1-3]: (1) the air flow is one-dimensional,

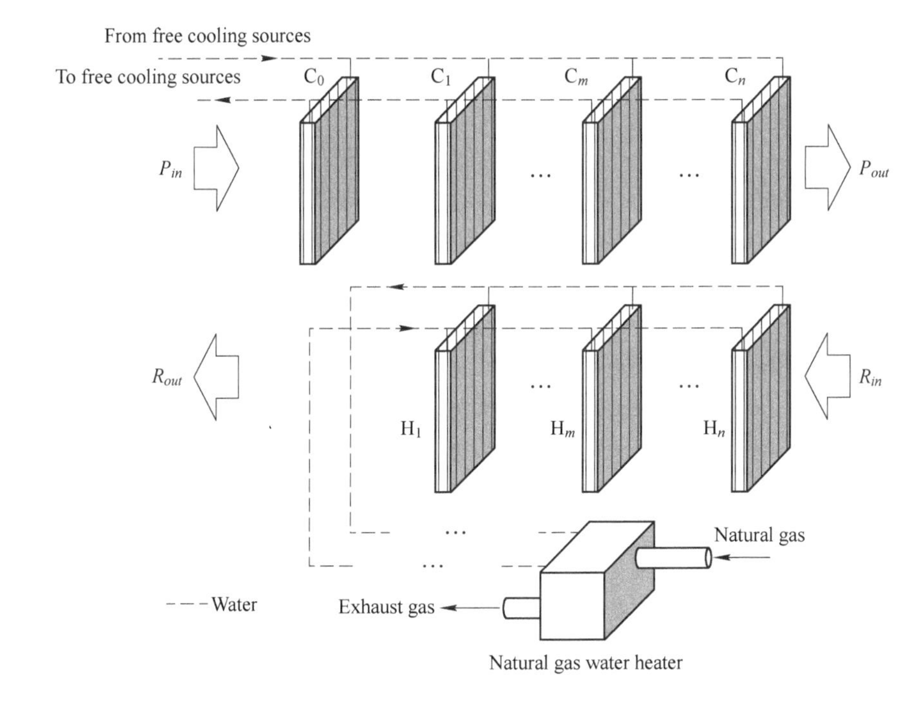

Fig. 9-5 Schematic of system C-2 using natural gas burner

and the axial heat conduction and mass diffusion in the fluid are neglected; (2) the channels are equally and uniformly distributed throughout the wheel; (3) the thermodynamic properties in the solid are constant and uniform; (4) the desiccant and substrate are of the same temperature in the wall thickness direction; and (5) air leakages between the two streams are negligible. The details about this model and experiment validations have been introduced in previous works[1-3]. In the present study, the desiccant wheel model is used to calculate temperature and humidity ratio of process air and regeneration air after each desiccant wheel.

Temperature of air after each cooler is set as t_s and that after each heater is t_{reg}. The required heating capacity (Q_h), cooling capacity (Q_c), and total heat change of process air (Q_t) can be calculated from Eqs. (9-1) ~ (9-3), and superscripts and subscripts of the parameters can be referred to Fig. 9-1.

$$Q_h = m_R \sum_{i=1}^{n} (h_R^{2i-1} - h_R^{2i}), \quad h_R^{2i} = h_{Rin} \text{ when } i = n \tag{9-1}$$

$$Q_c = m_P \left[(h_{Pin} - h_{P0}) + \sum_{i=1}^{n} (h_P^{2i-1} - h_P^{2i}) \right], \quad h_P^{2i} = h_{P_{out}} \text{ when } i = n \tag{9-2}$$

$$Q_t = m_P (h_{Pin} - h_{Pout}) \tag{9-3}$$

Cooling capacity of evaporators (Q_{evap}) and heat dissipated from condensers (Q_{cond}) can be calculated based on Q_h and Q_c for system A, system B and system C-1 from Eqs. (9-4) ~ (9-5).

For system A: $Q_{evap} = Q_c$, $Q_{cond} = Q_c(1 + 1/COP_{VCC})$ (9-4)

For system B and C-1: $Q_{cond} = Q_h$, $Q_{evap} = Q_h/(1 + 1/COP_{VCC})$ (9-5)

COP_{VCC} is the coefficient of performance of the VCC in Eq. (9-6)

$$COP_{VCC} = \frac{t_{evap} + 273.15}{t_{cond} - t_{evap}} \eta_{VCC} \tag{9-6}$$

where, η_{VCC} is the thermodynamic perfectness of the VCC, which is the ratio between COP_{VCC} and Carnot COP calculated through t_{evap} and t_{cond}.

t_{evap} and t_{cond} for the three systems can be simply calculated as follows:

For system A: $t_{evap} = t_s - \Delta t_{e-P}$, $t_{cond} = t_{reg} + \Delta t_{c-R}$ (9-7)

For system B: $t_{evap} = t_a - \Delta t_a - \Delta t_{e-a}$, $t_{cond} = t_{reg} + \Delta t_{c-R}$ (9-8)

For system C-1: $t_{evap} = t_a - \Delta t_a - \Delta t_{e-a}$, $t_{cond} = t_{reg} + \Delta t_{w-R} + \Delta t_{c-w}$ (9-9)

where, Δt_{e-P} is temperature difference between t_{evap} and the process air leaving each evaporator; Δt_{c-R} is temperature difference between t_{cond} and regeneration air leaving each condenser ($=t_{reg}$); t_a is the air temperature entering evaporators of system B and system C-1; Δt_a and Δt_{e-a} are temperature change of the air due to limited mass flow rate and temperature difference between t_{evap} and temperature of the air leaving evaporators of system B and system C-1; Δt_{w-R} is temperature difference between t_{reg} and temperature of the hot water entering each heater (t_{win}); and Δt_{c-w} is temperature difference between t_{cond} and t_{win}.

9.1.3 Performance evaluation indexes

Performances of these systems are analyzed and compared based on the same process air mass flow rate (m_P), inlet state ($t_{P_{in}}$, $\omega_{P_{in}}$) and supply air state ($t_{P_{out}}$, $\omega_{P_{out}}$). Regarding this, moisture removal capacity (MRC) defined in Eq. (9-10) is the same under the same working condition for each system.

$$MRC = m_P(\omega_{P_{in}} - \omega_{P_{out}}) \tag{9-10}$$

The regeneration air is essentially heated by thermal energy. Latent coefficient of performance (COP_{latent}) is defined as the latent heat removed from the process air divided by heating capacity of heaters (Q_h), shown as Eq. (9-11)[4]:

$$COP_{latent} = \frac{m_P(\omega_{P_{in}} - \omega_{P_{out}})h_v}{Q_h} \tag{9-11}$$

COP_{latent} is independent of heating systems. If heating and cooling systems are taken into consideration, the coefficient of performance of the whole system (COP_t) defined in Eq. (9-12) is used to evaluate the overall system performance.

$$COP_t = \frac{Q_t}{W} = \frac{m_P[c_{pa}(t_{P_{in}} - t_{P_{out}}) + (\omega_{P_{in}} - \omega_{P_{out}})h_v]}{W} \tag{9-12}$$

In Eq. (9-12), the total heat change of process air (Q_t) is considered. W is the power consumption of the heating and cooling system powered by electricity. For systems using vapor compression cycle, like system A, B and C-1, W_{comp} is calculated as:

$$W_{comp} = \frac{Q_{evap}}{COP_{VCC}} = \frac{Q_{cond}}{1 + COP_{VCC}} \tag{9-13}$$

where, Q_{cond} is the heat dissipated from condensers. For system A, the process air of which is cooled by evaporators, Q_{evap} equals to Q_c, as in Eq. (9-4). Q_{cond} is the sum of Q_{evap} and W_{comp}. Normally, Q_h takes a part of Q_{cond} and the rest is dissipated by H_0. Therefore, Q_{cond} is greater than Q_h. For system B and system C-1, Q_{cond} equals to Q_h.

For system C-2, if electrical heater is used, W of electrical heater is equal to Q_h. Different from the VCC and the electrical heater, which uses electricity to produce heat, Q_h is produced by burning natural gas for natural the gas burner. The volume of natural gas (V_{NG} in the unit of m^3/h) to produce Q_h can be calculated by Eq. (9-14).

$$V_{NG} = \frac{Q_h}{q_{NG}} \tag{9-14}$$

where, q_{NG} is the approximately 10.0kW · h/m^3 (heat per cubic meter of natural gas) considering an efficiency of 91%.

To make a fair energy consumption comparison between V_{NG}, which is the mass flow rate of natural gas, and W, which is electrical consumption, the energy consumption is converted to gram of standard coal (m_{ce} in the unit of gce/h) as Eq. (9-15):

$$m_{ce} = \Theta\alpha \tag{9-15}$$

where, Θ is the consumption of electricity (kW) or volume of natural gas (m^3/h); α is the conversion coefficient of electricity or natural gas to standard coal (gce) based on the power plant efficiency and lower heating value of primary fuels. Values of α are 319gce/(kW · h) electricity and 1330gce/m^3 natural gas[5] for following discussions.

The overall performances of the above systems can be calculated as Eq. (9-16).

$$\Lambda = \frac{m_{ce}}{Q_t} = \frac{\Theta\alpha}{Q_t} \tag{9-16}$$

Λ, in the unit of gce/(kW · h) (gram standard coal per kW · h total heat change of process air) is used to evaluate the efficiency of a system.

For systems using VCC, electrical heater and natural gas burner, Λ is calculated as Eqs. (9-17) ~ (9-19), respectively:

$$\Lambda_{VCC} = \frac{W_{comp}}{Q_t}\alpha_e = \frac{Q_{cond}}{Q_t} \cdot \frac{\alpha_E}{1 + COP_{VCC}} \quad (9\text{-}17)$$

$$\Lambda_{EH} = \frac{Q_h}{Q_t}\alpha_E \quad (9\text{-}18)$$

$$\Lambda_{NG} = \frac{V_{NG}}{Q_t}\alpha_e = \frac{Q_h}{Q_t} \cdot \frac{\alpha_{NG}}{q_{NG}} \quad (9\text{-}19)$$

9.2 Efficient configurations based on t_{reg} and Q_h or COP_{latent}

In this section, effects of SN and A_{ratio} on t_{reg} and Q_h or COP_{latent} are discussed without considering the types of cooling or heating systems. The optimal configurations for achieving the lowest t_{reg} and lowest Q_h or highest COP_{latent} are summarized.

9.2.1 Parameters for simulation analyses

Parameters about working conditions, desiccant wheels, VCC are listed in Table 9-1.

Table 9-1 Design conditions for the simulation analysis

Working conditions	Desiccant wheel	Cooling and heating system
Process air mass flow rate: 0.8kg/s BSC: 33℃, 19g/kg; WSC: 33℃, 14g/kg Regeneration air: 0.8/A_{ratio} kg/s, 26℃, 12g/kg Supply air: 0.8kg/s, 25℃, 9g/kg; t_s: 25℃; t_a: 26℃; Δt_a: 5℃	Radius: 0.5m Total thickness: 0.24m A_{ratio}: 1, 2 and 3 Air path: 2mm high and 2mm wide Nu: 2.463 RS: 15r/h SN: 1, 2, 3 and 4	Temperature after the cooling coils: t_c = 25℃ Temperature after the heating coils: t_{reg} High heat exchange efficiency: $\Delta t_{e-P} = \Delta t_{c-R} = \Delta t_{e-a} = \Delta t_{w-R} = \Delta t_{c-w}$ = 2℃ Low heat exchange efficiency: $\Delta t_{e-P} = \Delta t_{c-R} = \Delta t_{e-a} = \Delta t_{w-R} = \Delta t_{c-w}$ = 5℃ η_{VCC} = 0.4 and 0.6

The systems discussed in this chapter are used to process outdoor air and are regenerated by indoor air. The indoor air is at 26℃ and 12g/kg. Beijing summer conditions (BSC: 33℃ and 19g/kg) and Washington summer conditions (WSC: 33℃ and 14g/kg) are selected as the outdoor air to represent humid and mild climates. The supply air humidity ratio is fixed at 9g/kg, which meets the indoor latent heat removal requirement. t_s, which is temperature of process air after each cooler, is fixed at 25℃. Air flow rate of process air (m_P) is fixed at 0.8kg/s, and that of regeneration air (m_R) is equal to m_P/A_{ratio}. Temperature of air exchanging heat with the evaporator in systems B and C-1 is 26℃ and Δt_a is 5℃. The radius of desiccant wheel is 0.5m, and the total thickness of n-stage desiccant wheels equals to 0.24m and thickness of each desiccant wheel is 0.24/nm. *SN* varies from 1 to 4 and A_{ratio} equaling to 1, 2 and 3 are discussed in the paper. Scenarios of A_{ratio} smaller than 1 are not under discussion, so that m_P is larger than m_R to keep a positive pressure difference between the occupant room and the ambient environment. N_u of air path in desiccant wheels is selected to be 2.46, which relates only to width, height and shape of the air path cross section [6]. For the heating and cooling system, $\Delta t_{e-P} = \Delta t_{c-R} = \Delta t_{e-a} = \Delta t_{w-R} = \Delta t_{c-w} = 2℃$ is selected to represent high efficient heat exchange systems, while 5℃ is chosen to represent low efficient heat exchange systems. η_{VCC} equaling to 0.4 and 0.6 are discussed.

9.2.2 Effects of *SN* and A_{ratio} on t_{reg} and COP_{latent}

Air handling processes of single stage system and 4-stage system when $A_{ratio} = 1$ and 2 are shown in Fig. 9-6.

It can be seen from Fig. 9-6 that t_{reg} of 4-stage system is lower than that of single stage system. For desiccant wheels with higher A_{ratio} ($A_{ratio} = 2$), t_{reg} is higher than that of desiccant wheels with lower A_{ratio}($A_{ratio} = 1$).

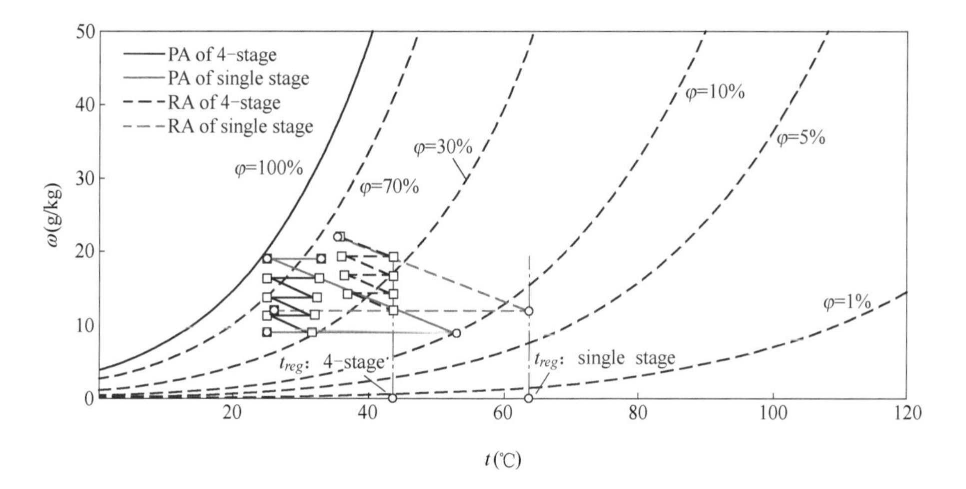

(a)

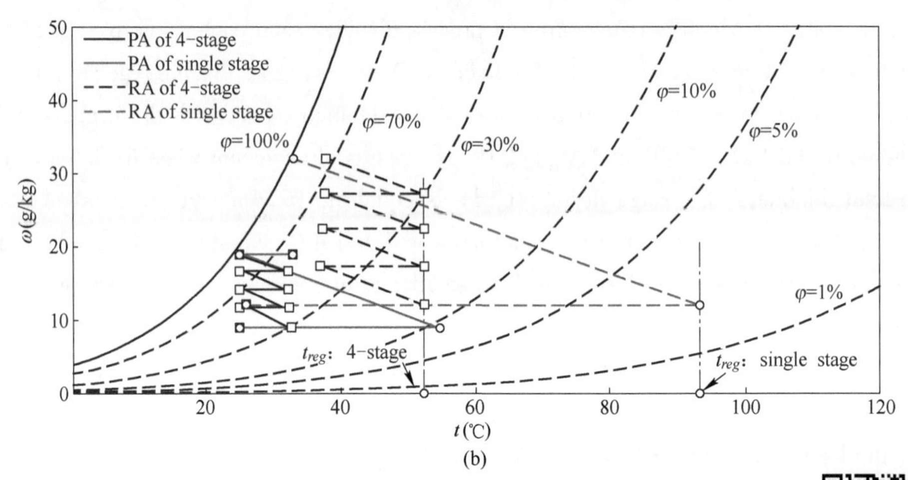

(b)

Fig. 9-6 Air handling processes of single stage and 4-stage configurations when $A_{ratio}=1$ and 2

(a) $A_{ratio}=1$; (b) $A_{ratio}=2$

t_{reg} of scenarios when SN changes from 1 to 4 and A_{ratio} changes from 1 to 3 under BSC and WSC are calculated using the desiccant wheel model. The results of t_{reg}, Q_h and COP_{latent} for different scenarios are shown in Fig. 9-7.

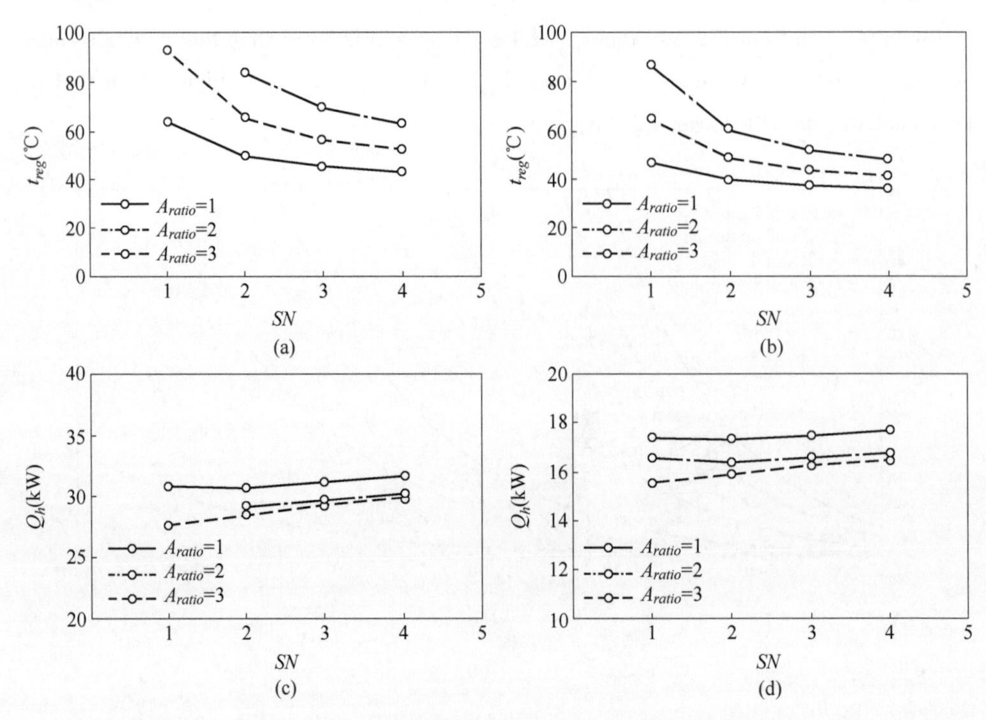

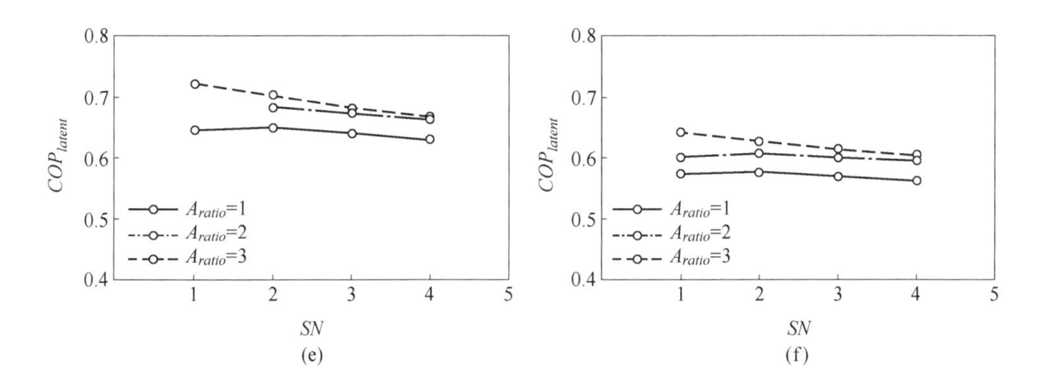

Fig. 9-7 Effects of A_{ratio} and SN on t_{reg}, Q_h and COP_{latent} under BSC and WSC

(a) t_{reg} under BSC; (b) t_{reg} under WSC; (c) Q_h under BSC;

(d) Q_h under WSC; (e) COP_{latent} under BSC; (f) COP_{latent} under WSC

The results show that under both mild and humid working conditions, t_{reg} reduces as the increase of SN and the decrease of A_{ratio}. However, the smallest Q_h is obtained when A_{ratio} is equal to 2. Q_h is lower for singe systems than scenarios with larger stage number. Since COP_{latent} is the latent heat change of process air divided by Q_h, single stage system with A_{ratio} equaling to 2 is better for higher COP_{latent}.

The suggested A_{ratio} and SN to achieve low t_{reg} and low Q_h or high COP_{latent} are summarized in Table 9-2.

It is shown in Table 9-2 that the suggested configurations of SN or A_{ratio} are different for achieving low t_{reg} and low Q_h or high COP_{latent}. To reduce t_{reg}, configurations of 3-stage~4-stage system with A_{ratio} equaling to 1 are preferred, whereas to reduce Q_h or increase COP_{latent}, single stage system with A_{ratio} equaling to 2 is preferred.

Table 9-2 Recommended configurations regarding t_{reg} and Q_h or COP_{latent}

Performance index	Suggested configuration
Low t_{reg}	$SN=3\sim4$ and $A_{ratio}=1$
Low Q_h or High COP_{latent}	$SN=1$ and $A_{ratio}=2$

9.3 Efficient configurations of A_{ratio} and SN for different heat sources

The previous analyses are based on t_{reg} and Q_h, which are independent of cooling and heating systems. Effects of t_{reg} and Q_h on energy consumption of different heat sources are not the same. For electrical heat or natural gas burners, the consumption of electricity or natural gas are influenced by Q_h but not by temperature. However, for the VCC, the power consumption of compressor is not only related to Q_h but also to temperature. The suggested configurations of *SN* or A_{ratio} for the above four systems are discussed next.

The adoption of heat recovery units in single stage system is analyzed. The suggested configuration with lower energy consumption for each system are summarized. Discussions are based on parameters in Table 9-1.

$\Delta t_{e-p} = \Delta t_{c-R} = \Delta t_{e-a} = \Delta t_{w-R} = \Delta t_{c-w} = 2℃$ and $\eta_{VCC} = 0.6$ are selected for discussion.

9.3.1 Recommended *SN* and A_{ratio} for the four systems without heat recovery

Systems A, B and C-1 use the VCC to heat the regeneration air. Effects of *SN* and A_{ratio} on performances of systems A, B and C-1 under BSC and WSC are shown in Figs. 9-8 and 9-9, respectively.

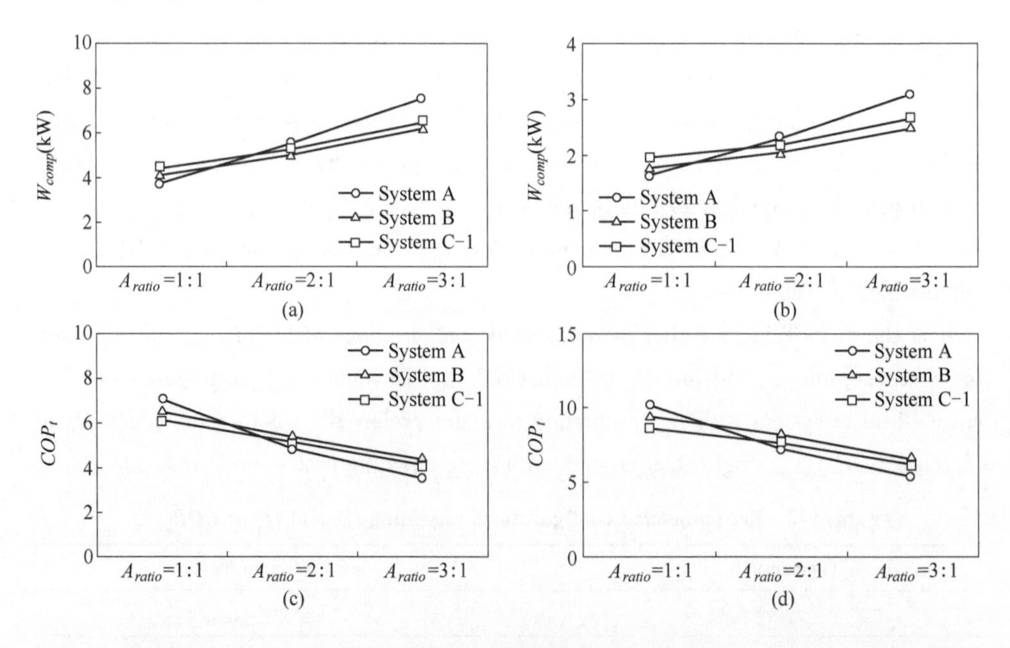

Fig. 9-8 Effects of A_{ratio} on W_{comp} and COP_t of systems A, B and C-1 in 4-stage configuration

(a) W_{comp} under BSC; (b) W_{comp} under WSC; (c) COP_t under BSC; (d) COP_t under WSC

Fig. 9-8 shows the effects of A_{ratio} when *SN* is 4. It shows that as A_{ratio} reduces, W_{comp} is reduced and higher COP_t is obtained. The recommended A_{ratio} is 1 : 1. Fig. 9-9 demonstrates the effects of *SN* on W_{comp} and COP_t, when A_{ratio} is 1.

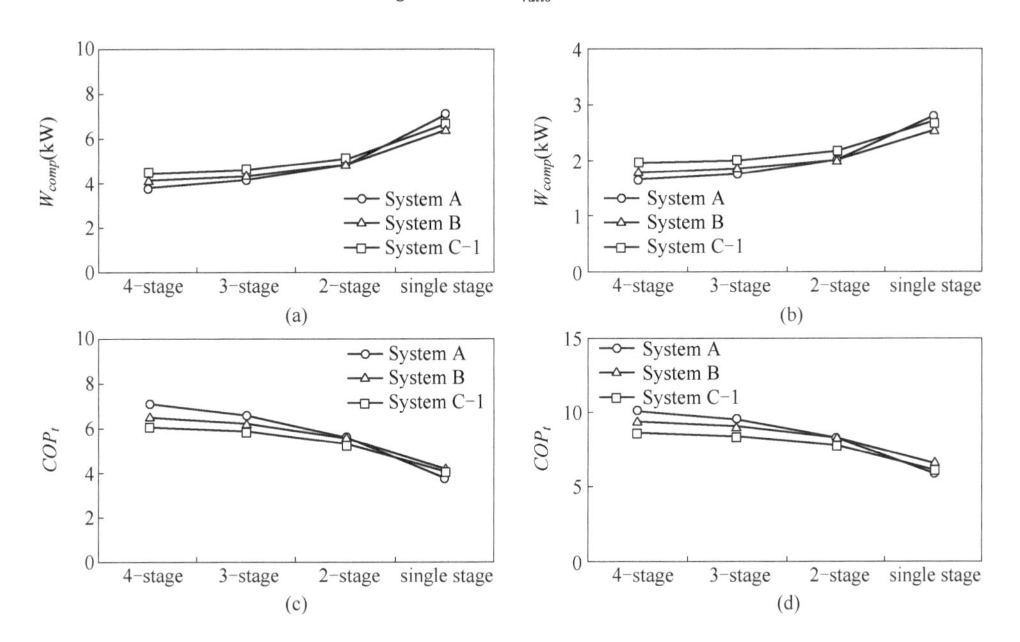

Fig. 9-9 Effect of SN on W_{comp} and COP_t of systems A, B and C-1 when $A_{ratio}=1$

(a) W_{comp} under BSC; (b) W_{comp} under WSC; (c) COP_t under BSC; (d) COP_t under WSC

It is common for the three systems that as the increase of SN, less energy is consumed by the compressor and higher COP_t can be obtained. The recommended SN is 3~4. As compared with Table 9-2, the recommended SN and A_{ratio} for the three systems are the same with those recommended for lower t_{reg}. Whereas, the results are different with those recommended for lower Q_h.

System C-2 adopts natural gas burner or electricity heater to heat the regeneration air. Fig. 9-10 is the results when natural gas burner is adopted.

Because the consumption of natural gas (V_{NG}) is related to Q_h, as shown in Eq. (9-14), the recommended SN and A_{ratio} are the same with those recommended for lower Q_h, but different with those recommended for lower t_{reg}, as shown in Table 9-2. Since energy consumption of electrical heater is only related to Q_h, the effects of SN and A_{ratio} on electricity consumption and the suggested SN and A_{ratio} for scenarios using electrical heater are the same with scenarios using natural gas burner. And scenarios with electrical heat will not be discussed in the following part.

In a summary, for heat pump driven systems, SN = 3 ~ 4 and A_{ratio} = 1 are recommended. For systems using gas burner or electrical heater to heat the regeneration air, multi-stage configurations are not suitable, and SN = 1 and A_{ratio} = 2 are recommended.

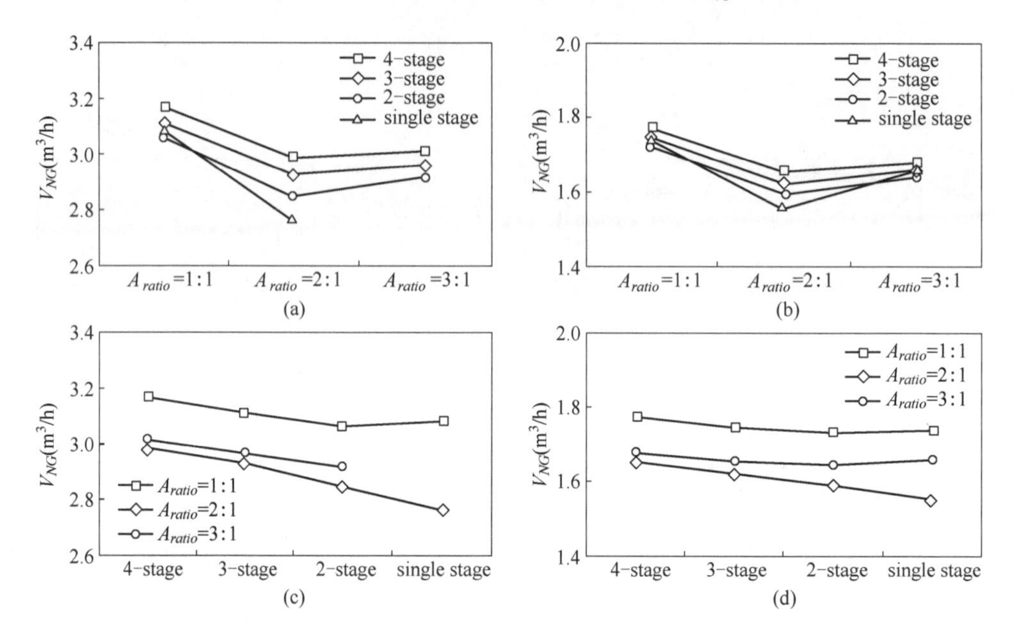

Fig. 9-10 Effect of *SN* and A_{ratio} on V_{NG} of system C-2 under BSC and WSC

(a) V_{NG} varying with A_{ratio} under BSC; (b) V_{NG} varying with A_{ratio} under WSC;

(c) V_{NG} varying with *SN* under BSC; (d) V_{NG} varying with *SN* under WSC

9.3.2 Single stage system with heat recovery

It is shown in Fig. 9-6 that for single stage system, there is large temperature difference between process air leaving desiccant wheel and regeneration air before entering the heater. There is huge energy conservation potential when heat recovery unit (HR) is adopted. The configuration of single stage with heat recovery configuration is shown in Fig. 9-11.

Energy conservation potential of HR is highly related to heat recovery efficiency (η_{HR}). Since the mass flow rate (m_P) and temperature leaving the desiccant wheel (t_{P_2}) of process air is constant under the same working condition, η_{HR} is defined as Eq. (9-20), representing the percentage of heat recovered by the process air, $m_P(t_{P_2} - t_{P_3})$, to the maximum heat that can be recovered by the process air, $m_P(t_{P_2} - t_{R_{in}})$, which is a constant value:

$$\eta_{HR} = \frac{t_{P_2} - t_{P_3}}{t_{P_2} - t_{R_{in}}} \tag{9-20}$$

η_{HR} is calculated as Eq. (9-21) for counter flow air-air heat exchanger:

$$\eta_{HR} = \frac{NTU}{NTU + 1}, \quad m_P/m_R = 1$$

$$\eta_{HR} = \frac{1 - \exp[-NTU(1 - m_P/Gm_R)]}{1 - (m_P/m_R)\exp[-NTU(1 - m_P/m_R)]}, \quad m_P/m_R \neq 1 \quad (9\text{-}21)$$

where, NTU is number of heat transfer unit of HR, which is defined as $KF/(m_P \cdot c_{pa})$. K, in the unit of $kW/(m^2 \cdot K)$, is overall heat transfer coefficient of HR; F, in the unit of m^2, is heat transfer area of HR. Since m_P is constant for all the scenarios and m_P/m_R equals to A_{ratio}, the effects of A_{ratio} on system performances can be evaluated under different NTU.

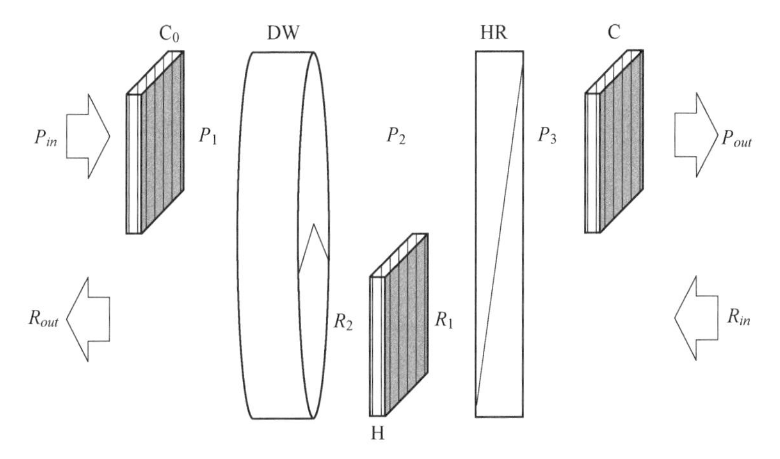

Fig. 9-11 Single stage with heat recovery configuration

Temperature of regeneration air leaving HR is calculated as Eq. (9-22).

$$t_{R_1} = t_{R_{in}} + \eta_{HR}(t_{P_2} - t_{R_{in}}) A_{ratio} \quad (9\text{-}22)$$

Heat capacity (Q_h) provided by the heaters, which are condensers of systems A and B, and the water-air heat exchanger of systems C-1 and C-2, can be calculated as:

$$Q_h = m_R c_{pa}(t_{reg} - t_{R_{in}}) - m_P c_{pa}\eta_{HR}(t_{P_2} - t_{R_{in}}) \quad (9\text{-}23)$$

The effects of A_{ratio} on W_{comp} and V_{NG} when NTU varies from 1 to 4 for systems A, B, C-1 and C-2 under BSC and WSC are shown in Figs. 9-12~9-13.

It is common for all the scenarios that W_{comp} or V_{NG} decreases at higher NTU. However, for higher A_{ratio} scenarios, the effect of NTU is weaker than scenarios with $A_{ratio} = 1$. This is because the increase rate of heat transfer efficiency with NTU slows down at higher m_P/m_R. As NTU increases from 1 to 4, η_{HR} increases from 0.5 to 0.8

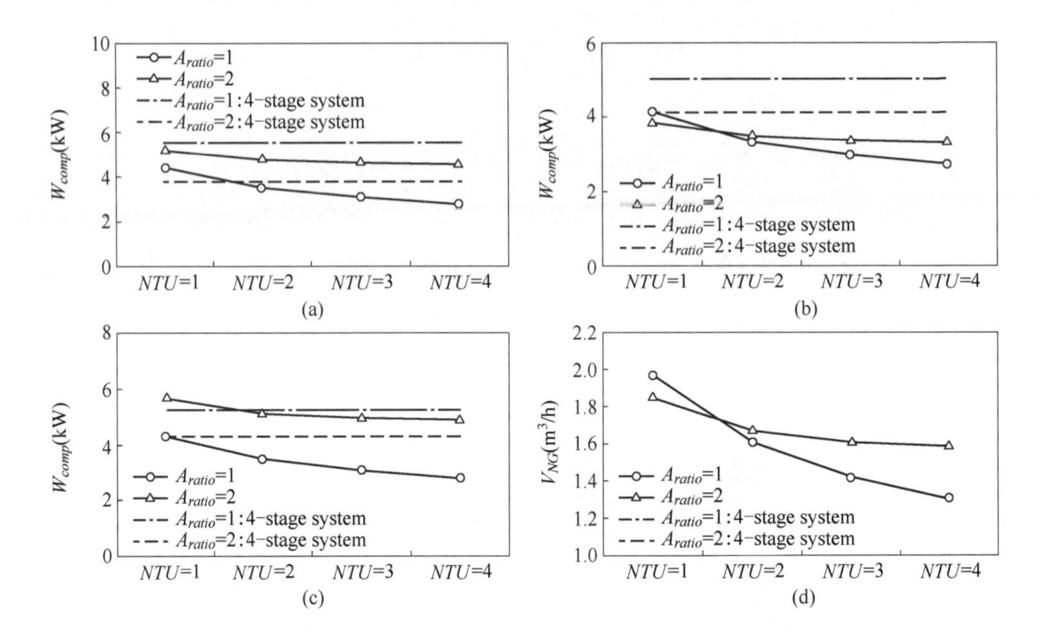

Fig. 9-12 Effect of *NTU* of heat recovery on energy consumption of the four systems under BSC

(a) System A; (b) System B; (c) System C-1; (d) System C-2

when m_P/m_R (or A_{ratio}) = 1, from 0. 387 to 0. 495 when $A_{ratio}=2$, and from 0. 302 to 0. 333 when $A_{ratio}=3$.

For systems A and C-1 under both working conditions, when *NTU* varies from 1 to 4, W_{comp} of single stage system with $A_{ratio}=1$ is constantly lower than that of single stage system with $A_{ratio}=2$. Whereas, for systems B and C-2, there is a cross point of the two lines, which is between $NTU=1$ and $NTU=2$. When *NTU* is smaller than the cross point, W_{comp} or V_{NG} for scenarios of $A_{ratio}=1$ is higher than scenarios of $A_{ratio}=2$. The results are reversed as *NTU* is higher than the cross point.

The results of single stage systems with heat recovery are compared with the corresponding 4-stage systems as shown in Figs. 9-12~9-13. When *NTU* increases above 2, the W_{comp} of systems with single stage and heat recovery configuration are constantly lower than that of systems with 4-stage configuration for $A_{ratio}=1$ or 2.

In a summary, heat recovery is an effective way to improve performances of single stage systems, which can be higher than the corresponding 4-stage counterparts. The contribution of *NTU* is more obvious when desiccant wheels with $A_{ratio}=1$ is used.

9. 3. 3 Discussion

The above analyses show that the recommended configurations, namely *SN*, A_{ratio}, are

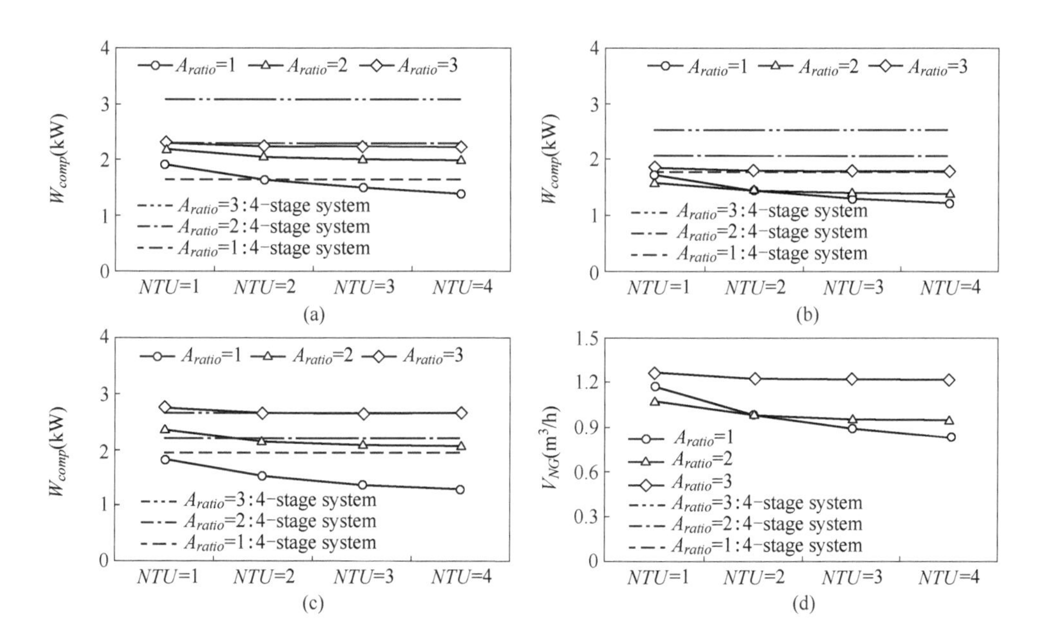

Fig. 9-13 Effect of *NTU* of heat recovery on energy consumption of the four systems under WSC

(a) System A; (b) System B; (c) System C-1; (d) System C-2

highly dependent on heat source types. The recommended system configurations for the above four systems with and without heat recovery unit are summarized in Table 9-3.

Table 9-3 Recommended configurations for the discussed systems

Energy source	System	Without heat recovery	Single stage with heat recovery
Condensing heat of electricity driven heat pump	System A	$SN = 3 \sim 4$ and $A_{ratio} = 1$	$A_{ratio} = 1$ and $NTU \geqslant 2$
	System B		
	System C-1		
Electrical heater or natural gas burner	System C-2	$SN = 1$ and $A_{ratio} = 2$	

Although multi-stage systems and $A_{ratio} = 1$ are beneficial for lowering t_{reg}, this is only meaningful for systems using the VCC. The recommended SN and A_{ratio} for systems using the VCC are the same with those recommended for lowering t_{reg}, as shown in Table 9-2. For electrical heater and gas burner, of which energy consumption are related to Q_h, the recommended SN and A_{ratio} are the same with those recommended for lowering Q_h, as shown in Table 9-2. Heat recovery has a great energy conservation potential for single stage systems. Since heat transfer efficiency is related to both NTU and m_P/m_R, which has the same value with A_{ratio}, the recommended configurations of the discussed systems are different with their counterparts without heat recovery. $A_{ratio} = 1$ and $NTU \geqslant 2$ are recommended for the four systems, which effectively reduces W_{comp} or V_{NG}, as compared with the counterparts without heat recovery under suggested SN and A_{ratio}.

9.4 Performance comparisons among different systems

In this part, performances of different systems are compared based on the recommended configurations listed in Table 9-3 with the aim of finding a better design for desiccant dehumidification systems with low primary energy consumption.

9.4.1 Comparison among heat pump driven systems

Performances of systems A, B and C-1 are discussed in this subsection, and the configuration of $SN = 4$ and $A_{ratio} = 1$ is selected. W_{comp} and COP_t under BSC and WSC for the three systems are shown in Fig. 9-14 (a) and (b).

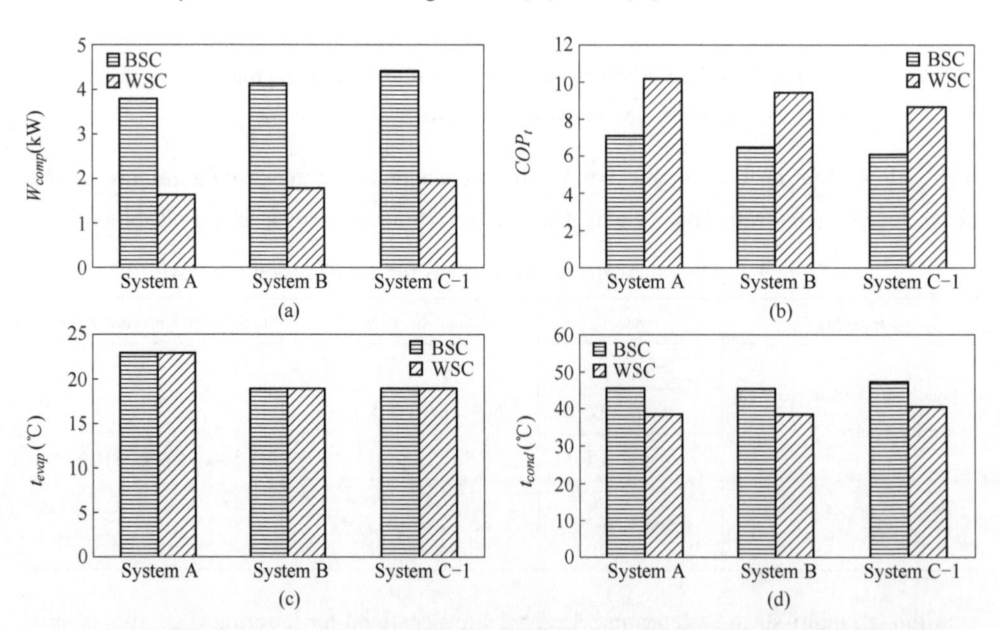

Fig. 9-14 Performances of the three systems using vapor compression cycle in the 4-stage configuration

(a) W_{comp}; (b) COP_t; (c) t_{evap}; (d) t_{cond}

It is shown that system A has the best performances among the three systems, while those for system C-1 are the worst.

As shown in Fig. 9-14 (c) and (d), t_{evap} of systems B and C-1 is lower than that of system A, while t_{cond} of systems C-1 is higher than those of systems A and B because of one more water flow loop between the condenser and the heaters of system C-1. COP_{VCC} of systems A, B and C-1 are 7.9, 6.6 and 6.2, respectively, under BSC, and are 11.4, 8.9 and 8.1, respectively, under WSC. Although Q_{cond} of system A is higher than those of systems B and C-1, which equals to Q_h, high COP_{VCC} of system A makes W_{comp} of system A lower than those of systems B and C-1. As for systems B and C-1, which are of the same Q_{cond} ($=Q_h$) and Q_t under the same working condition, higher COP_{VCC} of system B leads to lower W_{comp} and COP_t than those of system C-1, according to Eq. (9-12) and Eq. (9-13).

Fig. 9-15 is the results of systems in the configuration of single stage with heat recovery.

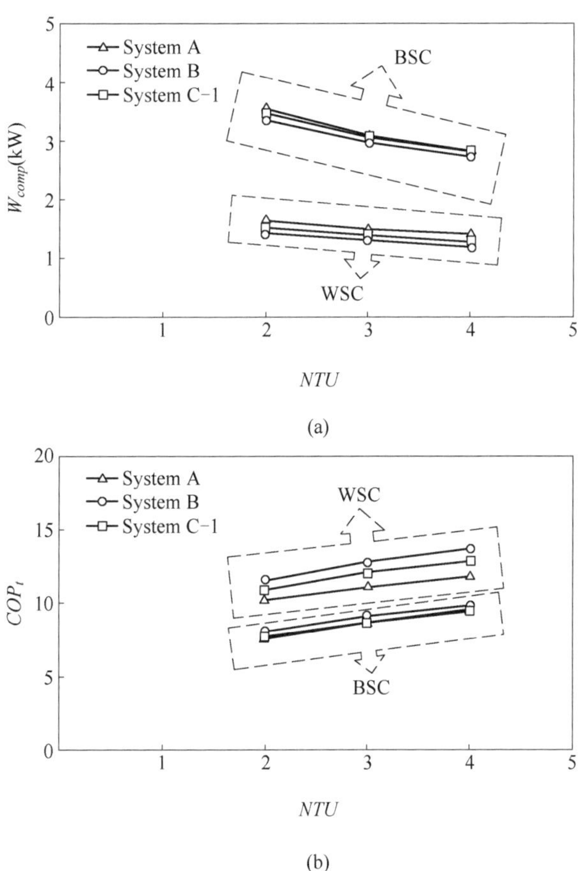

Fig. 9-15 Performances of the three systems using vapor compression cycle in the single stage and heat recovery configuration

(a) W_{comp}; (b) COP_t

NTU of HR varies from 2 to 4. Different from 4-stage scenarios, performances of system B are the best. However, the differences between the best and the worst (system A) are only within 5% and 15% under BSC and WSC, respectively. Considering energy consumption of cooling systems when free cooling is not available and the cooling system complexity for systems B and C-1, system A is highly recommended for the configuration of single system with heat recovery.

Based on the above analyses, systems A is recommended both for 4-stage configuration and single stage with heat recovery configuration. For systems using the VCC, it is not advised to add extra heat transfer loops, such as the water loop of system C-1 since the increase of t_{cond} or decrease of t_{evap} would reduce the VCC efficiency.

9.4.2 Performance comparison with different heat sources

In this subsection, performances of systems using the VCC are compared with system C-2 in the configuration of single stage with heat recovery and $A_{ratio} = 1$ from the consumption of standard coal per total heat change of process air (Λ_{VCC}). Since systems B and C-2 are of the same cooling system, 4-stage configuration and single stage with heat recovery configuration of system B are selected to represent systems using the VCC. As can be seen in Eq. (9-17), Λ_{VCC} is influenced by COP_{VCC}, which is related to t_{evap}, t_{cond} and η_{VCC}. Therefore, $\Delta t(\Delta t_{e-p} = \Delta t_{c-R} = \Delta t_{e-a} = \Delta t_{w-R} = \Delta t_{c-w}) = 2℃$ and 5℃, and $\eta_{VCC} = 0.6$ and 0.4 are selected for discussion. Δt_a is fixed at 5℃. α is 319gce/(kW · h) electricity based on the efficiency of power plant and 1330gce/m^3 natural gas, which is converted from lower heating values of natural gas and standard coal. The results are shown in Fig. 9-16.

It is shown as expected that Λ_{HP} reduces as the reduction of Δt and the increase of η_{VCC}. For single stage configuration with heat recovery, Λ_{VCC} and Λ_{NG} both reduces with the increase of *NTU* of HR. Λ_{VCC} of system B with HR is constantly lower than the 4-stage configuration and lower than Λ_{NG} of system C-2 with HR. However, Λ_{NG} of system C-2 with HR can be lower than Λ_{VCC} of system B with 4-stage configuration. Under BSC, Λ_{NG} of system C-2 when *NTU* varies from 2 to 4 is constantly lower than Λ_{VCC} of system B with 4-stage configuration when η_{VCC} is 0.4 and Δt is 5℃. Under WSC, Λ_{NG} of system C-2 with *NTU* of HR being 4 is just less than 10% higher than Λ_{VCC} of system B with 4-stage configuration when η_{VCC} is 0.4 and Δt is 5℃.

In conclusion, for systems using the VCC, the single stage and heat recovery configuration is simple and high efficient as compared with the multi-stage configuration, and is more efficient than systems using natural gas burners, from the point view of

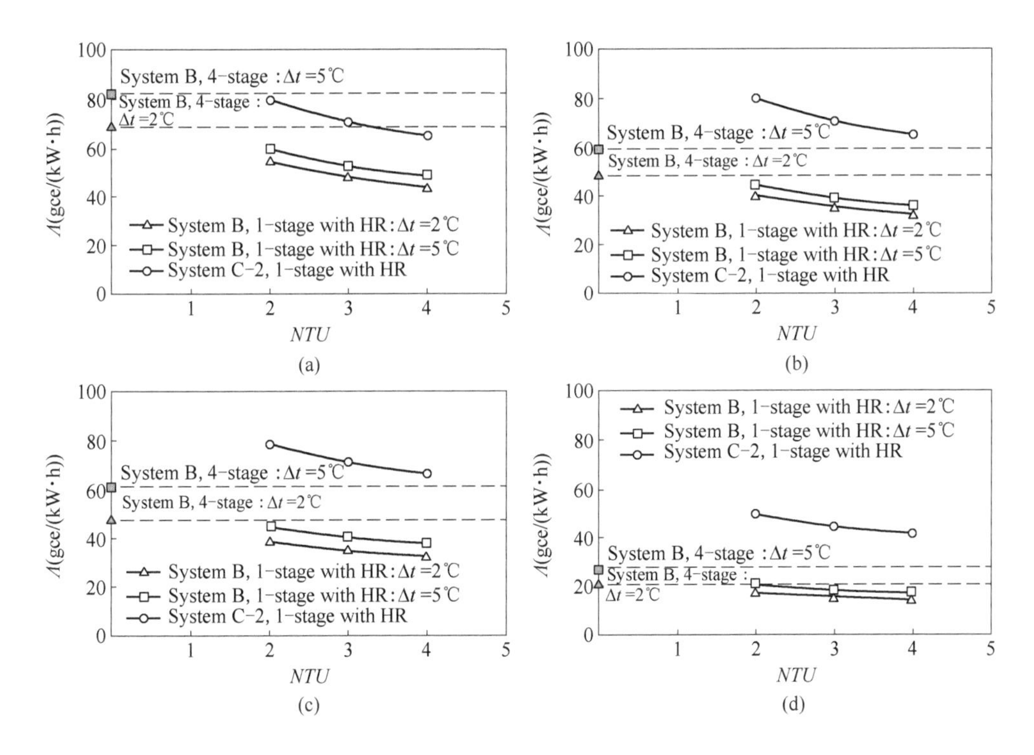

Fig. 9-16 Performances comparison between system B and system C-2

(a) $\eta_{HP}=0.4$, BSC; (b) $\eta_{HP}=0.6$, BSC; (c) $\eta_{HP}=0.4$, WSC; (d) $\eta_{HP}=0.6$, WSC

standard coal consumption. To keep better performances of 4-stage configuration of system B than system C-2 in the configuration of single stage with heat recovery, high efficient vapor compression cycle with η_{VCC} around 0.6 and Δt around 5℃ is necessary. Otherwise, the system C-2 in the configuration of single stage with heat recovery is suggested. The types of cooling and heating systems recommended for each configuration are summarized in Table 9-4.

Table 9-4 performance comparisons different systems and recommended systems for different configurations

Type of heat source	Configuration	
	4-stage and $A_{ratio}=1$	Single stage with heat recovery ($NTU \geqslant 2$) and $A_{ratio}=1$
Vapor compression cycle	System A has the best performance	Performance of the three systems are close, and system A is recommended; Performances are better than the 4-stage counterparts and systems using natural gas burner for regeneration
Natural gas burner	—	Performances are comparable with system B in the 4-stage configuration when η_{VCC} is 0.4 and Δt is 5℃

9.5 Conclusions

The efficient configurations of A_{ratio} and SN for desiccant wheel systems regenerated by heat sources such as condensation heat of vapor compression cycle, electrical heater and natural gas burner, are discussed in this chapter. Single stage with heat recovery configurations are compared with multi-stage configurations, too. The main conclusions are as follows:

(1) When electrical heater and natural gas burner are used, $SN=1$ and $A_{ratio}=2:1$ are suggested, which are the same with those for smaller heating capacity (Q_h). When vapor compression cycle is used, $SN=3\sim4$ and $A_{ratio}=1:1$ are suggested, which are the same with those for lower regeneration temperature. When heat recovery unit is applied to single stage systems, $A_{ratio}=1:1$ and $NTU\geqslant2$ are suggested for all the systems, and performances are higher than the corresponding 4-stage configurations.

(2) For systems with vapor compression cycle, it is suggested to directly heat the regeneration air with condenser like systems A and B. The system A is recommended both for 4-stage configuration and configuration of single stage with heat recovery.

(3) For systems with vapor compression cycle, the configuration of single stage with heat recovery is superior to their counterparts in multi-stage configuration and to systems using natural gas burners. To keep better performances of 4-stage systems using vapor compression cycle than natural gas burner regenerated systems in the configuration of single stage with heat recovery, high efficient vapor compression cycle with η_{VCC} around 0.6 and Δt around 5℃ is necessary.

Nomenclature

A_{ratio}	area ration of dehumidification to regeneration of a desiccant wheel
BSC	Beijing Summer Condition
C_i	number i cooler
COP	coefficient of performance
COP_{latent}	latent heat removed by desiccant wheel from process airdivided by Q_h
c_p	specific heat capacity, kJ/(kg · K)
DW	desiccant wheel
H_i	number i heater
h_v	specific heat of vaporization, kJ/kg
m	mass flow rate of air, kg/s

m_{ce}	gram of standard coal, gce/h
MRC	moisture remove capacity, g/s
NTU	number of transfer units of heat recovery unit
Nu	Nusselt number
Q	heating or cooling capacity, kW
q_{NG}	heat value of natural gas, (kW · h)/m^3
SN	number of stages
t	Celsius temperature, ℃
VCC	vapor compression cycle
V_{NG}	volume flow of natural gas, m^3/h
W	electricity consumption, kW
WSC	Washington airport Summer Condition

Greek symbols

Δ	difference
ω	humidity ratio, g/kg
Θ	consumption of electricity or volume of natural gas, kW or m^3/h
α	conversion coefficient of electricity or natural gas to standard coal, gce/kW · h or gce/m^3
Λ	efficiency, gce/(kW · h)
η	efficiency

Subscripts

a	air
c	cooling
$comp$	compressor
$cond$	condenser
$evap$	evaporator
h	heat
in	inlet
NG	natural gas
out	outlet

P process air

R regeneration air

reg regeneration

s supply

t total

w water

References

[1] R Tu, X H Liu, Y Jiang. Performance analysis of a two-stage desiccant cooling system [J]. Appl. Energy, 2014, 113 (1): 1562-1574.

[2] R Tu, X H Liu, Y Jiang. Lowering the regeneration temperature of a rotary wheel dehumidification system using exergy analysis [J]. Energy Conver. Manage, 2015, 89 (1): 162-174.

[3] R Tu, X H Liu, Y H Hwang, F Ma. Performance analysis of ventilation systems with desiccant wheel cooling based on exergy destruction [J]. Energy Convers. Manage, 2016, 123: 265-279.

[4] T Cao, H Lee, Y H Hwang, R Radermacher, H Chun. Experimental investigations on thin polymer desiccant wheel performance [J]. Int. J. Refrig, 2014, 44: 1-11.

[5] Building energy research center. 2017 Annual Report On China building Energy Efficiency [R]. 1st edition. Beijing: China Architecture and Building Press, 2017.

[6] L Z Zhang. Total Heat Recovery: Heat and Moisture Recovery from Ventilation Air [M]. New York: Nova Science Publisher, 2008.